Argumentation Schemes

This book provides a systematic analysis of many common argumentation schemes and a compendium of ninety-six schemes. The study of these schemes, or forms of argument that capture stereotypical patterns of human reasoning, is at the core of argumentation research. Surveying all aspects of argumentation schemes from the ground up, the book takes the reader from the elementary exposition in the first chapter to the current state of the art in the research efforts to formalize and classify the schemes, outlined in the last three chapters. It provides a systematic and comprehensive account, with notation suitable for computational applications that increasingly make use of argumentation schemes.

Douglas Walton is Distinguished Research Fellow of CRRAR (Centre for Research in Reasoning, Argumentation and Rhetoric), University of Windsor, and Assumption University Chair in Argumentation Studies (University of Windsor). The recipient of numerous fellowships, awards, and honors, he is the author of more than thirty books, most recently *Fundamentals of Critical Argumentation, Media Argumentation,* and *Witness Testimony Evidence.*

Chris Reed is senior lecturer and head of research at the School of Computing, University of Dundee. He is the head of the Argumentation Research Group at Dundee, which has been instrumental in the development of the Argument Interchange Format, an international standard for computational work in the area.

Fabrizio Macagno received his doctorate in linguistics from the Catholic University of Milan. He is coauthor with Douglas Walton of several papers published in international journals.

For Karen, Cathy, Franco, and Mirna, with love.

Argumentation Schemes

DOUGLAS WALTON
University of Winnipeg

CHRIS REED
University of Dundee

FABRIZIO MACAGNO
Catholic University of Milan

CAMBRIDGE
UNIVERSITY PRESS

CAMBRIDGE
UNIVERSITY PRESS

32 Avenue of the Americas, New York NY 10013-2473, USA

Cambridge University Press is part of the University of Cambridge.

It furthers the University's mission by disseminating knowledge in the pursuit of
education, learning, and research at the highest international levels of excellence.

www.cambridge.org
Information on this title: www.cambridge.org/9780521723749

First published 2008
6th printing 2013

A catalog record for this publication is available from the British Library.

Library of Congress Cataloging in Publication data
Walton, Douglas N.
Argumentation schemes / Douglas Walton, Chris Reed, Fabrizio Macagno.
    p.   cm.
Includes bibliographical references and index.
ISBN 978-0-521-89790-7 (hardback) – ISBN 978-0-521-72374-9 (pbk.)
1. Reasoning. I. Reed, Chris. II. Macagno, Fabrizio. III.Title.
BC177.W317   2008
168–dc22   2007045337

ISBN 978-0-521-89790-7 Hardback
ISBN 978-0-521-72374-9 Paperback

Contents

Acknowledgments

Chris Reed would like to thank the Leverhulme Trust, and Doug Walton would like to thank the Social Sciences and Humanities Research Council of Canada, for research grants to support a joint research project, Argumentation Schemes in Natural and Artificial Communication. The research project had the aims of developing a classification of schemes and building a corpus of natural language examples. Fabrizio Macagno would like to thank the Catholic University of Milan for a research grant that supported his work on the grant project while in residence at the University of Winnipeg in 2004 and the Social Sciences and Humanities Research Council of Canada for support for his second visit to Winnipeg in the summer of 2006.

Chris Reed and Doug Walton would like to thank the following journals and publishers for permission to reprint material from the following articles, in modified form: Chris Reed and Douglas Walton, "Diagramming, Argumentation Schemes and Critical Questions," in *Anyone Who Has a View: Theoretical Contributions to the Study of Argumentation*, ed. Frans H. van Eemeren, J. Anthony Blair, Charles A. Willard, and A. Francisca Snoek Henkemans (Dordrecht: Kluwer, 2003), 195–211. Material from pages 198–203 of this paper appears in Chapter 1, section 4; material from pages 196–207 appears in Chapter 1, section 6.

Chris Reed and Douglas Walton, "Towards a Formal and Implemented Model of Argumentation Schemes in Agent Communication," in *Argumentation in Multi-Agent Systems*, ed. Iyad Rahwan, Pavlos Moraitis, and Chris Reed (Berlin: Springer, 2005), 19–30. Chapter 10, section 5, is taken from parts of this paper; material from pages 26–27 of this paper appears in Chapter 11, section 2; material from pages 24–25 appears in Chapter 11, section 3.

Douglas Walton and Chris Reed, "Applications of Argumention Schemes," in *Argumentation and Its Applications: Proceedings of the 4th Conference of the Ontario Society for the Study of Argumentation (OSSA2001)*, ed. Hans V. Hansen, Christopher W. Tindale, J. Anthony Blair, Ralph H. Johnson, and Robert C. Pinto (Windsor, Canada: CD ROM, 2001). Material from pages 9–10 of this paper (the section "Pedagogy") appears in Chapter 1, section 4; material from pages 24–25 appears in Chapter 11, section 3; other material from this paper appears in Chapter 1, sections 1, 2, and 3.

Douglas Walton and Chris Reed, "Argumentation Schemes and Enthymemes," *Synthese: An International Journal for Epistemology, Methodology and Philosophy of Science* **145** (2005): 339–370. Chapter 8 is based on material from this paper, but it has been modified to suit the needs of the book.

Chris Reed and Douglas Walton, "Argumentation Schemes in Dialogue," in *Dissensus and the Search for Common Ground: Proceedings of the 7th Conference of the Ontario Society for the Study of Argumentation (OSSA2007)*, ed. Hans V. Hansen, Christopher W. Tindale, J. Anthony Blair, Ralph H. Johnson, and Robert C. Pinto (Windsor, Canada: CD ROM, 2001). Parts of this paper appear as sections 8–11 of Chapter 11.

Figure 10.1 in Chapter 10, section 2, originally appeared as Figure 6.1 in the book by Douglas Walton, *Ad Hominem Arguments* (Tuscaloosa: University of Alabama Press, 1998), 260.

Figure 10.2 in Chapter 10, section 3, originally appeared as Figure 6.2 in the book by Douglas Walton, *Ad Hominem Arguments* (Tuscaloosa: University of Alabama Press, 1998), 261.

Figure 12.9 originally appeared at the URL <http://compendium.open.ac.uk/conpendium-arg-schemes.html>.

Figure 12.11 originally appeared at <http://openlearn.open.ac.uk/file.php/2824/kmap/1183813181/(Many_Poineer_fund_G_1371082521411834697306 95_Outline.html>.

Figure 12.12 originally appeared at the URL <http://wiki.austhink.com/Walton>.

The authors would also like to thank Henry Prakken for stimulating discussions on closely related issues and Scott Jacobs for comments on Chapter 8 during discussions of the subject of argument refutation and rebuttal in January 2005 at the University of Arizona. They would like to thank David Godden for comments on Chapters 3 and 8. In particular, Godden's detailed and extensive commentary was enormously helpful to making revisions to Chapter 8. We would like to thank Rita Campbell for composing the index and Anahid Melikian for help with proofreading.

Introduction

Argumentation Schemes

The theory of argumentation is a rich interdisciplinary area of research spanning philosophy, communication studies, linguistics, computer science, and psychology. In the past few years, formal models of argumentation have been steadily gaining importance in artificial intelligence, where they have found a wide range of applications in specifying semantics for logic programs, generating natural language text, supporting legal reasoning, and facilitating multi-agent dialogue and negotiation on the Internet.[1] The most useful and widely used tool so far developed in argumentation theory is the set of argumentation schemes. Argumentation schemes are forms of argument (structures of inference) that represent structures of common types of arguments used in everyday discourse, as well as in special contexts like those of legal argumentation and scientific argumentation.[2] They include the deductive and inductive forms of argument that we are already so familiar with in logic. However, they also represent forms of argument that are neither deductive nor inductive, but that fall into a third category, sometimes called defeasible, presumptive, or abductive. Such an argument may not be very strong by itself, but may be strong enough to provide evidence to warrant rational acceptance of its conclusion, given that its premises are acceptable (Toulmin, 1958). Such an argument can rightly carry weight, or be a plausible basis

---

[1] Recent conferences and workshops dedicated to the theory of argumentation in artifical intelligence include the International Conference on Computational Models of Argument (COMMA 2006), the Computational Models of Natural Argument (CMNA) workshop series, and the Argumentation in Multi-Agent Systems (ArgMAS 04, 05, and 06) workshop series. In 2007, there has been a call for papers for a special issue of the IEEE journal *Intelligent Systems* on the topic of argumentation technology.

[2] Prakken (2005) has shown that because logic is too abstract to apply very effectively to legal argumentation, research in AI and law needs to be supplemented by an argumentation schemes approach.

for acceptance, on a balance of considerations in an investigation or discussion that is moving forward, as new evidence is being collected. The investigation can then move ahead, even under conditions of uncertainty and lack of knowledge, using the conclusion tentatively accepted.

To use a phrase from Anderson, Schum, and Twining (2005, p. 262), such presumptive arguments are necessary but dangerous. We need to use them as heuristics that provide rational grounds for accepting a conclusion tentatively even if it has not been conclusively proved, but we have to remain open-minded when we use such arguments, because they are fallible and inherently subject to default. A defeasible argument is one in which the conclusion can be accepted tentatively in relation to the evidence known so far in a case, but may need to be retracted as new evidence comes in. A typical case of a defeasible argument is one based on a generalization that is subject to qualifications. Should it come to be known that the present case is an exception to the generalization, the argument defaults, and its conclusion must be retracted. Defeasible arguments are especially prominent in legal and ethical reasoning, but they are everywhere, even in science, especially at the discovery stage of an investigation.

The recognition of the importance and legitimacy of defeasible argumentation has led to a recent paradigm shift in logic, artificial intelligence, and cognitive science. Common forms of defeasible arguments were long categorized as fallacious in logic textbooks. It is been only recently that, as these informal fallacies have been studied more intensively, more and more instances have been recognized where the forms of argument underlying them are reasonable, but inherently defeasible. For example, arguments based on expert opinion have long been categorized in logic textbooks under the heading of fallacious appeals to authority. However, it is clear that for practical purposes in everyday reasoning, and in many of our social and intellectual institutions, we could not get by without such arguments. Expert testimony, including ballistics evidence, DNA evidence, and many other forms of testimony by scientific experts, has become a dominant kind of evidence in the courts. It has become so dominant as evidence that it is on the verge of overwhelming our judicial system. Clearly, it is not helpful to condemn such evidence as inherently fallacious. Rather, the problem is to judge in specific cases when an argument from expert opinion can properly be judged to be strong, weak, or fallacious. Hence the importance of argumentation schemes has become readily apparent in recent years, as this paradigm shift about rational argumentation has affected many fields, including law, cognitive science, artificial intelligence, logic, philosophy of science, and indeed any field where standards of rational argument are centrally important.

There has emerged in recent years a considerable body of work on informal fallacies, collecting together a large corpus of examples, along with tools to identify, analyze, and evaluate the arguments in those examples. Clearly, this body of work provides a huge database, and repository of other materials, including many argumentation schemes, that are fundamental to any attempt to approach the project of providing a systematic overview of the current state of the art of research on argumentation schemes. The special advantage of the present book is that it builds on this previous research on fallacies, moving through the paradigm shift to the new idea of coping with the revolutionary notion that such "fallacies" are no longer fallacies.

Although this is the first book to bring together such a large number of schemes and to analyze and study them in such depth, even to the point of starting the project of classifying and formalizing schemes, prior works on schemes do exist. In a book on presumptive argumentation schemes by one of the authors (Walton, 1996), a list of twenty-six defeasible argumentation schemes was presented and analyzed. Among them are such common forms of argument as argument from sign, argument from example, argument from commitment, argument from position to know, argument from expert opinion, ad hominem argument, argument from analogy, argument from precedent, argument from gradualism, and several types of slippery slope argument. Each argument of this type is presented as providing only a defeasible support for its conclusion, subject to critical questioning in a context of dialogue. Matching each argumentation scheme is an appropriate set of critical questions. The method of studying defeasible argumentation schemes through the use of a set of matching critical questions can be credited to Hastings (1963). Arthur Hastings, in his innovative Ph.D. thesis at Northwestern University in 1963, set out a useful list of many of these schemes, with illustrative examples, and with a set of critical questions corresponding to each scheme. The method of evaluation of an argument fitting a scheme is that once the argument is put forward by a proponent, it may be defeated if the respondent asks an appropriate critical question that is not answered by the proponent. Hastings' approach seemed to have been ignored for many years, but as the field of argumentation studies developed, other researchers began to adopt his approach. For example, Kienpointner (1992) and Grennan (1997) produced comprehensive lists of schemes, stressing deductive and inductive forms.

This book takes a much more comprehensive and in-depth approach than any previous treatment of schemes. In Chapter 9, a compendium of schemes has been produced that presents sixty-five schemes. They are presented in a form that can easily be used by all of those who are interested in schemes or are working on them. Nearly all of the schemes in the compendium have been collected from the already existing literature, although there are a few new ones. Here, for the first time, they are brought together in one place. Chapter 1 introduces the beginning reader to schemes, and describes the basic tools of argumentation research needed to formulate the schemes more precisely and to understand how they work. All concepts in Chapter 1 are explained from the ground up, so that the beginning reader can understand the chapters that follow. The reader can next, in Chapter 2, gain further insight into how schemes work and how they are to be analyzed by examining the treatment of one of the most fundamental schemes, that for argument from analogy. Argument from analogy is especially important in law, notably in our Anglo-American justice system, where court decisions are arrived at by comparing a given case with a previously decided one. Thus Chapter 2 also reveals the importance of this particular argumentation scheme and the wide-ranging nature of the application of defeasible argumentation schemes.

Chapter 2 begins with a typical example of a legal decision by the courts that is based on argument from analogy. When a trained dog sniffs luggage in a public place and signals to the police that it contains drugs, should this event be classified as a search? The question is decided by comparing the case, by analogy, to previous cases that have already been decided by the courts. Previously, the logical literature on argument from analogy has tended to classify this form of argument as either deductive or inductive. We propose a new way of classifying it by treating it as a defeasible argumentation scheme that can hold tentatively on the balance of considerations, thus influencing future decisions without finally closing the issue one way or the other. By using tools developed in argumentation theory and artificial intelligence, we show how argument from analogy, as used in legal reasoning in typical cases, is closely associated with other argumentation schemes. Especially prominent, as shown by our analysis of these cases, are the argumentation schemes for verbal classification and argument from precedent. We are thus able to show, in a much deeper way than has been possible in the past, how argument from analogy, allied with these other argumentation schemes, provides new logical foundations for case-based reasoning in law.

Chapters 3, 4, and 5 describe many of the most common and important defeasible argumentation schemes, providing many examples to show how they work in everyday argumentation. Chapters 6 and 7 study two concepts that are not only fundamental to understanding how schemes work, but that also show how important schemes are as building blocks of the most common kinds of arguments. Chapter 6 shows how schemes can be used to help identify premises or conclusions that are implicitly assumed, but that have not been explicitly stated as part of an argument. Chapter 7 is about the notion of argument rebuttal. In other words, it is all about how one argument confronts and attacks, and possibly even defeats, another argument by adducing reasons that show that the other argument is not tenable. Some argumentation schemes have the specific purpose of functioning as rebuttals to other arguments. Although the notion of argument rebuttal is fundamental to the study of all rational argumentation, there are many controversies and disagreements about how it should precisely be defined, and the notion has never been clarified fully throughout the long history of the subject.

As we show in Chapter 8, the study of schemes has a long history going back to Aristotle's topics – common types of argument, often called commonplaces, that Aristotle saw as fundamental building blocks in a branch of logic he called dialectic. Aristotle also developed formal logic through his theory of the syllogism, and that approach to logic came to dominate the whole field, and indeed the intellectual scene generally, through the Middle Ages. As deductive logic became formalized in the twentieth century, the study of dialectic continued to be ignored. Although informal fallacies, as noted earlier, continued to be treated in the logic textbooks, the study of topics remained in a somewhat confused state, never gaining wide acceptance as a tool for the analysis of rational argumentation. Many had hoped that topics could be used as a tool for the discovery of new arguments, a technique for argument invention. This would be an extremely useful tool in many fields, but only with the advent of this book has it become a practical possibility.

The problem so far in modern argumentation studies is that the schemes have been developed in a rough-and-ready way. They have been meant to be practical tools to help students learn skills of argumentation and critical thinking by recognizing common forms of argument and by being able to criticize them by asking standard critical questions that probe the weak points of an argument. Such a practical tool has proved to be extremely useful, but if schemes are to be exploited by more exact fields like logic and artificial intelligence, they need to be defined and analyzed in a more precise and systematic manner. Indeed, as we show in Chapter 10, a systematic method of classifying schemes is a top priority, and the current work in artificial intelligence is developing methods for the formalization of schemes. These efforts, culminating in this book, represent the frontiers of the new research on schemes, aiming at the goal of developing tools for argument search in natural texts and for argument invention.

This volume surveys all aspects of argumentation schemes from the ground up, taking the reader from the elementary exposition of the first chapter to the latest state of the art in the research efforts to formalize and classify the schemes as outlined in the last three chapters. In Chapter 8, the history of schemes is surveyed, so that the reader can grasp how, even though their study was very much in the background for two millennia, there was active work on them during both the ancient and medieval periods. In Chapters 2 through 5 we pick out what we take to be the most important and common schemes, and analyze and discuss these schemes up to the present point of research on them. In Chapters 6 and 7 we discuss two underlying concepts, those of enthymeme and rebuttal, that are fundamentally important in helping us to understand the common structure that the schemes share and the promise that they hold as argumentation tools. Thus the whole book gives a panoramic survey of the state of the art of current research on schemes, from the ancient roots of the subject to recent research developments. It is a necessary tool for anyone interested in argumentation schemes, in the many fields that use them, and in the many other fields that will.

1

Basic Tools in the State of the Art

This chapter introduces the reader to argumentation schemes and explains, through the use of some examples, why they are important. Another aim of the chapter is to briefly review the literature on argumentation schemes, including the key works by Hastings, Walton, and Kienpointner, and to set it in a broader context, bringing out some characteristics of defeasible reasoning and argument evaluation that are fundamental to the study of schemes. Another is to introduce the beginning reader to some basic tools, like argument diagramming, that utilize schemes and need to be integrated with them. In this chapter we will introduce the reader to an automated system of argument diagramming called Araucaria. This technique is a box-and-arrow representation of the premises and conclusions of an argument, showing how one argument can be chained together with others to form a sequence of reasoning. This tool will be used in subsequent chapters, and so we need to introduce the reader to it now. One of our goals in the book is to show how argumentation schemes are in the process of being modeled by argument technology in the field of artificial intelligence (AI). However, we will reserve our fullest account of these developments for the last chapter of the book, even though, from time to time, we will mention aspects of them that impinge on our fundamental understanding of argumentation schemes as forms of reasoning.

Another aim of this chapter is to introduce the reader to the problem posed by the fact that many of the most important kinds of schemes are defeasible in nature, meaning that even after the argument has been accepted, it might later be defeated as new evidence enters into consideration. This factor of defeasibility raises the problem of how schemes are rationally binding. In deductive logic, if someone to whom an argument is directed accepts the premises of the argument, and the argument is deductively valid, that person must accept the conclusion. If he does not, he is in a position of inconsistency, a position that is logically untenable. However, defeasible schemes are not binding in this way, because it is open to the person to whom the argument is directed to ask critical questions about it before having to accept a conclusion. This feature, the attaching of critical questions to a scheme, turns out to be problematic in several respects. First of all, it challenges the traditional notion of argument cogency, whereby a cogent argument provides a sufficient reason to accept the conclusion. Second, it presents a problem in applying standard tools and techniques, like box-and-arrow diagrams, that model arguments as sets of propositions, called premises and conclusions, and inferential links between sets of them. It is not easy to see how critical questions can be analyzed as tools for argument evaluation within such a propositional model.

Schemes have recently been attracting more and more attention from those who are interested in exploiting the rich interdisciplinary area between argumentation and AI (Reed and Norman, 2003; Verheij, 2003). Of course, AI has long been interested in nondeductive forms of reasoning (for a good general review of the area, see Prakken and Vreeswijk, 2002). But schemes, as construed by argumentation theory, seem to provide a somewhat more fine-grained analysis than is typical within AI. One example lies in the granularity of classification of types: Kienpointner introduces over a dozen, Walton almost thirty, and Grennan over fifty, but none can claim exhaustivity. By comparison, AI systems are more typically built with a small handful. Pollock's (1995) OSCAR, for example, identifies fewer than ten – with an uneven amount of work spread between them. This profligacy in philosophical classification might be argued to be as much a problem as an advantage – explored further in Chapter 10 – but it serves to demonstrate that more detail is in some way being adduced. It is the contention of this book that those refined structures of reasoning yield nicely to a computational interpretation and can be implemented to useful effect. Eventually, in chapter 12, we will examine recent developments in computing that have the aim of formalizing schemes and building working systems for analyzing, evaluating, and constructing arguments using schemes.

1. INTRODUCING ARGUMENTATION SCHEMES

Perelman and Olbrechts-Tyteca, in *The New Rhetoric* (1969), in addition to the other authors mentioned in the Introduction, identified many of these defeasible types of arguments used to carry evidential weight in a dialogue, in a somewhat different style from that of Arthur Hastings' Ph.D. thesis (1963), where a more systematic analysis of many of the most common of these presumptive schemes is presented. The scheme itself, in Hastings' treatment, is specified by stating the form of premises and conclusion in each argument type. Hastings expresses one special premise in each scheme as a Toulmin warrant, which could be seen as a generalization or rule, linking the other premise or premises to the conclusion. Such a warrant is typically a defeasible generalization that is subject to qualifications, on the Toulim model. Along with each scheme, Hastings attaches a corresponding set of critical questions. These features set the basic pattern for argumentation schemes in the literature that followed.

Some argumentation schemes were used by van Eemeren and Groot-endorst (1984; 1992) in their work on critical discussion and fallacies. Kienpointner (1992) developed a comprehensive listing of argumenta-tion schemes that includes deductive and inductive forms in addition to presumptive ones. Walton (1996) identified some twenty-six (depend-ing on how you count them) argumentation schemes for common types of presumptive reasoning. Following Hastings' format, a set of critical questions attached to each scheme is the device for criticizing any argu-ment fitting the structure of the scheme. The asking of a question, along with the response to it, implies a kind of dialogue structure in which two parties interact with each other. If an argument put forward by a proponent meets the requirements of a scheme, and the premises are acceptable to the respondent, then the respondent is obliged to accept the conclusion. But such an acceptance – or commitment, as it is often called – is provisional in the dialogue. If the respondent asks one of the critical questions matching the scheme and the proponent fails to offer an adequate answer, the argument defaults. Thus we see that defeasibility is linked to a dialogue structure in which a burden can shift back and forth. The original weight of an argument, before it defaulted and had to be retracted, is restored only when the proponent gives a successful answer to the question.

An argumentation scheme that can be used as our first example is that for argument from sign. Let's take a case in which Helen and Bob are hiking along a trail in Banff, and Bob points out some tracks along the path, saying, "These look like bear tracks, so a bear must have passed along this trail." In the argumentation scheme that follows, taken from Walton (1996, p. 49), one premise is seen to function as a generalization.

Argument from Sign

Minor Premise: Given data represented as statement A is true in this situation.
Major (Generalization) Premise: Statement B is generally indicated as true when
   its sign, A, is true, in this kind of situation.
Conclusion: Therefore, B is true in this situation.

The major premise is a conditional stating that if A is true, then generally,
but subject to exceptions, B is also true. This generalization is defeasible.
The tracks could have been planted on the trail by tricksters. But in the
absence of evidence of some trickery, it is reasonable to provisionally
draw the conclusion that a bear passed along the trail. Argument from
sign is closely related to abductive inference, or inference to the best
explanation, since the best explanation of the existence of the observed
tracks is the hypothesis that a bear walked along the trail producing
the tracks. There could be other explanations, but in the absence of
additional evidence, the bear hypothesis could be plausible as a basis for
drawing a provisional conclusion.

Argumentation schemes include deductive forms of reasoning like
modus ponens, and inductive forms like arguing from a collected set of
data to a statistical conclusion drawn from the data. But they also include
forms of reasoning that are often necessary, but are more tentative in
nature and need to be judged circumspectly by reserving some doubts.
Such reasoning is presumptive and defeasible. This kind of reasoning
is only plausible and is often resorted to in conditions of uncertainty
and lack of knowledge. Presumptive reasoning supports inference under
conditions of incompleteness by allowing unknown data to be presumed.
Defeasible reasoning, as mentioned earlier, is of a sort in which the
conclusion can be withdrawn or modified if known (but uncertain) data
turn out to be flawed (Fox and Das, 2000). Walton (1996, p. 81) employs
the following example:

> A Ph.D. student, Susan, has spent more than five years trying to finish her
> thesis, but there are problems. Her advisers keep leaving town, and delays
> are continued. She contemplates going to law school, where you can get a
> degree in a definite period. But then she thinks: "Well, I have put so much
> work into this thing. It would be a pity to give up now."

This is not just an instance of a presumptive, defeasible argument, but an
instantiation of a particular pattern of reasoning, a particular scheme –
in this case, what is called the *argument from waste* by Perelman and
Olbrechts-Tyteca (1969), but what is more usually called the *argument*

from sunk costs. By taking this classificatory step, it becomes possible to probe in a straightforward, deterministic way for missing premises, and to consider a set of predetermined critical questions in carrying out an analysis and evaluation of the argument.

Argumentation schemes have recently been attracting increasing interest for several reasons. The first is their contribution to fallacy theory. As Walton has pointed out in a series of monographs, arguments that fit into traditional categories of fallacies seem, under the right circumstances, to be appropriate, acceptable, and persuasive. Walton (1996) posits that schemes offer one way of tackling this apparent contradiction. Argumentation schemes are also very attractive on pedagogical grounds. Schemes can be used as a way of providing students with additional structure and analytic tools with which to analyze natural arguments and to evaluate them critically.

Schemes also hold great potential for tackling a variety of problems in artificial intelligence (AI). The real world represents an immense challenge to artificial agents. Even if we focus only upon reasoning capabilities, and leave to one side the physical aspects of interacting with the world, an agent must deal with two fundamental problems: uncertainty and incompleteness. Not only will an agent not know everything, it cannot even be sure of the things that it does know – and this demands a complete shift from more traditional approaches to software design. Recent work (see Carbogim et al., 2000; Reed, 1997; and Walton, 2000 for reviews) has shown that argumentation offers a powerful means of tackling these problems by moving away from purely deductive, monotonic approaches to reasoning and toward presumptive, defeasible techniques. Typically, however, such reasoning systems will have to interact not only with the world, but also with humans. This places further demands on these systems: not only must this reasoning be carried out, it must also be presented, perhaps dialogically, in a form that is appropriate for human consumption. Once again, it has been demonstrated that dialogue-based theories of argumentation offer flexible, realistic, and, crucially, implementable techniques (Grasso et al., 2000; Reed, 1998).

Argumentation schemes are the forms of argument (structures of inference) that enable one to identify and evaluate common types of argumentation in everyday discourse. Walton (1996) identifies twenty-five argumentation schemes for presumptive reasoning. Matching each argumentation scheme, a set of critical questions is given. The two elements together, the argumentation scheme and the matching critical questions, are used to evaluate a given argument in a particular case, in relation to a context of dialogue in which the argument occurred. An argument used in a given case is evaluated by judging the weight of evidence on both sides at the point in the case where the argument was used. If all the premises are supported by some weight of evidence, then that weight of acceptability is shifted toward the conclusion, subject to rebuttal by the asking of appropriate critical questions.

In deductive logic, we are used to working with forms of argument. Deductively valid forms of argument like *modus ponens* and disjunctive syllogism are used as formal structures to analyze and evaluate arguments. In a comparable way, inductive forms of argumentation of various sorts can be used to model probabilistic argumentation. The new tool we need is a set of argumentation schemes (forms of argument) that can be used to model many kinds of plausibilistic (abductive, presumptive, defeasible) types of argument. But the modeling with respect to this third type of argument needs to be pragmatic in nature. The argument serves a probative function whereby probative weight is transferred from the premises to the conclusion. Probative weight is defeasible. Its function is to tilt a balance of considerations on an ultimate issue in a dialogue to one side or the other. An argumentation scheme is evaluated in a given case by light of appropriate critical questions in a dialogue. This approach sounds new and unusual to those traditionalists trained in deductive logic. But its utility is immediately apparent from a computational viewpoint, especially in relation to recent work on defeasible reasoning in AI.

The list of presumptive argumentation schemes given by Walton (1996) is not complete, but it identifies many of the most common forms of defeasible argumentation that should be the focus of research. Perelman and Olbrechts-Tyteca (1958) identified many distinctive kinds of arguments used to convince a respondent on a provisional basis. Arthur Hastings' Ph.D. thesis (1963) presented an even more systematic taxonomy by listing some of these schemes, along with useful examples of them. Recently, Kienpointner (1992) has produced an even more comprehensive outline of many argumentation schemes, stressing deductive and inductive forms. Among the presumptive argumentation schemes presented and analyzed by Walton (1996) are such familiar types of argumentation as argument from sign, argument from example, argument from commitment, argument from position to know, argument from expert opinion, argument from analogy, argument from precedent, argument from gradualism, and the slippery slope argument. Helpful examples of each type of argumentation are given and discussed. In other recent writings on argumentation (e.g., van Eemeren and Grootendorst, 1992), there is a good deal of stress laid on how important argumentation schemes are in any attempt to evaluate common arguments in everyday reasoning as correct or fallacious, acceptable or questionable. The existing formulations of the argumentation schemes are not very precise or systematic, perhaps because they have arisen out of practical concerns in dealing with real cases. New work is needed to refine, classify, and formalize these schemes. To provide a more detailed introduction to what argumentation schemes are and how they work, two examples will be presented. One is called argument from position to know, and the other is called argument from expert opinion.

2. ARGUMENT FROM POSITION TO KNOW
AND EXPERT OPINION

The argument from position to know is a type of argument based on the presumption by a proponent that a respondent is privy to some information or knowledge that can be extracted from him (her, it) by questioning. The classic example is the dialogue in which someone lost in a foreign city asks a stranger where the Central Station (or some other building or institution) is located. The questioner presumes, perhaps wrongly, that the person queried is familiar with the town. The following form of argument from position to know is that given by Walton (1996, p. 61).

Argument from Position to Know

Major Premise: Source *a* is in a position to know about things in a certain subject domain *S* containing proposition *A*.
Minor Premise: *a* asserts that *A* (in domain *S*) is true (false).
Conclusion: *A* is true (false).

Matching the argument from position to know, as indicated by Walton (1996), are the following three critical questions.

CQ1: Is *a* in a position to know whether *A* is true (false)?
CQ2: Is *a* an honest (trustworthy, reliable) source?
CQ3: Did *a* assert that *A* is true (false)?

As indicated earlier, argument from position to know shifts a probative weight from the premises to the conclusion, thus tilting the balance of considerations in a dialogue toward one side. But this outcome is only tentative, depending on what happens next in the dialogue. If an appropriate critical question is posed by the respondent, the probative weight shifts the balance of considerations to the other side. Only if the question is answered satisfactorily is the probative weight shifted back again.

Argument from expert opinion is a subtype of the more general argumentation scheme for argument from position to know, of a kind often used in an information-seeking dialogue. The special kind of information seeking in appeal to expert opinion arises from a situation where one party to the dialogue has information that the other lacks. The one party is an expert. The other is not. The expert has knowledge that the non expert wants to use in order to determine how to proceed with a problem or choice of actions. Argument from expert opinion is represented by the following argumentation scheme in the analysis given by Walton (1997, p. 210).

Argument from Expert Opinion

Major Premise: Source E is an expert in subject domain S containing proposition A.

Minor Premise: E asserts that proposition A (in domain S) is true (false).

Conclusion: A may plausibly be taken to be true (false).

As Walton (1997) makes clear, appeal to expert opinion should, in most typical cases at any rate, be seen as a defeasible form of argument. It is rarely wise to treat an expert as omniscient. However, there is quite a natural tendency to respect experts and to defer to them. Thus, for most of us, it is not easy to question the opinion of an expert. It verges on the impolite, and is best done in a careful way. But experts are often wrong, for many reasons. As a practical matter – for example, in matters of health and finance – you can do much better if you are prepared to critically question the advice of an expert in the right way. Thus, in principle, appeal to expert opinion as a form of argument is best seen as defeasible and as open to critical questioning. Appeal to expert opinion is a fallible form of argument that often carries probative weight. But, as the logic textbooks have rightly emphasized in the past, there is a tendency to defer to experts, sometimes too easily, and this tendency gives rise to fallacious appeals to expert opinion. Pressing ahead too aggressively, or "browbeating" the respondent, is associated with many cases of fallacious appeal to expert opinion. The sophistical tactic often used is for the proponent to try to trade on the respondent's respect for expert opinion by suppressing the respondent's legitimate critical questions in the dialogue.

Here are the six basic critical questions matching the appeal to expert opinion, as indicated by Walton (1997, p. 223):

1. *Expertise Question*: How credible is *E* as an expert source?
2. *Field Question*: Is *E* an expert in the field that *A* is in?
3. *Opinion Question*: What did *E* assert that implies *A*?
4. *Trustworthiness Question*: Is *E* personally reliable as a source?
5. *Consistency Question*: Is *A* consistent with what other experts assert?
6. *Backup Evidence Question*: Is *E*'s assertion based on evidence?

Someone attempting to analyze or critically evaluate a given argument can use the two devices of the argumentation scheme and the matching critical questions. The scheme identifies the form of the argument and its premises. Once the premises have been identified, they can then be questioned to see if there is support for them. The critical questions indicate other ways in which the argument can be questioned or criticized by indicating key assumptions that the worth of the argument depends on. The asking of a critical question throws doubt on the structural link between the premises and the conclusion.

The idea behind using critical questions to evaluate appeals to expert opinion is dialectical. The assumption is that the issue to be settled by argumentation in a dialogue hangs on a balance of considerations. Appeal to expert opinion can carry a small weight of presumption in the dialogue, even if by itself it is only a weak argument. If the given argument meets the requirements of the argumentation scheme and the premises are plausible (carry some weight as presumptions), that can give the conclusion some weight as a plausible assumption to go ahead with. But suppose the respondent asks one of the appropriate critical questions just indicated. The burden of proof shifts back to the proponent's side, defeating the argument temporarily until the critical question has been answered successfully.

3. CRITICAL QUESTIONS

One of the features of argumentation schemes that is key to evaluating whether an argument fitting a scheme should be judged strong or weak is the list of associated critical questions – questions that can be

asked (or assumptions that are held) by which a nondeductive argument based on a scheme might be judged to be (or presented as
being) good or fallacious. The critical questions form a vital part of
the definition of a scheme, and are one of the benefits of adopting a
scheme-based approach. One crucial aspect, then, of developing applications of argumentation schemes – computational or otherwise – is to
capture these critical questions in an appropriate way. The pattern of
most argumentation schemes is similar to *modus ponens*, typically with
something defeasible acting as the major premise (Walton, 2002b). In
a standard diagramming approach, a *modus ponens* argument would
be analysed as having a single conclusion supported by two linked
premises. That the premises are linked rather than convergent might
be demonstrated using Freeman's (1991) approach of considering a
reconstructed dialogue in which an imaginary interlocutor asks a specific
question after the presentation of the minor premise. So, in the following
example,

> (Ex1) *This computer does have an accumulator. It is built on the von Neumann
> architecture.*

the interlocutor might ask, "Why is that [the second sentence] relevant?,"
eliciting the major premise that *having a von Neumann architecture implies
having an accumulator.* (In a real dialogue, of course, a question such as
"Why is that relevant?" is more likely to elicit, through Gricean maxims,
not just the linked major premise, but also further arguments in support
of that premise.) It is Freeman's question of relevance, rather than of
ground adequacy ("Can you give me another reason?"), that moves the
argument from one premise to the other in this case, suggesting that
they are linked.

Viewing *modus ponens* as an example of a linked argument structure, it
is possible to see argumentation schemes in the same way: a conclusion
supported by two, or sometimes more, linked premises. Crucially, many
of these premises are often left implicit. In a *modus ponens* argument, it
is the usual practice to leave implicit the major premise – so usual, in
fact, that the enthymematic form has been analyzed as a separate argument form entirely, the *modus brevis* (Sadock, 1977). Including the major
premise of a *modus ponens* usually leads to hopelessly cumbersome text –
though in certain extreme situations (high levels of audience skepti-
cism or cognitive load), it may be appropriate to make all three components of the argument explicit (Reed, 1999). In argumentation schemes,

however, it is not just the (defeasible) major premises that are left implicit. In many cases there is a range of assumptions, all of which can be seen as acting as implicit linked premises. For example, recall the scheme capturing *argument from position to know* introduced earlier:

(P1) *a* is in a position to know whether *A* is true (false).

(P2) *a* asserts that *A* is true (false).

(C) Therefore, *A* is true (false).

(CQ1) Is *a* in a position to know whether *A* is true (false)?

(CQ2) Is *a* an honest (trustworthy, reliable) source?

(CQ3) Did *a* assert that *A* is true (false)?

In a canonical use of this scheme, the second premise, P2, is asserted explicitly, as is the conclusion. Premise P1 is left implicit (and, as Walton points out, is probably assumed by the hearer by Grice's Principle of Charity, by which an assumption of honesty and relevance is made). The argument thus has its conclusion C, supported by the two linked premises P1 and P2 (if either premise fails, then the argument falls down, just as with the minor and major premises of a *modus ponens*).

In addition, however, the propositional content of the critical questions forms necessary assumptions if the argument scheme is to successfully carry the burden of proof. Thus, for an argument from position to know to be successful, it is necessary for an audience to accept that *a* is honest – in addition to the premises P1 and P2. Furthermore, this additional premise is also linked: if *a* were believed to be dishonest, then the entire argument would fall down.

The complete set of linked premises employed in a scheme is thus the union of those given as premises and (the propositional content of) those listed as critical questions. In the current fluidity of active work on argumentation schemes, the distinction between premises and critical questions may be unclear. For example, in argument from position to know, CQ1 and P1 are closely related and might be characterized as a single premise; CQ2 is clearly distinct and forms a second premise; but although P2 and CQ3 are similar, it might be argued that the critical question is subtly different, aiming for the very words that *a* spoke, as opposed to a paraphrasing or interpretation. It is not the aim here to resolve such potential disputes, but rather to show that any particular interpretation of a list of premises and critical questions can be adequately characterised as a set of linked premises, many of which may be left implicit in an actual text.

4. ENTHYMEMES, SCHEMES, AND CRITICAL QUESTIONS

The term 'enthymeme' is standardly used in logic to refer to an argument in which one or more statements that are part of the argument
are not explicitly stated. Enthymemes are sometimes loosely referred to
as arguments with "missing premises," but sometimes the missing statement is the conclusion. There are many problems with enthymemes that
make the notion a difficult one to capture by means of some mechanical
process. Attributing unstated assumptions to an arguer is a perilous kind
of inference to draw, for it depends on interpreting what the arguer presumably meant to say. Any argument expressed in a natural language text
of discourse is notoriously difficult to interpret. First of all, vagueness and
ambiguity are common. But even worse, arguers sometimes achieve plausible deniability by exploiting innuendo and concealed meaning. When
a meaning is attributed to him, the arguer may deny it, even alleging
that the other party has committed the straw man fallacy. This fallacy is
the tactic of exaggerating or distorting an opponent's argument to make
it more vulnerable to refutation (Scriven, 1976, pp. 85–86). One might
think that the problem of enthymemes could be solved by attributing
arguments to someone else only if the argument comes out as deductively valid. But here an even worse problem lurks (Burke, 1985; Gough
and Tindale, 1985; Hitchcock, 1985). Making the argument valid may
not represent what the arguer really meant to say. Maybe the argument
he intended to put forward is invalid. At any rate, it is not too hard
to appreciate that the problem of enthymemes is far from trivial, and
that it would be extremely difficult to find some algorithm that could
mechanically plug in the right missing statements.

Parenthetically, it might be noted that even the term 'enthymeme'
itself seems to be a historical misnomer. Burnyeat (1994) has examined
the textual evidence of Aristotle's manuscripts and the early commentators on them. In the *Prior Analytics* (70a10), Aristotle wrote that an
enthymeme is an incomplete (*ateles*) *sullogismos* from plausibilities or
signs. But Burnyeat has cast doubt on whether Aristotle wrote the word
ateles in the original manuscript. It seems more likely that it was inserted
by one of the earliest commentators and then kept in. According to
Burnyeat's analysis, what Aristotle really meant by 'enthymeme' is a
plausibilistic argument of the kind he treated in the *Topics* and the
Rhetoric. Such an argument is syllogistic in appearance, but based on
a warrant that is defeasible, or true only "for the most part" (to use
Burnyeat's translation of Aristotle's phrase). If Burnyeat's interpretation

is right, the outcome is significant for argumentation theory. It means that 'enthymeme', in the original Aristotelian meaning, refers to presumptive argumentation schemes, not to incomplete arguments.

Now let's go on to discuss the general question of how the critical questions are related to missing premises. To pose this question more effectively, we need to consider a reformulation of appeal to expert opinion as an argumentation scheme. Various ways of setting up the schemes by formulating explicit premises can be considered. The scheme representing appeal to expert opinion as a form of argument quoted earlier, from Walton (1997, p. 210), could be called version I. In an alternate version, a conditional premise that links the major to the minor premise has been added.

Appeal to Expert Opinion (Version II)

Major Premise: Source *E* is an expert in subject domain *S* containing proposition *A*.
Minor Premise: *E* asserts that proposition *A* (in domain *S*) is true (false).
Conditional Premise: If source *E* is an expert in a subject domain *S* containing proposition *A*, and *E* asserts that proposition *A* is true (false), then *A* may plausibly be taken to be true (false).
Conclusion: *A* may plausibly be taken to be true (false).

Version II has taken the old argumentation scheme and added a premise that expresses the Toulmin warrant that gives the argument its backing. What version II reveals is that the argument has a *modus ponens* structure as an inference. But it is not a deductively valid *modus ponens* argument. It has the form we could call defeasible *modus ponens*. For example, in a given case, an argument having the form of version II could throw weight on the conclusion that a proposition *A* is plausible. But then it might be pointed out that *E* is not a credible expert, for some reason. This information would defeat the appeal to expert opinion, undermining the previous grounds for accepting *A*.

Now the question arises whether version II could be made even more explicit. Could it be done by building the critical questions into the argumentation scheme? According to this proposal, the new scheme would have the following form.

Appeal to Expert Opinion (Version III)

Major Premise: Source *E* is an expert in subject domain *S* containing proposition *A*.

Minor Premise: *E* asserts that proposition *A* (in domain *S*) is true (false).

Conditional Premise: If source *E* is an expert in a subject domain *S* containing proposition *A*, and *E* asserts that proposition *A* is true (false), and *E* is credible as an expert source, and *E* is an expert in the field *A* is in, and *E* asserted *A*, or a statement that implies *A*, and *E* is personally reliable as a source, and *A* is consistent with what other experts assert, and *E*'s assertion is based on evidence, then *A* may plausibly be taken to be true (false).

Conclusion: *A* may plausibly be taken to be true (false).

Version III makes the conditional premise seem cumbersome and hard to remember. Another way to accomplish the same result would be to add the content of each of the critical questions as a separate premise. This yields version IV.

Appeal to Expert Opinion (Version IV)

Major Premise: Source *E* is an expert in subject domain *S* containing proposition *A*.

Minor Premise: *E* asserts that proposition *A* (in domain *S*) is true (false).

Conditional Premise: If source *E* is an expert in a subject domain *S* containing proposition *A*, and *E* asserts that proposition *A* is true (false), then *A* may plausibly be taken to be true (false).

Expertise Premise: *E* is credible as an expert source.

Field Premise: *E* is an expert in the field that *A* is in.

Opinion Premise: *E* asserted *A*, or made a statement that implies *A*.

Trustworthiness Premise: *E* is personally reliable as a source.

Consistency Premise: *A* is consistent with what other experts assert.

Backup Evidence Premise: *E*'s assertion is based on evidence.

Conclusion: *A* may plausibly be taken to be true (false).

In version IV, all the critical questions have been built in as premises. Now the argumentation scheme is complete by itself, and we don't need the device of critical questions any longer, or so it would seem.

Technically speaking, version III or version IV would work as well as version II, with accompanying critical questions, as a format for analyzing and evaluating appeals to expert opinion as arguments. It doesn't really matter that much which version you use. The advantage of version II is that it strikes a nice balance. It shows you what you basically need as the core of the appeal to expert opinion. It indicates to a user what essential elements give this form of argument the weight that it can carry to command rational assent in a case by shifting a presumption from one side of a dialogue to the other. But then the critical questions offer the user (interlocutor, analyst, evaluator, student) a choice among strategies for probing the weak points in such an argument. Like a traditional topic, they function as a memory device. We tend to defer to an expert, and may be hard-pressed to think of the right question to ask. To open the discussion up, a user can cast around among the list of standard critical questions and find one that best expresses his doubts or his failure to make sense of what the expert has said. Thus version II is a good way in which to express the form of appeal to expert opinion.

Version II is also the most attractive option for diagramming. Having to include in a diagram all the implicit premises of version IV introduces unnecessary complexity, while diagramming the extra, convoluted warrant of version III fails to elucidate the structure of the argument. Instead, marking the general, typical structure with a scheme, and then allowing access to that scheme's critical questions during the analysis process, allows the analyst the flexibility to include the critical question premises where they are included in the original text, but to leave them out of the diagram where they are not required. The full set of critical questions is retained in the definition of the scheme to remind the analyst of the assumptions that are being made, and to aid in the process of evaluation. This is the approach adopted in Araucaria.

5. PEDAGOGY

One of the advantages of adopting argumentation schemes as a component of argument analysis is in teaching critical thinking skills. In the first place, the structure provided by a scheme narrows the options, and can serve as an aid to the student in identifying missing premises. Thus if a *circumstantial argument against the person* (Walton, 1996, p. 58) is a strong candidate for explaining a particular stretch of text, the analyst might look in more detail for the appropriate premises – that the speaker claims that everyone should act in a particular way, and so on. The analyst is then led to identify premises specified by the scheme but not present in the text – such as the premise usually left implicit in circumstantial ad hominem arguments, that if any speaker claims that something should hold for everyone, then it should hold for that speaker too.

The critical questions associated with each scheme have several roles to play. First, they aid in the initial identification – if critical questions are inappropriate in a given case, the chances are that the argument scheme too is inappropriate. Of course, the critical questions are also crucial in guiding an analyst toward an evaluation of the argument, supporting in some detail a critical approach to the argumentation presented. In more formal approaches, the only tools that a student has available are soundness and validity; critical questions offer rich contextual prompts to support the analysis process. The use of argumentation schemes, and their close relation to fallacies, also introduces another key advantage: flexibility. By accepting the fact that there may be a scale from good to bad argument, it becomes possible to equip learners with the flexibility necessary for handling real – rather than logic-textbook – arguments. The scale thus encompasses deductively strong arguments and downright fallacies (in the broadest sense, including examples such as affirming the consequent, which is not only 'bad' but also fails by most standards to appear 'good') – and as well as examples lying somewhere in between, where support might be 'substantial' or 'slight'.

Two further features of argumentation schemes in the context of pedagogy are the ease with which they can be integrated into traditional diagramming techniques and the possibility of supporting such diagramming with software tools. Every text has an argumentative purpose, that is, it aims at supporting a conclusion with the use of inferential passages proceeding from premises. Its logical structure can be represented as a tree, constituted by arguments, connected or convergent, dependent on assumptions or other statements, directed toward the end of the communicative move. For this reason, every text is, at the argumentative level, a hierarchical structure of links and dependencies, a complex map of relations. Diagrams are visualizations of this structure; they are means to represent and identify the specific role of each subsequence and each implicit premise. Diagramming, consequently, is an indispensable pedagogic tool: by inquiring into the fundaments and the function of every discursive step, the analyzer can reconstruct and understand the implicit basis of every statement, how it is connected with the rest of the move, and identify the inferential link that warrants the link itself. Argumentation schemes are related to the latter point of this analysis. They are the inferential patterns, the quasi-logical models used to support a conclusion. They motivate the connections, represent how the premises act as reasons, and assign to them a specific function. For these reasons, diagramming is an effective means to concretely apply topical models; on the other hand, argumentation schemes are indispensable to analyzing a text under the argumentative point of view, the purpose of diagrams themselves.

A diagram is constituted by a set of propositions (premises or conclusions) and a set of arrows representing the inferential steps. The technique of visualizing the reasoning structure of a text was used first in law by Wigmore (1931), in order to facilitate evidential assessment, and developed by David Schum (1994) with the introduction of generalizations, introduced to show the nature and defeasibility of the inference. The concept of generalization, used to analyze the reasoning, is based on Toulmin's model of argument, composed of premises, conclusions, and warrants. The warrant is the reason why a determinate conclusion can be drawn from the given starting points. In Walton's discussion (1996, Chapter 6), the warrant, or generalization, is examined in terms of schemes that can be traced back to basic patterns, differently instantiated in particular arguments. Inferential steps, for this reason, are organized, in diagrams, in a finite set of sequences connecting sets of points by argumentation schemes. Argument evaluation follows Rescher's rules of practical reasoning, distinguishing between linked and convergent arguments. While in the former the strength of the inference to the conclusion depends on the weaker premise, in the latter it is the strongest premise that influences the evidential strength of the inference.

While the importance of diagrams has been recognized in law, as we have seen, and in computer science, where they are used to structure evidential reasoning, their application to teaching philosophy is fairly recent (Rowe et al., 2006). The arguments a philosopher uses as persuasive means to prove a thesis using rational argumentation can be reconstructed, and the interpretation displayed on an argument diagram. Diagrams are instruments for textual analysis: by examining every sequence of reasoning, visualizing the relations between its points, and showing the inferential pattern and the implicit assumptions, the analyst can assess and judge the fundamental reasons and assumptions found in philosophical theories.

Diagrams are useful tools for both teachers and students. By means of visualizations, the principal steps of reasoning supporting a thesis may be represented in an organized structure, showing the position of every subargument. By recognizing inferential steps in a chain of reasoning, it is possible to reconstruct missing premises necessary to support the conclusion as reasonably following from the starting points. In this way, explaining the basic arguments and their critical points of weakness to students becomes much more effective. By means of applying such diagrams, the user can rapidly develop critical skills. In order to draw a diagram of a text, it is necessary, first of all, to identify premises, ultimate and final conclusions, and to connect them in linked or convergent arguments, according to their function. By developing the capability for dividing a text into argumentative sequences and organizing them, the user can improve his or her logical thinking skills. For these reasons, diagramming is a useful instrument in teaching, and an efficient means for developing the basic skills for the critical study of philosophical, literary, and political discourses, and arguments in every kind of communicative text.

6. INTRODUCING ARAUCARIA

Current collaborative research is under way at the University of Dundee to build a software tool that integrates traditional argument reconstruction and diagramming with the specification of argument schemes. A key feature of this software is that selecting schemes allows students to view critical questions and supply appropriate missing premises. In practical terms, the software is also designed for portability; it runs on Windows, Mac, and UNIX, and it saves arguments using XML, a common interlingua that is easily converted into web pages.

Araucaria is a software tool for supporting the process of constructing an argument diagram (Reed and Rowe, 2001). It is available free on the web at ⟨www.computing.dundee.ac.uk/staff/creed/araucaria⟩. It supports argumentation schemes, and it has an online repository of analyzed arguments. Once an argument has been analyzed, it can be saved in a format called AML (the Argument Markup Language) that can then be used for many purposes – for example, in a database or on a web page. An example is shown in Figure 1.1. Work is currently under way to provide web access to the online database of argument analyses independently of the Araucaria application. Araucaria has been designed for use by teachers and students in critical thinking courses, or in courses with a critical thinking aspect. But because it is a powerful tool in certain respects, its most important application may be to research problems in the field of argumentation.

Araucaria is similar to a software tool called Reason!Able devised by Tim van Gelder of the Department of Philosophy of the University of Melbourne, which has been well tested and is very simple and easy to use. Where Araucaria is aimed at argument analysis, for researchers and undergraduate teaching, Reason!Able is aimed at argument construction, for more introductory teaching earlier in the curriculum. The two thus complement each other.

Applying Araucaria to many basic problems of argumentation and informal logic has just begun. We will use some simple examples to discuss

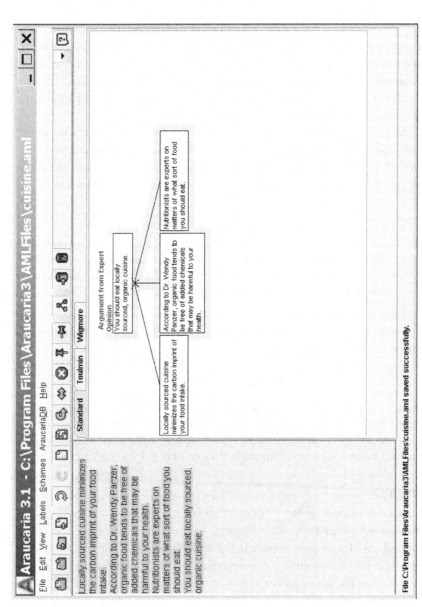

FIGURE 1.1. Araucaria main window.

25

some of the more basic points. In this discussion, we concentrate on our current joint research project that has the aim of developing a more sophisticated analysis, classification, and formalization of argumentation schemes. To begin, some introduction to schemes is presented. But to keep the discussion within reasonable limits, the scheme for appeal to expert opinion is taken as a case in point.

As an illustration of how this scheme is applied in Araucaria, we select two cases for analysis. Both are from the leading logic textbook (Hurley, 2000). Both are presented by Hurley (p. 139) as examples of the fallacy of "appeal to unqualified authority" or *argumentum ad verecundiam*.

The Bradshaw Example

Dr. Bradshaw, our family physician, has stated that the creation of muonic atoms of deuterium and tritium holds the key to producing a sustained nuclear fusion reaction at room temperature. In view of Dr. Bradshaw's expertise as a physician, we must conclude that this is indeed true.

The basic problem of fallaciousness in the Bradshaw example arises from the field critical question. As Hurley puts it, "The conclusion deals with nuclear physics, and the authority is a family physician" (p. 139).

The Tobacco Example

James W. Johnston, chairman of R. J. Reynolds Tobacco Company, tes- tified before Congress that tobacco is not an addictive substance and that smoking cigarettes does not produce any addiction. Therefore, we should believe him and conclude that smoking does not in fact lead to any addiction.

The basic problem of fallaciousness in the tobacco example arises from subquestion one of the trustworthiness critical question. Even if one should take him to be an authority, Johnson may be presumed to be biased. As Hurley puts it (p. 139), Johnston had a "clear motive to lie," for if he had admitted that tobacco is addictive, government regulations could have put his company out of business.

Let's consider how these two examples would be processed by Arau- caria, or indeed by any comparable system for argument analysis and diagramming. The two premises and the conclusion in the Bradshaw example can be highlighted, and the linked argument diagram can be constructed. If the argument were cleaned up a little before being inserted into Araucaria as text, it might come out something like this.

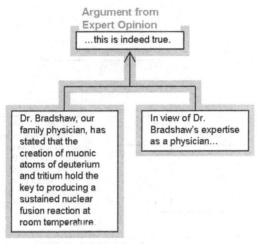

FIGURE 1.2. Argument diagram of the Bradshaw example.

Cleaned-Up Version of the Bradshaw Example

Dr. Bradshaw says that the muonic atoms are crucial to nuclear fusion (etc.).

Dr. Bradshaw is an expert in the field of medicine.

Therefore, (C) the claim that muonic atoms are crucial to nuclear fusion (etc.) may plausibly be taken to be true.

(In Figure 1.2, the shaded area around the three boxes shows the putative use of the argumentation scheme Appeal to Expert Opinion).

The problem is that the subject domain containing the proposition displayed in Figure 1.2 is not medicine, but physics. Therefore, this argument doesn't even get off the ground. The domain variable, S, in the major premise stands for medicine, while S in the minor premise stands for physics. The problem seems to be one of equivocation, or perhaps one of the argument not fitting the argumentation scheme at all (although it may superficially appear to, in the view of the uncritical thinker).

And yet there is another way of diagnosing the problem or fallacy in the argument in the Bradshaw example. If the field critical question is asked, the answer is "No, E is not an expert in the field that A is in." So here we seem to have a kind of duplication. The fault is diagnosed twice. Is this really necessary or desirable? Should the scheme and critical questions for Appeal to Expert Opinion be reformulated to eliminate this redundancy? That is the problem, anyhow.

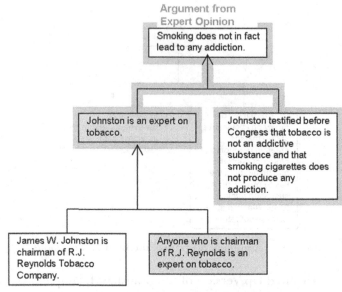

FIGURE 1.3. Argument diagram of the tobacco example.

Now consider the tobacco example. Like the previous one, this argument could perhaps be cleaned up a little in order to make it more visibly match the scheme.

Cleaned-Up Version of the Tobacco Example

(A) Johnston is chairman of R. J. Reynolds.
(B) Anyone who is chairman of R. J. Reynolds is an expert on tobacco.
(C) Johnston is an expert on tobacco.
(D) Johnston says that tobacco is not addictive (etc.)
Therefore, (E) "Tobacco is not addictive" may plausibly be taken to be true.

(In Figure 1.3, the scheme is again shown as a shaded outline; the greyed boxes indicate reconstructed claims that have been introduced during the analysis rather than being present in the original.)

One problem here is the slight dubiousness of the second premise, (B), as a generalization. This premise is true in certain respects, meaning that such a person is an expert on certain aspects of tobacco, like its manufacturing. But it is false in other respects, because such a person is not necessarily, or as far as we know, a medical or scientific expert on addiction or on the properties of addictive substances. But this is not the major problem with the argument as an *ad verecundiam*, judging by Hurley's diagnosis. The main problem is that the example triggers the bias sub-question of the trustworthiness critical question (to put it in our terms).

It should be mentioned that both these cases are relatively simple examples of the *ad verecundiam* fallacy taken from a logic textbook. In the textbook, they are used pedagogically to introduce students to the most simple or obvious kind of case that the students will see as falla-cious right away. In more complex examples, the mistake or blunder is not so obvious. And indeed, in the kind of case emphasized by Walton (1997), the proponent adopts a strategy of blocking progress in a dia-logue by trying to prevent the respondent (in advance) from raising the appropriate critical questions. One leading example in Walton's book is a case where the parents of a sick child are prevented from asking ques-tions about how to help their child by physicians who dismiss their claims as "anecdotal," suggesting that the parents do not really have a right to discuss questions of medical treatment with them. But this kind of case raises a problem identified by Jovicic (2002, p. 29). In the case she postu-lates, a proponent advances an argument from expert opinion using the appropriate argumentation scheme, and the premises are presumptively strong. But he is an arrogant person who blocks off the attempts of the nonspecialist audience to ask appropriate critical questions. Thus, by the Walton criterion, his argument commits the *ad verecundiam* fallacy. And yet, suppose that the argument, when presented to an audience of spe-cialists, who do not even need to ask these critical questions, is based on evidence in the field, making it presumptively strong. It may be too early to tell what the best solution to this problem is. But it does suggest that the argumentation scheme for appeal to expert opinion, even with the matching critical questions, may be only part of the answer when dealing with the *ad verecundiam* fallacy. Somehow the argumentation scheme, the critical questions, and the profile of dialogue (Krabbe, 1999) may all need to be taken into account in the big picture. The problems with the tobacco case and the Bradshaw case are just the beginning.

In the tobacco case, then, the argument has some problems fitting the form of the appeal to expert opinion. But once it gets past this snag, its underlying problem is deeper. So in the tobacco case, as contrasted with the Bradshaw case, the deeper problem that is the basis for judging the argument to be an *ad verecundiam* fallacy comes out only when the right critical question is asked. And even then, the precise diagnosis of the fault is pinpointed exactly only when the level of the critical subquestions is reached.

The problem, then, is one of finding a uniform method of balancing off the format of the argumentation scheme in relation to the critical questions so that the processes of argument analysis and evaluation are most user-friendly. Maybe a little bit of redundancy is OK, as long as all the bases are touched at least once.

7. PROBLEMS TO BE SOLVED

A general problem is how an argumentation scheme can have normative bite in a dialogue if the respondent can continue the dialogue by asking critical questions or by otherwise challenging the argument. If these arguments are defeasible, how can they ever be used to pin down a respondent's commitments? One tool that can be used to deal with this problem is the profile of dialogue (Krabbe, 1999). A profile of dialogue is a sequence of moves that represents only a small part of a longer sequence of dialogue. For example, it might represent a question, a reply to that question, and then a next move or two. Profiles are not just descriptive tools for identifying common patterns of moves in examples of argumentation. They can also be used in a normative way to represent how an ideal sequence of dialogue should go, or to diagnose faults, errors, or fallacies. The argumentation scheme for appeal to expert opinion, along with the set of matching critical questions, can easily be used to set up a normative profile for the typical kind of case in which appeal to expert opinion is used to support a claim. The first point in the profile will be an argument or question put forward by the respondent. The next point will be the appeal to expert opinion put forward by the proponent in reply to this move. At the next point, the respondent's set of allowed options can be represented by eight branches in a tree diagram. The respondent can (a) ask a critical question, (b) challenge one of the premises of the appeal to expert opinion, or (c) accept the conclusion of the argument as a commitment. Thus the profile of dialogue shows how the argument has normative bite when used in a dialogue.

Another problem concerns enthymemes. Can the critical questions be used, in addition to the argumentation scheme, to specify additional missing premises that can be added to a given argument? Because the critical questions are formulated in advance, it seems possible that they could be used as part of an automated device to pick out missing premises in enthymemes. But this problem leads back to the earlier one. It could be called the completeness problem for critical questions. Once the respondent has run through the list of critical questions matching a scheme, can he go on to ask even more specific critical questions raised by the previous answers? The problem is one of how argumentation schemes are binding on a respondent. Presumptive schemes are defeasible. They are not deductively valid. The question, then, is how long the process of asking critical questions can continue before the argument must finally be accepted as binding the respondent to acceptance of the conclusion, if he has accepted the premises.

As an example, let's go back to appeal to expert opinion, where the basic critical questions are known to have subquestions falling under each of them. For example, three critical subquestions have been cited (Walton, 1997, p. 217) as falling under the trustworthiness critical question.

Subquestions for the Trustworthiness Question

Subquestion 1: Is *E* biased?
Subquestion 2: Is *E* honest?
Subquestion 3: Is *E* conscientious?

Bias means failure to represent both sides of an issue in a balanced way. Bias is not always bad, because advocacy is sometimes quite appropriate in argumentation. Still, bias can be important in judging the worth of an argument based on appeal to expert opinion. Honesty means telling the truth, or whatever is perceived as being the truth of a matter. Conscientiousness means care in collecting sufficient information. Thus the subquestions just listed represent more specific ways in which the trustworthiness of an expert can be questioned.

Using this scheme, the completeness problem can be posed. Suppose the proponent has answered all of the six basic critical questions posed by the respondent in prior dialogue exchanges. Is the respondent obliged at that point to accept the appeal to expert opinion as reasonable? If he accepts the premises, is he now obliged to accept the conclusion as a commitment in the dialogue? Or can he carry on asking more specific critical subquestions? The danger is that the dialogue could go on indefinitely. What burden of proof is appropriate for the proponent? At what point can he stop the process and say that his appeal to expert opinion should now carry weight?

There are two issues that are combined by the foregoing considerations. One is the issue of the critical questions, and whether there should be some kind of burden of proof attached to asking them. The other is the issue of how arguments should be attacked or criticized generally.

This second issue is often phrased in terms of Pollock's (1987) distinctions between defeaters and undercutters, but this terminology can be a bit confusing itself. Let's begin with the idea that there are two ways to attack (criticize, refute) an argument. One is to use a counterargument. A counterargument is an argument with a conclusion that is the opposite (negation) of the original argument that was attacked by it. The other way is to attack the premises of the argument, either by questioning them or by arguing that one or more of them is false. This seems simple enough, but it applies only to deductive, or perhaps to inductive arguments. With defeasible arguments, the situation is more complex, because an opponent can attack the inference rule, the warrant or generalization the argument is based on, by citing an exception to the rule.

Thus, in general, a defeasible argument can be attacked in only three ways, by an attack on a premise, by a counterargument with an opposite conclusion, or by an argument attacking the inference rule. But some see the inference rule as really just acting as another premise. You can attack it, or you can attack any other premise of the argument. Thus, from this point of view, there are just two ways of attacking (and defeating) any argument. You can attack the premises or you can attack the argument by presenting a counterargument with the opposite conclusion. Let's call the latter form of attack a rebuttal.

Next there is the issue of where and how critical questions fit in as a form of attack on a defeasible argument, or something similar to an attack. One possible theory is that the critical questions represent additional premises that are additional assumptions of the argument at a deeper level. They are like unstated premises. This is all controversial, however. If the critical questions can be treated as implicit premises, that supposition has implications for any attempt to formally model argumentation. Another possible theory is that some critical questions function as implicit premises, while others function as starting points for finding rebuttals. The crucial difference is that the latter have a burden of proof attached for the questioner, while the former do not.

To take a hard look at one argumentation scheme to see how these two approaches will differ, let us return once again to the critical questions matching the appeal to expert opinion, and examine them individually.

1. *Expertise Question*: How credible (knowledgeable) is E as an expert source?
2. *Field Question*: Is E an expert in the field that A is in?

3. *Opinion Question*: What did *E* assert that implies *A*?
4. *Trustworthiness Question*: Is *E* personally reliable as a source – for example, is *E* biased?
5. *Consistency Question*: Is *A* consistent with what other experts assert?
6. *Backup Evidence Question*: Is *E*'s assertion based on evidence?

On the surface, it looks plausible that all these critical questions, except for (possibly) four and five, can be seen as implicit premises. Let's look at them one at a time. (1) When you put forward an appeal to expert opinion, you assume, as part of the argument, that the source is credible, or has knowledge in some field. (2) You assume that the expert is an expert in the field of the claim made. (3) You assume that the expert has said something, made some pronouncement, from which the claim can be extracted by inference or, in some cases, even by direct quoting. (6) You assume that the expert's assertion was based on some evidence within the field of his or her expertise. The argument doesn't make much sense, or hold up as a plausible appeal, without these assumptions being part of it. But four and five seem to be different. Consider four first. If the expert turns out to be biased, or to be dishonest, then if there is evidence for such claims, that evidence attacks the argument. The reason is that a finding of bias or dishonesty attacks the credibility of the source, potentially destroying the whole core of the argument from expert opinion. But to mount such an attack, the critic has to produce some fairly substantial evidence. Otherwise, the question is merely an innuendo. Next consider five. If the claim can be shown to be inconsistent with what other experts in the field say, then that is an argument against the claim. It is a rebuttal, especially if what the other experts say represents the generally accepted opinion in the field. But that needs to be shown by telling us what the other experts have in fact said, and by showing how these statements conflict with what our expert has said. So once again, asking this kind of critical question puts a burden of proof on the questioner.

The key difference is one of burden of proof. The trustworthiness and consistency critical questions seem to have a positive burden of proof attached to the side of the questioner. The other critical questions can just be asked out of the blue, so to speak. Once asked, this type of critical question must be given an appropriate answer, or the original argument falls down. With these critical questions, the burden of proof remains on the side of the proponent of the appeal to expert opinion. Merely asking the question makes the original argument default. Asking the trustworthiness or consistency critical questions is a harder task. If you want to get the original argument to default by asking the critical question, you have to back it up with reasons.

8. HOW ARE SCHEMES BINDING?

According to the standard approach to argument evaluation in informal logic, a given argument needs to be evaluated in light of three factors – acceptability of the premises, relevance of premises to the conclusion, and whether the premises provide a sufficient reason to accept the conclusion. Johnson and Blair (1994) use the term 'cogency' to define the standard for an argument to be evaluated as successful. According to their standard, an argument is cogent if and only if its premises are acceptable, relevant to the conclusion, and sufficient for acceptance of the conclusion. Using basically the same kind of criterion, Govier (2005, pp. 63–76) states that acceptability, relevance, and good grounds are the conditions of argument cogency. It is possible to add a fourth criterion, that there are no known better reasons for the opposite conclusion. However, it is also possible to subsume this fourth criterion under the third criterion of sufficiency. It all depends on how you define 'sufficiency'.

The reason that the notion of sufficiency is up for debate is that the main types of schemes important for the study of everyday argumentation are nonmonotonic. This means that the argument fitting the scheme always needs to be seen as subject to defeat as new premises are added by new information that becomes relevant. In such an argument, then, acceptance of the premises, along with the argument that links them to the conclusion, is not absolutely sufficient for the acceptance of the conclusion. It is only relatively sufficient, meaning that the conclusion needs to be only tentatively accepted, in light of revisions that may later be needed. In other words, such an argument is cogent only in the sense that it provides sufficient evidence to accept the conclusion at some particular point in an investigation that may later lead to the rejection of this conclusion. This defeasible notion of sufficiency incorporates an argument from lack of knowledge. In the absence of known evidence so far showing that the conclusion may not be acceptable, the argument provides sufficient grounds for provisionally accepting the conclusion.

It appears then that when you try to apply the three-part definition of a cogent argument to presumptive argumentation schemes, the notion of cogency needs to be redefined, or retailored to fit the reality of defeasible argumentation schemes. Some might argue that an argument based on a defeasible argumentation scheme is simply not cogent, because

acceptance of the premises is not sufficient for acceptance of the conclusion, at least not sufficient in any absolute sense. Others might argue
that the notion of cogency, and with it the notion of sufficiency, can be
stretched to fit the requirements of defeasible argumentation schemes.

However, when it is so stretched, there is a question of how such an
argument based on a defeasible argumentation scheme is binding. If
acceptance of the premises doesn't absolutely commit the respondent to
acceptance of the conclusion, so that he can now or later retract or reject
that conclusion, how is such an argument binding on the respondent?

One way to approach the problem of showing how presumptive argumentation schemes can be binding on a respondent is to utilize the
notion of burden of proof. According to this approach, when an argument fitting one of the schemes is put forward in a given case, it shifts the
burden of proof to the side of the respondent in a dialogue. One could
say that, by putting forward such an argument, the proponent seizes the
initiative, putting the respondent on the defensive until he either attacks
the argument successfully or provides some grounds for doubting it. This
approach could be called the shifting burden of proof theory of the binding nature of argumentation schemes. However, notice that the notions
of cogency and argument sufficiency have been stretched here to fit a dialogue format in which arguments can be put forward by one side as offering the other side reasons to accept a conclusion. There are two sides to
a dialogue. In such a dialogue, the other side might choose to accept the
conclusion, based on the argument just given, or not. Thus a cogent argument does not have to lead to the respondent's accepting the conclusion,
even though he might accept the premises, without having other avenues
open for continuing the dialogue. The respondent might attack the
argument, by putting forward counterarguments, or may simply express
doubts about the conclusion based on the evidence that he has.

Still, it would seem that a cogent presumptive argument that meets all
three criteria in some form or other (especially depending on what is
meant by the term 'sufficient') should put some pressure on the respondent to either accept the conclusion or give some good reason why he
is not prepared to accept it. Recall here that when we talk about cogent
arguments in this sense, we're invoking a normative notion of cogency.
We are talking about an argument that is successful, meaning that a
rational arguer should accept the conclusion, given that he accepts the
premises, and given that the other two requirements of cogency are met.
In this light, the shifting burden of proof theory need not imply that
a real person in a real case of argumentation has to, or should, accept

the conclusion of such an argument. The theory implies only that the shifting burden of proof requires that a respondent who does accept the premises of the argument, in a case where the other two requirements also hold, but does not accept the conclusion, is being illogical. This meaning of 'illogical' implies that he is not meeting the procedural rules of the dialogue that require certain normative standards of rational argumentation.

One approach to this bindingness problem can be defined by a warrant called a 'relation of conveyance' between the premises and the conclusion (Katzav and Reed, 2004, 2005). According to Katzav and Reed (2004, p. 5), premises represent conveying facts, whereas conclusions represent conveyed facts. Warrants represent the relationship between the conveying facts and the conveyed facts, and take the form of conditionals. So conceived, the scheme is an artifact that makes explicit the relation of conveyance represented by the warrant. This approach conceives of the binding nature of a scheme in terms of the relation of conveyance between the premises and the conclusion. But what exactly does 'conveyance' mean? It can easily be defined in relation to deductive validity and inductive strength, but as applied to presumptive schemes, it brings us back again to this notion of shifting burden of proof in a dialogue. A presumptive argumentation scheme imposes a relation of conveyance on the respondent such that if he accepts the premises, and if the scheme is applicable, and if all the requirements of the scheme are met, the conclusion is conveyed to him by these factors. That doesn't mean he has to accept it, or cannot now present reasons for doubting it; but it does mean that he has now been given a cogent reason for accepting it, and he has to deal with that. Otherwise, he is somehow being illogical or unreasonable, or failing to follow procedural requirements for engaging in rational argumentation.

The bottom line is that the binding nature of a presumptive scheme needs to be sought in the relationship between the scheme and the critical questions that match it. When a proponent presents a cogent presumptive argument that fits one of the schemes to the respondent, he needs to respond to it in some appropriate way. The obvious way is for him to either accept the argument or ask one of the set of appropriate critical questions. This way of approaching the problem is quite attractive, as it allows for defeasibility of presumptive schemes. Sufficiency doesn't mean that the respondent has to accept the conclusion of the argument, period. It means only that he needs to respond to it in an appropriate manner, which could include expressing doubts about it by asking appropriate critical questions. Such an argument is still cogent in that it performs a function of conveyance by putting pressure on the respondent to acknowledge the argument and respond to it in an appropriate way. However, although one appropriate way is simply to accept the conclusion, other ways are available. One is to question the premises. Another is to bring in new information that adds to the premises, leading to a different conclusion, or so the respondent may argue. Another is to bring forward a counterargument, an argument with a different conclusion from the one previously brought forward by the proponent. Still another option is the asking of critical questions other than those questioning the original premises put forward by the proponent. Such critical questions could express doubts about implicit presumptions that the argument was based on, or could cite exceptions to the general rule that is the warrant of the argument in virtue of which the premises are conveyed to the conclusion.

The problem of defining sufficiency rears up again, once we move on to accept the notion that a cogent argument fitting a presumptive scheme is binding only in the sense that it puts pressure on a respondent to behave in certain ways rather than forcing him to accept the conclusion of the argument. According to the new requirements for sufficiency of schemes proposed here, the respondent has a number of options. Were he to exercise these options, the dialogue could go on and on. Each of the questions he asks, the counterarguments he puts forward, or other legitimate kinds of moves he might make, could extend the dialogue. Such a chain of extended argumentation could be relevant. Another factor is that each of the critical questions corresponding to a scheme can have some subquestions (Walton, 1997). Presumably, the respondent has the right to ask these subquestions, if he so chooses, rather than immediately accepting the conclusion of the proponent's argument. Here the notion of sufficiency seems to be stretched very thin, because it would seem that the respondent has many ways out, and could continue the dialogue indefinitely, without ever accepting the conclusion of the proponent's cogent argument. It could even seem that there is the possibility of an infinite regress, except that the dialogue will eventually be closed off at the concluding stage. But it could be that, even at that stage, the respondent still has not accepted the conclusion of the proponent's argument. It may seem, in the face of such a possibility, that the notion of sufficiency has been watered down to the point where it means very little. Maybe the respondent can go on and on arguing and asking critical questions, and in the end never be forced to accept the conclusion of the proponent's argument even though all the requirements of cogency have been met (possibly except for sufficiency, whatever that means, since it is now at issue). This question of whether there should be a rule that puts an end to the process of critical questioning has been called the completeness problem for presumptive argumentation schemes (Walton, 2001, p. 160; Walton and Reed, 2003, p. 203).

9. DIRECTIONS FOR AI

The importance and potential utility of argumentation schemes within AI and related areas has been briefly explored, but perhaps the starkest conclusion is the dearth of research upon which AI can build. Even after the work of Hastings, Kienpointner, Walton and others, argumentation schemes are still poorly understood, with many issues remaining to be addressed. Researchers in AI are eager to get to work building upon the results of argumentation theory. So, as a contribution towards stimulating work in the area, this section presents a four-point list of desiderata of issues that AI needs to see addressed. The list is not intended to be exhaustive; rather, it focuses on key issues motivated by the areas discussed earlier.

First, although the work presented here, and the research upon which it builds, has attempted to elucidate somewhat the internal structure of argumentation schemes, a fuller and more principled solution is still required. The critical questions matching a scheme can be seen as representing additional relevant factors that might cause an argument to default. A question then arises: Could the critical questions be reformulated as additional premises in the argumentation scheme itself? Are all the premises of a scheme always linked? Or are critical questions nothing more than rephrasings of premises, or annotations for the student? The relationship of critical questions to missing premises that might be implicit in an argument is a problem that needs to be explored.

Second, the problem of relevance is as substantial and pressing in philosophy as it is in AI. To what extent do argumentation schemes structure or contribute to a definition of relevance that is computationally tractable? Third, how are schemes organized? Many relationships between schemes seem to suggest a taxonomic hierarchy, but how rigid is this hierarchy, and how are properties of schemes lower down in the hierarchy inherited from those higher up? Along how many dimensions does the hierarchy run? Finally, perhaps the single most important task – which has as prerequisites at least the three just mentioned – is to pin down definitions of argumentation schemes. The characterization needs to be formal enough to support computational implementation, while retaining the unique advantage possessed by schemes of having the flexibility requisite for untidy, unsanitized real-world argumentation.

The formulation of the set of presumptive schemes by Walton (1996) was rough and ready. The variables and constants used in the schemes are quite a varied bunch, and have not all been incorporated into any single overarching formal structure. Only the most rudimentary attempt was made to classify the schemes by a tree structure exhibiting how some fall under others. In many cases, the organization of the premises of the scheme and the matching critical questions was obviously clumsy. In some instances, for example, it seemed that the critical question merely asked whether one of the premises was true or acceptable. Thus it looked as if either the premise or the critical question was redundant. These same problems were perhaps even more evident in Hastings' (1963) initial attempt to introduce a comprehensive set of schemes with matching critical questions.

Now that we have a new software system for argumentation diagramming that can accommodate argumentation schemes, many of these technical issues of how to clean up the schemes appear more pressing. Previously, they may have seemed relatively minor matters of detail to the working argumentation theorist or teacher of critical thinking. But now they demand our attention. In this presentation, some of the most elementary of these technical questions of formalization of schemes are raised. As a means of arranging these questions, let us lay out our aims as desiderata for a theory of argumentation schemes. Such a theory should be

- *rich* and sufficiently exhaustive to cover a large proportion of naturally occurring argument.
- *simple*, so that it can be taught in the classroom and applied by students.
- *fine-grained*, so that it can be usefully employed as both a normative and an evaluative system.
- *rigorous*, and fully specified, so that it might be represented in a computational language such as XML.
- *clear*, so that it can be integrated into traditional diagramming technique.

This is a challenging list to tackle, not least because some of these desiderata are at odds with one another: the more fine-grained our theory is, for example, the less likely it is to be at all exhaustive. Similarly, rigorous specification is crucial for computational representation, but a significant barrier for application to the real world. Happily, some of the aims do hang together: simplicity, for example, works to support not only computer representation, but also diagramming and classroom teaching. This, then, is where we are headed.

10. WHERE WE GO FROM HERE

In this chapter we have introduced some of the basic tools needed to analyze schemes, like argument diagramming, and have indicated in a general way how schemes have become important in computing, particularly in multi-agent computing of the kind currently used for Internet communication. Now that we have an idea of what schemes are, and how they can be used in different areas of application, we can turn to the study of the individual argumentation schemes themselves. The first scheme we choose for consideration is argument from analogy. Of all the defeasible schemes, it represents the form of argument that has gained the most attention. There is currently a huge literature on argument from analogy, not only in argumentation studies, but also in linguistics, law, cognitive science, and artificial intelligence. This literature seems so huge that it is intimidating to try to say anything new about argument from analogy. In addition, when one tries to formulate an argumentation scheme for argument from analogy, the project is initially made extremely difficult by the fundamental disagreements that deadlock the literature on how to approach this type of argument. There is a widely held view that it should be seen as a form of deductive reasoning, and many of those more pragmatic commentators who do not see it as deductive see it as so inherently fuzzy and unamenable to precise analysis that they place it in the category of the mystical.

Despite these apparent obstacles, we will begin with argument from analogy, for several reasons. One is that it is very easy for readers to appreciate how this form of argument works by comparing one case or situation to another. Another is that it is easy for readers to see how fundamentally important this type of argument is. For example, it is fundamental to our whole Anglo-American justice system, which is based on the principle of taking a court ruling as a precedent and then judging the case at trial by how similar the new case is to the previously decided one. This process of case-based reasoning by precedent has argument from analogy at its central core. Basically, the system works by comparing one case to another, and then judging the new case in relation to the outcome of the prior comparable case.

Arguments from analogy are so common and familiar that we are very skilled at drawing conclusions based on analogies we perceive. Probing more deeply into the logical structure of arguments from analogy, however, poses problems. Although argument from analogy has traditionally been included in logic textbooks as an informal fallacy, the paradigm shift produces the problem of diagnosing the precise differences between the reasonable cases and the fallacious ones. How can we begin by defining what similarity is? Research on analogy in artificial intelligence has shown that judgment of similarity depends on a perception of an event, case, or image that is formed from uncountably many small details. Such details of an image comprise not only its shape but also the lines, colors, and textures that make it up. If two people compare two photographs of a common scene, there are indefinitely many ways in which the two images could be judged to be similar to each other or different from each other. The ability to recognize patterns and draw analogies based on them is a fundamental skill of animal as well as human reasoning. Ant colonies recognize patterns of food distribution in order to develop routines for foraging and allocating collective tasks (Mitchell, 2002). Our immune system even perceives patterns related to the shapes of viruses or organisms that invade the system, and bases its response on perceptions of similarity to previous infections (Mitchell, 2001). Thus it is not surprising that argument from analogy has been an intense subject of research in artificial intelligence in recent years.

What is called the "see" function (Wooldridge, 2002, p. 34) represents the agent's ability to bring in perceptions from its environment. Multi-agent reasoning commonly takes place in a situation where an agent's drawing a conclusion is based on its perception that one situation is similar to another. A problem is that the two situations might represent two different cases, even though one is similar to the other. Therefore, although they are similar in many respects, inevitably they are also different in many respects. Any conclusion is therefore drawn by a defeasible inference based on an initial perception of a situation that may need to be updated and corrected as new information comes in. In such cases, the fallacy may not be the initial leap to the conclusion, but rather the refusal to update or correct this reasoning as new evidence comes in.

Prakken (2005) has shown that artificial intelligence and law is one area of research that stands to benefit considerably from adopting an approach that takes argumentation schemes in legal reasoning into account. He has shown that the more traditional approach to legal logic, based on deductive logic and models of inductive reasoning, needs to be supplemented with an approach using argumentation schemes. He has identified certain schemes in particular representing legal forms of argument of a common and centrally important kind. The schemes he cites include argument from position to know, argument from consequences, argument from appearance, and schemes for rule application, including precedent-based reasoning and argument from analogy. Each of these schemes will be treated later in this book, and they can all be found in the compendium (Chapter 9). We agree with Prakken that legal reasoning is an excellent testing ground for applying argumentation schemes, and in Chapter 2 we take the case of argument from analogy as a paradigm that can be used to show the promise of the schemes approach as a research program for studying AI and law.

2

Argumentation Schemes for Arguments from Analogy, Classification, and Precedent

Argument from analogy is one of the fundamental forms of argumentation on which many other forms of argument – argument from precedent in law, for example – are based (Gordon, 1995; Weinreb, 2005). Our system of Anglo-American law is based on *ratio decidendi*, the principle that if a case has been decided by a court in a certain way, then a new case that is similar to it should be decided in the same way. Indeed, argument from analogy is the foundation of all case-based reasoning (CBR) in which the argumentation turns on a comparison of one case to another (Ashley and Rissland, 2003). CBR not only compares one case to another as similar, but also compares cases as more similar to others with respect to a given case, depending on the description of the problem posed in the given case. Thus argument from analogy is an extremely important and fundamental species of argumentation. So much has been written on it, in so many fields, including philosophy, cognitive science, artificial intelligence, linguistics, psychology, law, and computing, that we can barely scratch the surface here. Our more limited aim is to develop tools that can be used to identify the precise form of arguments from analogy, allowing us to better understand its close relationships with other important schemes, especially those representing argument from verbal classification and argument from precedent.

Those who have written on the subject of argument from analogy in law tend to fall into two camps, "skeptics" and "mystics," both of which leave the justification of argument from analogy as a rational form of argument twisting in the wind. The skeptics think either that it is a form of deductive reasoning, or that it does not represent any kind of form of rational argument at all (Schauer, 1995, p. 187; Brewer, 1996).[1] The mystics think that reasoning by analogy from precedents can be justified as having procedural benefits, even though it is sometimes wrong, and that it lacks rational force even when it is right (Sherwin, 1999). Both views assume that reasoning from analogy does not have "rational force" as a distinctive form of argument used to justify inferring a conclusion from a set of premises, apart from criteria of deductive or inductive reasoning, or purely pragmatic criteria, which can be applied to it in a piecemeal way. In this chapter, we will challenge this central assumption of both sides by taking positive steps to map out an argumentation scheme that represents a distinctive logical form of reasoning from analogy. This scheme, along with others related to it, is used to show how arguments from analogy in legal reasoning can be analyzed and evaluated in a way different from conventional methods used to evaluate deductive or inductive reasoning.

We consider a classical case from the legal literature on argument from analogy concerning the issue of whether a police officer's use of a drug-sniffing dog constitutes a search for legal purposes (Brewer, 1996). Applying recently developed tools in argumentation and artificial intelligence, we show that the legal reasoning in this case can be identified, analyzed, and evaluated using argumentation schemes. We show not only that argument from analogy is involved, but also that several closely related argumentation schemes, like those for argument from classification, are combined with argument from analogy in interesting ways. We consider how sets of critical questions matching each of these schemes can be used as devices to aid in the evaluation of typical legal arguments like those in the case of the drug-sniffing dog. We show that artificial intelligence systems based on argument from analogy, of a kind fundamentally important in the analysis of legal reasoning (Chorley and Bench-Capon, 2004), reveal that there are two sides to a case, each of which sets up a chain of defeasible reasoning, leading to a conclusion that represents its claim or hypothesis in the case. We survey standard treatments of argument from analogy in logic, and show how these treatments help to bring out key factors used to specify the form of argument from analogy and related argument forms. Most importantly, we show how typical instances of argument from analogy in legal reasoning are

---

[1] In a survey of work on argument from analogy, Juthe (2005, p. 1) points out that philosophers tend to think that arguments from analogy must be reformulated into deductive or inductive arguments in order to be valid arguments because their method of arguing is to find counterexamples to a universal premise.

based on argument from classification, which is in turn based on other argumentation schemes like the scheme for argument from an established rule and the scheme for argument from precedent. We suggest that a consideration of how all these schemes fit together and apply to typical cases of legal reasoning, like that of the drug-sniffing dog, should suggest the need for a new approach to the role of argument from analogy in analyzing the structure of such arguments. Finally, we propose a hypothesis about how argument from analogy works as a fallacy, and pose some unsolved problems about evaluating arguments from analogy.

1. THE CASE OF THE DRUG-SNIFFING DOG

Brewer (1996) describes the process of analogical reasoning as having three stages. At the first stage, the analogical reasoner is uncertain about how to classify something with respect to a matter in question. He offers the following question as an example. If a trained dog sniffs luggage left in a public place and signals to the police that it contains drugs, should this event be classified as a search according to the Fourth Amendment? This decision about how to classify the event has legal consequences. If it can be classified as a search, information obtained as a result of the dog's sniffing the luggage is not admissible as evidence in a case against the owner of the luggage. If the sniffing is not classified as a search, the information so obtained is admissible as evidence. On Brewer's analysis, this first classificatory stage of reasoning by analogy leads to a later evaluative stage in which the given event is compared to other cases that have already been classified legally as being searches or as not being searches.

How argument from classification works in legal reasoning has been illustrated by the classic illustration of Hart (1957–58) concerning a rule that vehicles are not allowed in a public park. According to this rule, clearly a car is a vehicle, whereas a bicycle is not classified as a vehicle. We then come to the undecided question of whether a skateboard can be classified as a vehicle. This decision, of course, involves analogy to some extent, for it has to be considered whether a skateboard is more like a car or more like a bicycle, in respect to the question of whether it is a vehicle. In this example, various kinds of arguments are involved. The type of argument initially is one of classification, but it later gives rise to an argument from analogy. Argument from precedent is also involved. Suppose it has already been decided in court that a bicycle is not a vehicle. That decision can later provide grounds for supporting the argument that a skateboard is not a vehicle, assuming that a skateboard is similar to a bicycle, insofar as the question of whether it should be classified as a vehicle is concerned.

Consider once again Brewer's case of the drug-sniffing dog. Suppose the following three general rules (Weinreb, 2005, p. 24) have already been established by prior kinds of cases decided by a court.

> *Rule 1*: If a police officer sees something in plain view in a public place, the information collected is not classified as a search.
>
> *Rule 2*: If a police officer opens luggage and then observes something inside the luggage, the information collected is classified as a search.
>
> *Rule 3*: If a police officer listens surreptitiously to a conversation in a private place, it is classified as a search.

These are three specific rules that are accepted based on specific kinds of cases that might previously have been decided by the courts.

There is a more general kind of rule involved as well, called an 'analogy-warranting rule' by Brewer. In the case of the drug-sniffing dog, the analogy-warranting rule (AWR) can be stated as follows (Weinreb, 2005, p. 24).

> AWR: If a police officer obtains information about a person or thing in a public place without intrusion on the person or taking possession of or interfering with the use of the thing, it is not a search for purposes of the Fourth Amendment.

Brewer claims that the analogy-warranting rule provides a generalization that enables a conclusion to be drawn on the basis of the facts in a particular case by deductive reasoning. Weinreb (p. 25) supports Brewer's claim that this rule as stated is sufficient to resolve the issue deductively in the case of the drug- sniffing dog. However, he adds the qualification that it may have to be refined in another case with different facts, citing the following question (ibid.): "if a police officer were to use an X-ray machine that allowed him to see what is inside a container without opening it or taking possession of it would the information he collected be a search?" Weinreb (ibid.) adds that in order to deal with this new kind of case one may want to add the qualification "and without using a technological device" to the generalization (or one may not want to). The issue raised by this kind of question is whether or not the generalization should be qualified.

According to the account given by Weinreb (2005, p. 31), the initial version of the analogy-warranting rule in a case like that of the drug-sniffing dog is only a preliminary step in the process of evaluating an argument from analogy. As the argumentation moves on, there is what is described as a process of "reciprocal reflective adjustment" that arrives at a formulation of the rule that fits the particular issue to be decided in an individual case. For example, one such rule might be (p. 26): if a dog is made to sniff luggage, without interfering with it or its owner, for the purpose of revealing whether it contains drugs, there is not a search to which the requirements of the Fourth Amendment apply. This rule bridges the specific case being decided to the general rule found in the Fourth Amendment. Here we have yet another kind of rule involved in the chain of argumentation. The sequence of argumentation appears to go back and forth from the specific case to the statute that is the basis of the general rule, combining facts and rules with various other rules of wider or narrower generality.

As already noted, Brewer sees the sequence of argumentation in the drug-sniffing dog example as an argument from analogy that is part of a wider three-stage process in which it is embedded. As Brewer describes the whole process, the reasoner formulates a rule covering the phenomenon that provoked the inquiry, applies it to the particular case, and then, by means of a valid deductive inference, reaches the appropriate conclusion. Weinreb (2005, p. 27) comments that although Brewer refers to the three-stage process as an argument from analogy, the work of the analogy is really completed by the end of the first stage. On Brewer's account, the role that the argument from analogy plays in the process is not logical or a matter of justfication, but psychological or epistemological in nature (p. 27). On his account, it is merely a discovery process that explains how a lawyer or judge happened to hit on that particular rule. Once having formulated the rule, the judge or lawyer proceeds by a process of deductive reasoning in which the analogy has no more work to do. Brewer is inclined to see the persuasive force of argument from analogy as being "mystical," or impossible to explain or clearly formulate by any precise model of logical reasoning. His reasoning is that once the three stages of argumentation from analogy have been completed in a particular case, the argument depends on the deductive relationship between the premises and the conclusion, and its rational force is no greater with the analogy than without it (p. 30). Thus Brewer tends to be dismissive of argumentation from analogy, seeing it as a psychological or mystical process that, once completed, no longer plays any important role in reconstructing the process of logical reasoning used as rational justification for the legal conclusion finally arrived at.

The interesting question raised by the case of the drug-sniffing dog is how argument from classification and argument from analogy can be combined with other forms of legal reasoning, like argument from precedent, by a logical process of fitting rules to facts. Part of this task is to untangle how each type of argument is related to the others in a typical case of the kind Brewer cites. There may be various grounds supporting argument from classification, and, as noted earlier, some of them may be based on argument from analogy. Argument from precedent and argument from an established rule also fit into the sequence of argumentation that makes up the process according to which an ultimate conclusion is drawn. Our analysis will help to untangle the individual steps in the process. It will define argument from analogy as a specific argumentation scheme, distinguishing it from argument from classification, defined as a different argumentation scheme. By this means, it will be shown that argument from analogy is not merely a mystical device of argument invention, and that it has a form that can be precisely defined as a structure of reasoning that can be used to justify a claim by giving reasons to support it that can be criticized and evaluated.

2. ARGUMENT FROM ANALOGY AS TREATED IN LOGIC TEXTBOOKS

Aristotle was also very well aware of the power of argument from analogy and presented a number of interesting examples of its use. This one is based on the assumption that choosing magistrates for office by lot (lottery) is a bad practice, just as wrestlers chosen by lot do not perform well, because they are not able to contend.

> Were one to say that it is not fitting the magistrates chosen by lot should be in office; for it is just the same thing as though one were to pick out wrestlers by lot; not taking such as are able to contend, but those on whom the lot may fall. (*Rhetoric*, 1851, II, xx, 4, p. 166)

This argument from analogy is based on the alleged similarity between wrestlers chosen by lot, who do not perform well, and magistrates chosen by lot. The conclusion drawn is that they do not perform well either. The ultimate conclusion of the argument is that magistrates should not be chosen by lot. Such a policy is a bad practice, because, like wrestlers chosen the same way, they will not perform well.

Twentieth-century logic textbooks tend to see argument from analogy as dangerous and misleading. Some even classify it as fallacy, based on the assumption that it is, generally speaking, a fallacious form of argumentation. For example, in an early influential text (Beardsley, 1950, p. 107), the argument from analogy is considered a fallacy. Analogies, on his view, can be used to illustrate, or may lead to a tentative hypothesis, but when they are used as arguments, as reasons to support a claim, the reasoning is not valid. His attack on argument from analogy derives from the following formalization of the argumentation scheme (ibid.):

X has certain characteristics *a, b, c* . . .
Y has the characteristics *a, b, c* . . .
But Y also has the characteristics *x, y, z* . . .
Therefore, X has the characteristics *x, y, z.*

Arguments fitting this scheme are held to be fallacious, because "no matter how many characteristics a pair of things have in common, there may be any number of other ways in which they are different" (Beardsley, 1950, p. 108). In other words, the warranting premise underlying the argument from analogy, " if X and Y have a number of characteristics in common, then any further characteristics found in Y will also be found in X" (Beardsley, 1956, p. 65), is not valid. On Beardsley's theory, the strength that an argument from analogy might have stems from the number of similarities between the things being compared, and not from an underlying principle that explains the similarities. On this view, arguments from analogy are not based on a generalization connecting the observed similarities and the attributed property.

Beardsley's theory is based on a principle of isomorphism or matching of two cases that is reminiscent of what is now called pattern recognition in computing. Like a map of the geographic features of an area, the two terms of an analogy must share the highest number of characteristics in order for the argument to be strong. The problem seems to be that argument from analogy is a plausible form of argument only when it is used for guessing; it is not good enough to be used to prove a claim. It can be confused with a deductive argument, where the two terms of comparison are related by a generalization that may or may not hold up. If not, the argument may default, and thus prove to be fallacious. For instance, the difference between the following two examples lies in the different nature of the generalization. In the first example, the generalization "All people whose parents both have blue eyes are people with blue eyes" is based on genetic laws, and thus the generalization could warrant the conclusion. The second case, however, is based on a generalization apparently, but wrongly, transforming the argument into a valid deduction (Barker, 1989, p. 189).

> John's parents both have blue eyes, and so do Jim's; John has blue eyes; therefore Jim must have blue eyes.
> John's parents both read Greek, and so do Jim's; John likes horseradish; therefore Jim must like horseradish. (Beardsley, 1956, p. 66)

In the second case, no accepted warranting principle or generalization can be found as the basis of the two comparisons. Hence this case is an instance of fallacious argument from analogy.

The development of modern treatments of argument from analogy could be described as a series of attempts to overcome the problem that Beardsley has identified.[2] Beardsley's criticism of argument from analogy is interesting because it points out important features of the scheme. The key factor is not the number of common features that are the basis of the analogy, but the class under which the terms are categorized. In some cases, elements that don't fit the classification are suppressed (Barker, 1989, p. 192). On the other hand, the idea of a generalization warranting the conclusion, based on the principle just outlined, is different from the idea of comparison of terms. A classification is different from a generalization. A resemblance must be distinguished from common characteristics (p. 190). The generalization in the example just given, for instance, must be distinguished from the classification of John and Jim as genetically similar from a particular point of view.

In *Within Reason* (1990), Burbridge describes the argument from analogy. He defines the subject of the conclusion as the Primary Subject (PS), the predicate referred to the latter as the Targeted Predicate (TP), the object compared to it as the Analogue (A), and the features connecting the analogue with the primary subject as Similarities (S). The structure of the topic assumes the following form (pp. 12, 13):

Scheme	Example
Premise 1: PS is like A in $S^1 \ldots S^n$	The universe and a machine are similar in that both are divided into an intricate pattern of parts and subparts.
Premise 2: A has TP.	A machine has a maker.
Conclusion: So, PS has TP.	Therefore, the universe has a maker.

---

[2] Juthe (2005, pp. 6–8) has provided other examples of argument from analogy from logic textbooks, and some examples from the history of philosophy.

Burbridge argues that the soundness of the reasoning is based on the relations between the targeted predicate and the similarities between the two compared terms. If they are closely connected, the argument provides good grounds for the conclusion. If, on the contrary, they are more closely related to the dissimilarities between the analogue and the subject, the inference is unsound (p. 19). The objection that can be raised, showing the unacceptability of the reasoning from analogy, takes the following form (p. 31).

> *Premise 1*: The Objector has assumed that because PS is the same as A, it shares the targeted predicate TP.
> *Premise 2*: PS and A are not identical but analogues, different though similar.
> *Premise 3*: The difference between PS and A is more closely connected to TP than the similarity is.
> *Premise 4*: Therefore, TP does not apply to PS.

Burbridge's evaluation of argument from analogy by the device of raising objections is comparable to the method of critical questions outlined earlier.

A comparable account can be found in Johnson and Blair (1983). On their analysis of the fallacy of faulty analogy, the criterion used to evaluate the argument is the similarity in the respect required to support the conclusion (p. 100). In Copi and Jackson (1992), the relation between the properties is defined in terms of relevance: "an analogy is relevant to establishing the presence of a given attribute provided it is drawn with respect to other circumstances affecting it. One attribute or circumstance is relevant to another, for purposes of analogical argument, if the first affects the second, that is, if it has a causal or determining effect on the other" (p. 199). On this analysis, the strength of the argument depends on the relation cause-effect or effect-cause. Analogy is claimed to depend on relevance, defined in causal terms. More than failure of relevance seems to be involved, however, in cases where an argument from analogy seems to be the most serious kind of fallacy. This kind of case occurs when the argument from analogy is so strongly persuasive at first sight that it overwhelms all resistance, but later, subject to critical questioning, it deflates.

An excellent example of just how argument from analogy can be over-rated in this way is furnished by examining the legal doctrine of striking similarity. According to Patry (2006), this doctrine was referred to as early as the nineteenth century, but took on a renewed life in the often-cited opinion of Judge Frank in the 1946 case of *Arnstein v. Porter*. This

case concerned an allegation of copyright violation based on the striking similarity between two songs, certain portions of which were shown
to be almost completely identical. It is part of the burden of proof for
copyright infringement that both (1) access to the material and (2) copying of it must be proved (Patry, 2006). Patry (personal communication,
2006) summarizes the components that need to be established in order
to prove infringement of copyright.

> The substantive law of copyright says that to win, defendant must have
> (1) copied a copyrighted work owned by plaintiff, and have copied a (2)
> material amount of material that is copyrightable. To copy, one must have
> had access to the work as a practical and legal matter. At that first stage of
> the inquiry it matters not what or how much defendant copied; it is the
> mere fact of copying that has to be established. If you can't establish access,
> the case will be dismissed because defendant can't have copied. Access is
> defined as not just as proof defendant had the work in hand, but more
> liberally as a reasonable opportunity to have been exposed to the work, at
> least for works that are widely distributed.

Similarity – for example, a similarity between two melodies – can also be
part of the evidence. In some cases, however, the similarity seems to be
so striking that it can overwhelm the need to prove other elements of the
evidence, convincing a judge or jury all by itself.

An interesting comment on this kind of case was made by Judge Alex
Kozinski in a paper called "How I Narrowly Escaped Insanity" (2001).
Judge Kozinski (2001, p. 1301) relates that around 1980 he wrote a
science fiction novel about extracting a person's mind and implanting
it into another person's head. When the movie *Total Recall* came out in
1990, he found the similarities between the movie and his unpublished
novel uncanny. The story lines appeared to him to be identical, but
the clincher was a scene near the end of the movie, where the villain
kicks over a fish tank and the camera shows the fish squirming on the
floor. The identical scene had appeared in his novel. Because of this
striking similarity to his own novel, he was convinced that his novel had
to have been pirated. But later he started to have second thoughts. He
had never finished the novel, nor had he ever sent it to anyone to read.
He concluded, in the end, that despite the similarities, it was just a
coincidence. Nevertheless, for a time he found the striking similarity
such convincing evidence that he found it very hard to resist the feeling
that his ideas had been stolen.

3. IS ARGUMENT FROM ANALOGY DEDUCTIVE OR INDUCTIVE?

There is currently a controversy in logic on whether argument from analogy is a deductive or inductive form of reasoning. In the following discussion, we will adopt the strategy of slipping between the horns of this dilemma by proposing a third alternative. But before getting to that point, some examples illustrating deductive and inductive arguments from analogy need to be considered. We begin with one of the most famous arguments from analogy used in philosophy, the violinist example of Thomson (1971).

The Violinist Example

You awake to discover that you have been connected by a machine to a famous, gravely ill violinist who has suffered acute kidney failure. The machine cleanses his blood; poisons in his bloodstream will kill him unless he is connected to you by this machine. You are the only perfect blood match for this violinist, and if you are detached from the machine, the violinist will die. Construction of a new machine that can cleanse the violinist's blood, so that you can be unhooked from him, will take about nine months.

When confronted with this example, most people would say that it is wrong for you to be compelled to remain attached to the violinist. Thompson's argument is that by analogy, it is wrong to compel a woman who is pregnant as the result of rape to retain the fetus. The argument is that if you have the right to detach yourself from the violinist, then a woman who is pregnant as a result of rape has a right to an abortion.

According to Waller (2001, p. 202) the violinist example is a deductive argument from analogy because it is based on the following general principle: a person does not have an obligation to save or sustain a life when he or she has done nothing to take on that obligation. This principle is taken to imply, as an instance, that a woman who is pregnant as a result of rape does not have an obligation to save or sustain the life of the fetus. Waller's analysis of the violinist example parallels Brewer's analysis of the drug-sniffing dog example. Both are analyzed as deductive arguments from analogy based on drawing a conclusion from a universal generalization as applied to a specific case that fits the requirements stated in the generalization.

According to Waller (2001, p. 201), deductive arguments by analogy have the following form.

We both agree with case *a*.

The most plausible reason for believing *a* is the acceptance of principle *C*.

C implies *b* (*b* is a case that fits under principle *C*).

Therefore, consistency requires the acceptance of *b*.

Waller (p. 202) classifies this form of argument, as exemplified by the argument from analogy in the violin example, as deductive, not inductive. His reason is that it does not argue that because people generally believe they have a right to detach themselves from violinists, they are likely to hold that rape victims have the right to detach themselves from fetuses. Instead of classifying it as an inductive argument, Waller classifies it as what he calls a deduction from principle.

In contrast with deductive arguments from analogy, Waller offers an example of an inductive argument from analogy (2001, p. 202). Alice likes novels by a certain author. She concludes that since she and Barbara have similar tastes in novels, Barbara will like novels by the same author. According to Waller, this example is not an instance of a deductive argument from analogy, because there is no implication that Barbara must like novels by this author in order to be logically consistent. Waller offers (ibid.) the following structure to represent the form of inductive arguments from analogy.

D has characteristics *e*, *f*, *g*, and *h*.

E also has characteristics *e*, *f*, *g*, and *h*.

D also has characteristic *k*.

Having characteristics *e*, *f*, *g*, and *h* is relevant to having characteristic *k*.

Therefore, *E* will probably also have characteristic *k*.

According to Waller (p. 203), important differences are obscured if inductive arguments from analogy are analyzed under the pattern used for deductive arguments from analogy. In using a nondeductive argument from analogy, we're trying to determine why we believe something by formulating the general principle on which our belief is based. In using an inductive argument from analogy, a different sort of investigation is required to determine the relevant characteristics that are shared by two cases.

The type of argument Waller calls a deductive argument from analogy is called an a priori nondeductive argument from analogy in Govier's classification system. Govier classifies such analogical arguments as deductive because they require discovery of an underlying universal principle that may not be known in a given case, and that may be obscure. She also sees such arguments from analogy as not properly being classifiable as deductive because they are persuasive in nature, and lose their persuasive force when recast into a deductively valid form. Guarini (2004) agrees with Govier that the principle on which an argument from analogy is based is uncovered only after much reflection and analysis. Our view of how to classify such arguments will be even stronger than Govier's, as opposed to Waller's, because we will argue that even the final version of the argument from analogy, once this process of reflection and analysis has been completed, is not (in typical cases) deductively valid. We will claim that the generalization, or what Guarini calls the principle of the case, is defeasible. Hence, even once formulated and applied, it typically can be defeated by contrary instances in the future, of a kind that cannot be anticipated.

4. THE SCHEMES FOR ARGUMENT FROM ANALOGY

Argumentation schemes represent typical patterns of reasoning used in everyday conversational reasoning in persuasion dialogue, as well as in special contexts like legal and scientific reasoning. They are the historical descendants of Aristotle's topics, recently studied by Hastings (1963), Perelman and Olbrechts-Tyteca (1969), Kienpointner (1992), Pollock (1995), Walton (1996), and Grennan (1997). Recently argumentation schemes have been applied and developed in a formal manner in artificial intelligence, where they are recognized as having potential as tools for modeling the reasoning of artificial agents (Prakken and Vreeswik, 2002; Reed and Norman, 2003; Verheij, 2003; Bex, Prakken, and Walton, 2003; Prakken, 2006). The movement to apply argumentation tools to AI, and then to use these tools to model reasoning used to derive conclusions in legal evidence, is part of the new evidence scholarship (NES) that has grown up in recent years (Tillers, 2002). Of special importance for the modeling of legal reasoning in AI is argument from analogy, given its central role in CBR (Ashley and Rissland, 2003).

The argumentation scheme for argument from analogy (Walton, 1989, p. 258) is the following. It is based on an assumption fundamental to CBR, namely, the principle that two cases can be judged to be similar to each other. It is a fact about human cognition that we very commonly make a judgment that one case is similar to another in drawing conclusions about what to do in daily life. How we draw such comparisons is a subject for psychology. But according to a branch of AI called pattern recognition, we recognize patterns in what we perceive and identify as an object or scene, and we are skilled at matching such patterns, seeing one thing as similar to another. However, we can easily make mistakes in such judgments, due to camouflage, or because of our expectations about what we would normally encounter in a familiar situation. Even so, judgments of similarity of one case to another are often made, and they are used to provide data in the major premise of an argument from analogy. According to the argumentation scheme for argument from analogy (Walton, 1989, p. 258), this form of argument reasons from a source case, C_1, to a target case, C_2.

Argument from Analogy: Version 1

Major Premise: Generally, case C_1 is similar to case C_2.
Minor Premise: Proposition A is true (false) in case C_1.
Conclusion: Proposition A is true (false) in case C_2.

The following example of argument from analogy, which could be called the shoe example, is quoted from Walton (1989, p. 258):

I infer that a new pair of shoes will wear well on the grounds that I got good wear from other shoes previously purchased from the same store.

In this case, the inductive argument from analogy is quite a weak one, because this store might carry a variety of different kinds of shoes, and some of them might wear well, while other might not. The analogy does not seem to fail altogether, because the new shoes are similar to the others in a relevant respect, coming from the same store. There could be some reason why this factor is relevant. Perhaps, for example, this store is run by an owner who selects better-quality shoes that tend to wear well.

It is often asserted that there should be a requirement of relevant similarity that acts as a premise for argument from analogy (Weinreb, 2005, p. 32). For example, consider the argument: this apple is red and tastes good; this ball is red; therefore, it will taste good. Here the argument from analogy fails because the observed similarity between the source and the target is not "relevant to the further similarity that is in question" (ibid.). The observed similarity of being red is not relevant to the further similarity of tasting good. What is important here, according to Waller (2001), is to determine relevance for the inductive argument. For example, in the novels case, the author of the book is a relevant characteristic for determining whether Barbara is likely to enjoy it, while the color of the dust jacket is not a relevant characteristic for this purpose. On Waller's analysis, inductive arguments from analogy need to be based on relevant characteristics in a way that deductive arguments from analogy do not. However, it seems that not all arguments from analogy are based on relevant factors shared by the two cases.

Guarini (2004, pp. 164–166) cites a very interesting example that can be used to explain how argument from analogy is used very successfully in a court of law, the case of *DeGraffenreid v. General Motors* (Crenshaw, 1998). In this case, a group of black women sued General Motors for discriminatory hiring practices, arguing that GM had not hired enough people who were both female and black. GM had already hired many white women and black men. The argument in this case was specifically that they had not hired enough people who were both female and black. Initially, the case was dismissed on the grounds that the statute covering such cases did not explicitly identify compound classes, (e.g., race combined with gender) as grounds for a discrimination suit. When the suit was brought forward again, a comparison was made to a common previous type of case in which white men had sued for reverse discrimination. The basis of this previous argument was the contention that there had been discrimination against people in the compound class of being both male and white. The argument in favor of black women was based on the assumption that the two types of cases should be treated in the same way. Guarini (2004) classifies this case as an instance of deductive argument from analogy that depends on a principle but does not require relevance, contrasting it with other examples of argument from analogy that do depend on relevance.

Guarini distinguishes between two schemes for argument from analogy. The first one, which he calls the core scheme, has the following form (Guarini, 2004, p. 161).

a has features $f_1, f_2, \ldots, f_n$.
b has features $f_1, f_2, \ldots, f_n$.
a and b should be treated or classified in the same way with respect to
$f_1, f_2, \ldots, f_n$.

Guarini (p. 162) cites the argument from analogy in the case of *DeGraffenreid v. General Motors* as an example that fits the core scheme.

Guarini argues that the core scheme can be extended in various ways. He postulates a second scheme for argument from analogy by adding two further steps to the core scheme (2004, p. 162):

a is *X* in virtue of $f_1, f_2, \ldots, f_n$.
Therefore, *b* is *X*.

Guarini takes the similarities $f_1, f_2, \ldots, f_n$ in this scheme to be "relevant similarities" (ibid.), but adds that he does not include the term 'relevant' in the scheme because "it appears to be common practice not to include relevance claims in argument reconstruction" (ibid.). On his analysis, all five steps are necessary to model the argument in Thomson's violin example, but only the first three are necessary to model the argument in the case of *DeGraffenreid v. General Motors.* In the latter case, he writes (ibid.), there is no need for the last two steps, because the argument "depends simply on the view that the cases in question should be treated in the same way, without arguing for some specific classification of those cases." In the violin example, however, the similarities cited have to be relevant if the argument is to be persuasive.

We agree with Guarini that the argument from analogy used in the violin example is fundamentally different from the one used in the case of *DeGraffenreid v. General Motors.* To take this difference into account, we propose that an additional premise be added to version one of the scheme to the effect that the observed similarity between the source and the target is relevant to the further similarity that is in question. This approach leads to the postulation of a new version of the argumentation scheme for argument from analogy.

Argument from Analogy: Version 2

Major Premise: Generally, case C1 is similar to case C2.
Relevant Similarity Premise: The similarity between C1 and C2 observed so far is relevant to the further similarity that is in question.
Minor Premise: Proposition *A* is true (false) in case C1.
Conclusion: Proposition *A* is true (false) in case C2.

In the following gold example, cited by Copi and Cohen (1983, p. 101), two situations, prospecting for gold and scientific research, are presented as similar in relevant respects. The reason Copi and Cohen give (p. 101) is that both fall under the category of "quest," constituted by difficulty, training, and fortune.

> As in prospecting for gold, a scientist may dig with skill, courage, energy and intelligence just a few feet away from a rich vein – but always unsuccessfully.

Consequently in scientific research the rewards for industry, perseverance, imagination and intelligence are highly uncertain. (Kubie, 1954, p. 111, quoted in Walton, 1996, p. 78)

These are held to be relevant similarities, and the argument from analogy in the gold example is held to be a strong one. But what is meant by 'relevant' when the expression 'relevant similarity' is used? The answer given is that properties shared by the two objects fall under the category of "quest." Isn't this answer about how the two objects are defined or classified as common knowledge? That may not be a matter of relevance so much as a matter of making explicit further supporting premises about how each object is classified. This matter of classification will be further studied.

Is failure of the requirement of relevant similarity the reason why the argument in the red ball case fails? Maybe not. The reason may be simply the failure of the major premise, requiring the ball and the apple to be generally similar. They share some properties, like being red and being round, but they do not seem to be generally similar. Maybe, as in the shoe case, there are some other factors in the background that defeat or undercut the argument. For example, we know that a ball is a manufactured item made out of some substance like rubber, while we know that an apple is a fruit, a common food item. Just as in the shoe case, the argument stands or fails based on unstated but generally known premises that support or undermine the explicit ones. The explanation of whether an argument is strong or weak may turn on these factors, and not on some mysterious notion of relevant similarity that seems hard to define or explain. These considerations suggest that we need to look into the factors behind why an argument from analogy fails.

We can see from the examples we have cited that argument from analogy can be quite reasonable in many instances. Why, then, is it so often categorized as fallacious in the logic textbooks? There is a difference between a weak argument and a fallacious one, the difference being that a fallacy has a powerful tendency to deceive. An excellent example of just how compelling an argument from analogy can be is provided by examining the legal doctrine of striking similarity, described in section 2 of Chapter 2. According to Patry (2006), this doctrine was referred to as early as the nineteenth century, but took on a renewed life in the often-cited opinion of Judge Frank in the 1946 case of *Arnstein v. Porter.* This case concerned an allegation of copyright violation based on the striking similarity between two songs, certain portions of which showed an almost complete identity. Part of the burden of proof for copyright infringement is that both access to the material and copying of it must be proved (Patry, 2006). Similarity – for example, a similarity between two melodies in a pair of songs – can also be part of the evidence. In some cases, however, the similarity seems to be so striking that it can overwhelm any need to prove other elements of the evidence, convincing a judge or jury virtually by itself.

5. ARGUMENT FROM ANALOGY AS A DEFEASIBLE FORM OF ARGUMENT

One very nice thing about Guarini's discussion of the violinist example is that he puts the argumentation sequence in the form of a dialogue, as follows (2004, pp. 155–156):

Jack: I think abortions are always immoral, except when a woman's life is in immediate physical danger.

Jill: What about cases where a life was made dependent on a woman through force, like in cases of rape?

Jack: That is profoundly unfortunate, but while rape is morally objectionable, the right to life of the fetus outweighs the concern that the life of the fetus was made dependent on the woman through force.

Jill: Well, imagine that you were kidnapped, knocked unconscious, and when you woke up you discovered you were connected to a world famous violinist. Your kidneys are filtering his blood. You can disconnect yourself, but if you do so the violinist dies. The only way to keep the violinist alive is to stay hooked up until a suitable kidney is found for transplant, and that is expected to take about nine months. Isn't it okay to disconnect yourself?

Jack: I think so

Jill: Shouldn't you then say that abortions are morally acceptable in cases of rape?

In describing how the argument from analogy works in this famous example, Waller (2001, p. 156) uses the term "persuasion" to suggest that the analogical principle used in this reasoning is defeasible. He writes that in the dialogue, Jack is persuaded to believe the principle, meaning that he did not believe it prior to the argumentation in the dialogue. As Guarini puts the point, "you cannot be persuaded of something you already believe"(2004, p. 156). The conclusion he draws is that the analogical argumentation used in the dialogue does not trade on what he calls an "exceptionless" principle, but on what we would call a defeasible principle, one that is inherently subject to exceptions. Guarini (ibid.) reinforces this point by describing how he has taught Thomson's paper many times in university classes. According to his experience, those who are persuaded by the argument from analogy tend to fall into two classes. There are those who are persuaded by the principle they initially rejected, but now come to accept, and there are those who are persuaded to change their view even though they are not prepared to endorse any principle. According to Guarini, the ones who fall into the first group are like Jack in his constructed dialogue. Those who fall into the second group are so "drained" by the casuistry of cases (as Guarini describes the process) that they lose interest in generating and testing moral principles. This group begins to suspect that any moral principle they can state on the subject of abortion will be found to have exceptions. This observation is interesting, from our viewpoint, because it suggests the defeasible nature of the generalizations used in ethical reasoning in typical casuistic disputes about ethical dilemmas. The point applies very well to generalizations in legal reasoning (Bex, et al., 2003; Anderson, Schum, and Twining, 2005).

In our opinion, Guarini's observations and discussion are very revealing, because they offer a lot of evidence showing that the argument from analogy in the violinist example is neither deductive nor inductive in its logical form. It is quite accurately described, in our opinion, as being defeasible in nature. Its logic is brought out very well by the dialogue that Guarini constructs to illustrate how one party uses the argument to persuade the other to accept some contested proposition that he did not previously believe. In our opinion, Guarini is quite right to classify this kind of argumentation as occurring in the framework of a persuasion dialogue. Moreover, it is quite revealing that he describes the argumentation in the case as fitting a casuistry of cases in which any moral principle brought forward by one party to persuade the other party to accept some conclusion is subject to exceptions that may be found by the other party. This structure is known in the literature on AI and law as a persuasion dialogue (Bench-Capon, 2003; Bench-Capon and Prakken, 2005).

The earlier examples and list of critical questions suggest that argument from analogy is best seen as a defeasible argumentation scheme that is inherently weak and subject to failure, but that can still be reasonable if used properly to support a conclusion. Thus it should not be impaled on the dilemma of being only a deductive or inductive form of argument.[3] We propose that it be seen as a defeasible kind of argument that should be evaluated as strong or weak in specific respects by

---

[3] On the importance of defeasibility as a property of legal argumentation, see Prakken (1997, 2001, 2002), Prakken and Sartor (1996, 1997), and Prakken and Vreeswijk (2002).

subjecting it to critical questioning in a dialogue. The following critical questions matching version one of the scheme for argument from analogy are given by Walton (1989, p. 258).

Critical Questions for Argument from Analogy

CQ1: Is A true (false) in C1?
CQ2: Are C1 and C2 similar, in the respects cited?
CQ3: Are there important differences (dissimilarities) between C1 and C2?
CQ4: Is there some other case C3 that is also similar to C1 except that A is false (true) in C3?

In this case, CQ4 is especially appropriate, as applied to the shoes example, because there might be quite a difference between the two pairs of shoes, for example, even though both were bought at the same store. These four critical questions represent what is meant to be a practical approach to evaluating arguments from analogy. They are important to keep in mind to provide a device to help students who are learning critical thinking, when confronted by an argument from analogy, to scan quickly to look for key points of weakness in the argument in order to challenge its applicability and strength in a given case.

The problem of how to represent the critical questions as ways of rebutting or undercutting arguments that fit an argumentation scheme is beginning to be investigated in the AI literature. Verheij (2003) showed that critical questions can have four roles.

1. They can be used to question whether a premise of a scheme holds.
2. They can point to exceptional situations in which a scheme should not be used.
3. They can set conditions for the proper use of a scheme.
4. They can point to other arguments that might be used to attack the scheme.

Verheij argues that critical questions of kind one, which merely question whether a given premise in a scheme is true, are redundant and can be ignored. The reason is that it is a condition of the use of any argument that the premises are true, or at least are acceptable, and pointing to the need for this assumption is merely restating the premise of the argumentation scheme. Verheij and others (Reed and Walton, 2005) have also begun to investigate whether the remaining kinds of critical questions could be reformulated as implicit premises of a scheme, or as kinds of counterarguments that might be used to attack or defeat the scheme.

Thus critical questions have become linked to the more general problem, much discussed in AI, of how defeasible arguments are to be evaluated. Any first pass through the literature on defeasibility begins with the distinction between undercutters and defeaters (Pollock, 1995, pp. 40–41). Pollock defines a defeater, sometimes called a rebuttal or refutation of an argument, as another argument that has the opposite (negation) of the original conclusion as its conclusion.[4] Pollock defines an undercutter as a counterargument that attacks the inferential link between the premises and the conclusion in the original argument. Pollock (1995, p. 41) offers the following example.

> For instance, suppose *x* looks red to me, but I know that *x* is illuminated by red lights and red lights can make objects look red when they are not. Knowing this defeats the prima facie reason but it is not a reason for thinking that *x* is *not* red. After all, red objects look red in red light too. This is an *undercutting defeater*. (Pollock's italics in both instances)

The undercutter is the weaker form of attack, one that only raises questions on whether the original argument supports its conclusion, leaving room for doubt. The defeater is the stronger form of attack.

Looking at the list of critical questions just given, it would seem that CQ1 and CQ2 are redundant, while CQ4 looks more like a defeater. CQ3 could be either an undercutter or a defeater, perhaps depending on how important the differences are. The strategy for attacking an argument from analogy suggested by CQ4 is to produce a counteranalogy. For instance, in the shoe example, I might argue that somebody else bought a pair of shoes from this same store that did not wear well at all. This attack would seem to be a defeater. At any rate, the issue currently being investigated in AI is whether the critical questions can be reconfigured as undercutters, defeaters, or attacking arguments of some sort, which can be modeled as arguments made up of propositions, as opposed to being seen as questions. For questions are much harder to model, using current AI techniques, than arguments made up of propositions that are premises or conclusions.

Sophisticated automated systems have been developed in computing to provide a more exact basis for reasoning from analogy on the basis of similar cases. One is the CATO system, which teaches law students how to create case-based arguments. CATO has templates for argument

---

[4] Pollock (1995, p. 40) calls a "rebutting defeater" a reason for denying the conclusion of an argument where the premises offer a prima facie (defeasible) reason for the conclusion.

moves, like argument from analogy, and it also has rules for distinguishing between cases that are not similar, which can show how to attack a rule (Ashley and Rissland, 2003, p. 41). Another is the HYPO system, in trade secrets law, which produces arguments from analogy and counterattacks to them. HYPO argues for one side by making a legal point, and then argues for the other side by responding with a counterpoint (Rissland, 1990, p. 1971). Such arguments are based on a given or "old" case that has been decided and a "new" one that has not. The new case is called the problem case, while the old case is called the base case. One way to evaluate arguments from analogy commonly used in AI is to identify so-called factors that are the basis of comparison between one case and another. This technique can be explained by considering a legal example that was presented by Chorley and Bench-Capon (2004, P. 94).

The Wild Animals Example

In all three cases, the plaintiff (P) was chasing wild animals, and the defendant (D) interrupted the chase, preventing P from capturing those animals. The issue to be decided is whether or not P has a legal remedy (a right to be compensated for the loss of the game) against D. In the fox case, *Pierson v. Post*, P was hunting a fox on open land in the traditional manner, using horse and hound, when D killed and carried off the fox. In this case, P was held to have no right to the fox because he had gained no possession of it. In the ducks case, *Keeble v. Hickeringill*, P owned a pond and made his living by luring wild ducks there with decoys, shooting them, and selling them for food. Out of malice, D used guns to scare the ducks away from the pond. Here, P won. In the fish case, *Young v. Hitchins*, both parties were commercial fishermen. While P was closing his nets, D sped into the gap, spread his own net and caught the fish. In this case, D won.

The fish case (*Young*) is the problem case; the fox case (*Pierson*) and the ducks case (*Keeble*) are the base cases that provide theories that can arguably be used for the purpose of deciding the problem case. A hypothesis is formed from the base case by matching it by similarity or difference to the problem case. The system developed by Bench-Capon and Chorley, called AGATHA, presents arguments for and against a hypothesis as a dialogue between the plaintiff's side and the defendant's side, using argument from analogy. The evaluation of a hypothesis is based on four factors, representing points of similarity or difference between cases, and three values, which represent goals or rules used for judging the worth of factors. In the wild animals case, Chorley and Bench-Capon (2004, p. 94) cite four factors and three values.

Four Factors

pNposs: The plaintiff did not have possession of the animal.
pLand: The plaintiff owned the land.
pLiv: The plaintiff was pursuing his livelihood.
dLiv: The defendant was pursuing his livelihood.

Three Values

LLit: the value of reducing litigation.
MSec: the value to secure enjoyment of property rights.
MProd: the value to promote productive activity.

The factors correspond to critical questions CQ2 and CQ3. They represent the respects in which one case can be held to be similar to or different from another. Indeed, once the factors have been identified, they can easily be compared, and we can calculate the respects in which the one case is similar to or different from the other. The argument from analogy can then be used to set up two sides of a dialogue, with arguments on each side based on the factors. Each side has a hypothesis, a conclusion it tries to argue for, using an argument from analogy. For example, the defendant could set up his hypothesis as a conclusion by arguing that the fish case (the problem case) is similar to the fox case, the base case in which it was judged that the plaintiff had no right to the fox. The plaintiff could then set up an opposite hypothesis by arguing that the ducks case is similar to the fish case, because in both cases the plaintiff was pursuing his livelihood. Thus an extended sequence of arguments from analogy can be carried on, back and forth, using the factors to set up hypotheses. These arguments can be evaluated as strong or weak, using the values. Once factors and values are stated precisely in a given case, the final outcome of the argument from analogy can be computed so that one can calculate which side has the stronger argument.

The best way to approach any argument from analogy is to begin by identifying the premises and conclusion of the argument in the given case. Having done that, one can probe into the case a bit further by asking one or more of the critical questions. The questions, along with the replies given to them, form a dialogue. If the case is from a written text, the analyst has to supply both sides. If the arguer is present, she and the questioner can have an actual dialogue. In either event, as the dialogue proceeds, the argument can be analyzed in more depth by identifying the factors and values. Of course, the participants may also dispute what the factors and values are. But once the analysis gets to the point where factors and values are being identified, a better body of evidence that can be used to support or refute the hypotheses on both sides can be developed. By this means, an argument from analogy can be evaluated as strong or weak, based on the evidence in a case.

We close this section with one further observation. In arguments from analogy, the conclusion is inferred based on the properties presented as common to the two (or more) compared cases. As in the scheme described earlier, the elements are classified under a verbal category, and certain characteristics are pointed out. A classification, founded on these features, is then applied to the analogous or similar case, and the classification of the similar case follows consequently. The main feature of the argument that needs to be scrutinized is the relation between the common properties and the property predicated in the conclusion.

6. ARGUMENTS FROM CLASSIFICATION

Arguments from analogy are often closely related to arguments from classification, and indeed it will be a prevailing theme of this chapter that we need to talk of argument from analogy in terms of argument from classification. To introduce argument from classification, let us begin by stating a deductive form of it, where the variable a is a thing or individual, and F and G represent classes of things.

Deductive Form of Argument from Classification

All F's can be classified as G's.
a is an F.
Therefore, a is a G.

This form of argument is deductively valid, and it can therefore be contrasted with argument from analogy, best seen as not being deductively valid in the kinds of cases that will be studied here. However, arguments from analogy are often so closely related to, and dependent on, arguments from classification that there is a tendency to mix them up, or to fail to evaluate a given case as involving both kinds of argument. It may be at least in part for this reason that arguments from analogy have often been described as fallacious by twentieth-century logic textbooks. But as we will show, they can be reasonable under the right conditions, even though they should often be critically questioned and examined very carefully before giving them much weight. They are often weak arguments taken by themselves, but can be pivotal, on a balance of considerations, in a larger mass of evidence of which they are a part.

Arguments from classification are based on two main components: the description, or presentation, of the facts or events, and their classification, proceeding from properties presented in the definition itself. The classification may derive from a semantic aspect of the words used to describe the event: if *x* is classified as a terrorist, the implicit premise that terrorists are classified as enemies, stemming from the accepted meaning of the word 'terrorist', automatically leads to the conclusion that *x* is an enemy. Such verbal categorization is a speaker's choice, but it may also be backed by common or shared knowledge. For instance, if we claim that terrorists are dangerous fanatics who need to be locked up, we may base our categorization on the common opinion shared by many people. If this classification can be taken for granted, it will then apply to our interlocutor as well, and may be taken for granted as an implicit premise. We don't even need to argue explicitly that terrorists are dangerous to society, because this assumption proceeds from common knowledge that is already accepted. The classification is inherent, already implied by the properties the terms are categorized as having. Classification is, therefore, a matter of choice of words that support or attack a viewpoint. A subject of a classification is tacitly included under a point of view with different characteristics that are relevant to an issue. For this reason, a classification of a term based on a definition always implies the exclusion of relevant opposed perspectives. Definitions, especially of ethical terms, are rarely, if ever, argumentatively neutral. Defining almost always means advancing an opinion about a situation, often one that is not explicitly formulated. Argument from verbal classification proceeds from semantic, endoxic, or shared properties of a definition. For this reason, the main problem concerns the acceptance of the definition itself, and this aspect is the key to grasping the fallacious uses of arguments from classification.

A famous example is that of a religious discussion where the interlocutor's position is labelled as "heresy." Based merely on this verbal classification, it is concluded that the position itself is wrong, on the implicit basis of the generalization that every heresy is wrong. The concealed assumption is the loaded nature of the term 'heresy', a matter of definition. The opponent's standpoint, in fact, may be heretical on the grounds that everyone, the authorities, or its opponents have classified it under this term. On the other hand, the proponent might have used that definition without any backing warranting that the heresy of the position is known or accepted by the hearer. In this second case, the risk is the fallacy of begging the question. Here, a premise supporting the conclusion can itself be supported only by using the conclusion as a premise. If the latter, based on a classification or on the way a term is defined, is the only means to prove the former, the fallacy of question-begging epithet is said to have been committed. This kind of argument is a device used to avoid the requirement of fulfilling a burden of proof (Walton, 1996, p. 54).

As we have just shown, some arguments from verbal classification are based on universal generalizations of a kind that are not subject to exceptions. However, many arguments of this kind are based on a classification that can be defeasible, and subject to arguable exceptions. To cover such cases, the following scheme for argument from verbal classification (Walton, 1996, p. 54) has a major premise that can represent a defeasible generalization in some instances.

Scheme for Defeasible Argument from a Verbal Classification

Major Premise: If some particular thing *a* can be classified as falling under verbal category *C*, then *a* has property *F* (in virtue of such a classification).
Minor Premise: *a* can be classified as falling under verbal category *C*.
Conclusion: *a* has property *F*.

If there is a problem concerning the definition of a key term, the classification can be questioned, and the argument may be weakened or even defeated. This can be done by attacking the generalization or by questioning whether it fits the particular thing at issue. The two critical questions focus on the application of the generalization to the concrete case, and on its strength.

CQ1: Does *a* definitely have *F*, or is there room for doubt?
CQ2: Can the verbal classification (in the second premise) be said to hold strongly, or is it one of those weak classifications that is subject to doubt?

In many instances, the argument would not be plausible unless the property *F* had an argumentative value based on a commonly accepted meaning of a term. In other words, the plausibility of the argument resides in the assumption that the classification, implicitly or explicitly, leads by inference to the conclusion claimed. In the heresy case, the fact that heresy is generally considered wrong is the implicit classification that is the concealed basis of the plausibility of the argument. The value of the move is based on the communicative implicature that the classification implies.

Finally, in this section we argue that part of the argument cited by Brewer as being the central argument from analogy, based on the analogy-warranting rule (AWR), is an instance of argument from classification. This argument can be put in the following form.

The Warranting Rule Argument in the Drug-Sniffing Dog Example

If a police officer obtains information about a person or thing in a public place without intruding on the person or taking possession of or interfering with the use of the thing, it is not a search for purposes of the Fourth Amendment.

In the case of the drug-sniffing dog, the police officer obtained information about a person or thing in a public place without intruding on the person or taking possession of or interfering with the use of the thing.

The case of the drug-sniffing dog is not classified as a search for purposes of the Fourth Amendment.

The first premise is a generalization, of the kind called a conditional in logic and a rule in AI (Bex et al., 2003). The second premise specifies a fact about a specific case that fits the rule. The conclusion classifies the specific case under the definition of the term 'search' set out in the rule. Thus we classify this central argument in the drug-sniffing dog case as an argument from classification, and not as an argument from analogy. Clearly, this thesis is controversial, judging from the way Brewer and Weinreb treat the argument in the drug-sniffing dog case. However, rather than continuing to dispute the issue at this juncture, we move on to make another relevant point. It is the point that argument from analogy can be combined with argument from classification.

Govier, in her textbook *A Practical Study of Argument* (1992), shows how the structure of a set of compared terms can be seen as the basis for analogical thinking. She brings this lesson out by connecting argument from analogy with argument from classification. On her analysis, argument from classification is based on determining features and the commonality of two compared terms. The following form of argument specified by Govier (1992, p. 274) can be viewed as stating a special argumentation scheme that combines argument from analogy with argument from classification.

Argument from Analogy Based on Classification

The analogue has features *a*, *b*, and *c*.

The primary subject has features *a*, *b*, and *c*.

It is by virtue of features *a*, *b*, and *c* that the analogue is properly classified as *W*.

So, the primary subject ought to be classified as *W*.

This scheme shows how argument from analogy is based on classification and, in many typical instances of the kind we have examined, is combined with argument from classification. They are related by the fact that they are based on a classification of the elements proceeding from properties pointed out in their description. Argument from classification leads to the conclusion that a particular case or thing has a determinate property because it may be classified as generally having that property. On the other hand, argument from analogy proceeds from the similarity of two facts under a particular point of view to their categorization under the same class. Thus there are many close connections between argument from analogy and argument from classification.

7. ARGUMENTS BASED ON RULES AND CLASSIFICATIONS

The pervasiveness of argument from classification and argument from analogy in the most common kinds of arguments in ethics and law can be brought out by considering the common kind of case in which an argument, and the reply to it, are based on applying a rule to a case. Such arguments are fundamental to case-based reasoning. In these arguments, argument from an established rule is the pragmatic counterpart to argument from verbal classification. To begin with, the situation is categorized as an instance of an established rule, and, therefore, the course of action prescribed in it supposedly has to be applied to the specific case. The argument is derived from recognizing the case as an instance of a general class in a rule, and from this classification of it the conclusion follows that the respondent should act in accordance with the rule. For example, in the following case, the conclusion has a concrete effect on the student's action. It is a prohibition flowing from the application of a rule that leads to the refutation of an argument.

Student: I don't think I will able to get my essay in on Tuesday. Would it be OK if I handed it in next week?

Professor: We all agreed at the beginning of the year that Tuesday is the deadline. That is the rule. (Walton, 1995, p. 91)

This is the scheme for argument from an established rule (Walton, 1996, p. 92).

Scheme for Argument from an Established Rule

Major Premise: If carrying out types of actions including the state of affairs A is the established rule for x, then (unless the case is an exception), x must carry out A.

Minor Premise: Carrying out types of actions including the state of affairs A is the established rule for a.

Conclusion: Therefore, a must carry out A.

It can be seen that the pattern of this scheme is similar to that of argument from verbal classification. But it differs from the latter because it belongs to a different kind of dialogue, that of deliberation, and because the missing premise itself, the generalization, is differently supported. The critical questions are directed against the generalization in the rule.

Critical Questions for Argument from an Established Rule

CQ1: Does the rule require carrying out types of actions that include A as an instance?

CQ2: Are there other established rules that might conflict with, or override, this one?

CQ3: Is this case an exceptional one, that is, could there be extenuating circumstances or an excuse for noncompliance ?

The last critical question concerns the problem of the exceptional case. It is the problem of the classification of the situation as falling under the law or rule. The counterargument is a kind of rebuttal of the implicit assumption that the case in question represents an instance of the rule. In the following example, the argument is rebutted by denying the classification of the student under the class "students obligated to get the essay in on Tuesday." The rule, defining the conditions under which the subjects are considered or not considered elements of the class, excludes the student from being obliged to follow the rule.

Student: I don't think I will be able to get my essay in on Tuesday. Would it be OK if I handed it in next week?

Professor: We all agreed at the beginning of the year that Tuesday is the deadline. That is the rule.

Student: But I had a bad case of the flu last week, and I have a note from my physician to prove it. (Walton, 1995, p.93)

This reply represents the argumentation scheme known as argument from an exceptional case (Walton, 1996, p. 92).

Scheme for Argument from an Exceptional Case

Major Premise: If the case of *x* is an exception, then the established rule can be waived in the case of *x*.
Minor Premise: The case of *a* is an exception.
Conclusion: Therefore, the established rule can be waived in the case of *a*.

This form of argument is based on argument from verbal classification. The classification of *x* as an exception allows the application to it of the property "to be waived." The critical questions stem from the classification.

Critical Questions for Argument from an Exceptional Case

CQ1: Is the case of *a* a recognized type of exception?
CQ2: If it is not a recognized case, can evidence why the established rule does not apply to it be given?
CQ3: If it is a borderline case, can comparable cases be cited ?

The last critical question indicates how argument from an exceptional case and its rebuttals are based on argument from analogy, as one case is compared to another. This connection will emerge even more dramatically when we consider arguments from precedent.

8. ARGUMENT FROM PRECEDENT AND PRACTICAL ARGUMENT FROM ANALOGY

Argument from an exceptional case focuses on the problem of the classification of a case as an instance of the class the norm deals with. Argument from precedent points out another aspect of the normative counterpart to the argument from verbal classification, namely, the link between the two properties. This scheme may be described as an extension of the previously described scheme in which the law or rule establishes that the classification of a case as having a property *F* leads to the conclusion that the property *G* applies to it as well. In a legal type of dialogue, where plausibility value and probative weight are judged differently than they are in ordinary conversational argument, precedent is judged by similar cases decided in previous trials (*ratio decidendi*). The criteria of classification can be clearly established in legal cases by classifications and definitions that are codified in law. Argument from precedent can be used to apply a rule to a case, but it can also be used to attack a prior argument that a rule should be applied to a case. The following refutation scheme represents this common form of attack. It is based on the capability of recognizing and regulating the possibility of exceptions to a rule. In law, as well as in everyday argumentation, argument from precedent shifts the burden of proof to the side of the respondent. The move itself may need to be defended, or the major premise may be undercut and the argument refuted.

Student: I don't think I will able to get my essay in on Tuesday. Would it be OK if I handed it in next week?

Professor: We all agreed at the beginning of the year that Tuesday is the deadline. That is the rule.

Student: Yes, but you said to Ms. Reasoner that she could hand her essay in a week late because she has another assignment due this week. I have another assignment due, too. So I should be able to hand mine in a week late, too. (Walton, 1995, p. 94)

The scheme for argument from precedent fitting this example is the following (Walton, 1996, p. 94). It leads to a conclusion that the rule must be modified.

Refutation Scheme for Argument from Precedent

Major Premise: Generally, according to the established rule, if x has property F, then x also has property G.

Minor Premise: In this legitimate case, a has property F but does not have property G.

Conclusion: Therefore, an exception to the rule must be recognized, and the rule must be appropriately modified or qualified.

The following critical questions represent strategies to defend the argument from established rule by attacking the opponent's argument.

Critical Questions for Argument from Precedent

CQ1: Does the established rule really apply to this case?

CQ2: Is the case cited legitimate, or can it be explained as only an apparent violation of the rule?

CQ3: Can the case cited be dealt with under an already recognized category of exception that does not require a change in the rule?

These critical questions represent possible moves to defeat an argument from precedent.

The practical counterpart to the scheme for argument from analogy is practical reasoning from analogy. The similarity of two situations in some respects is considered a reason to act in a similar way. For instance, in the example below, the rebels are compared by Reagan to the American patriots, while they are presented by his opponent as similar to the Vietcong. "Patriot" and "rebel" are the two categories used to describe the same situation. The force of this argument lies in the persuasive force of the definitions. Patriots are, by definition, people who fight for their freedom and, therefore, are "good," while rebels are people who do not respect the law and social order and, for this reason, are "bad." The course of action of helping good or bad people is, therefore, classified as good or bad. In this argument, similarities are fundamental to justification of the action.

> President Reagan, in a speech for congressional funds to aid the Contra rebels in Nicaragua, compares the Contras to the American patriots who fought in the War of Independence. A speaker in Congress opposed to sending aid to the Contras compares the situation in Nicaragua to the war in Vietnam. (Walton, 1989, p. 256)

The schemes for this type of practical inference are the following (Walton, 1989, p. 257):

Positive Scheme for Practical Argument from Analogy

The right thing to do in $S1$ was to carry out A.
$S2$ is similar to $S1$.
Therefore, the right thing to do in $S2$ is carry out A.

Negative Scheme for Practical Argument from Analogy

The wrong thing to do in $S1$ was to carry out A.
$S2$ is similar to $S1$.
Therefore, the wrong thing to do in $S2$ is carry out A.

Practical argument from analogy is the basis of another type of argumentation scheme, the slippery slope and its variants. The most important premise is the similarity between the case beyond contention and another, a_{k-1}, closely related, differing only in a negligible detail. The two cases are presented as almost the same: therefore, following a reasoning step comparable to the practical argument from analogy, they are classified in the same way. The same scheme is successively applied to a_{k-1}, and the analogical reasoning is followed until the controversial case a_i is reached. The chain of preceding steps proves that the latter is analogous to the premise beyond contention and, therefore, that it should be classified in the same manner.

9. THE CASE OF THE DRUG-SNIFFING DOG AGAIN

Returning to the case of the drug-sniffing dog, we can see how much of what was considered to be argument from analogy, as that case was treated, is really based on either argument from classification or argument from precedent. For example, consider once again the three general rules cited as having been established by prior cases.

> *Rule 1*: If a police officer sees something in plain view in a public place, the information collected is not classified as a search.
> *Rule 2*: If a police officer opens luggage and then observes something inside the luggage, the information collected is classified as a search.
> *Rule 3*: If a police officer listens surreptitiously to a conversation in a private place, it is classified as a search.

These rules set down classifications based on how the previously classified open-textured term was defined in previous court rulings. The analogy-warranting rule, formulated by Brewer as the essential principle on which arguments from analogy are based, is better classified as an argument from classification. Brewer claims, as noted in section one, that this rule as stated is sufficient to resolve the issue in the case of the drug-sniffing dog. But really, this rule is not part of an argument from analogy, nor does it belong to the argumentation scheme for the argument from analogy. Instead, it is an argument from classification, supposedly based on evidence from the Fourth Amendment. Thus it is better seen as an argument based on classification, and perhaps also on statutory interpretation and precedent, that can either support or attack the basic argument from analogy in the drug-sniffing dog case. We take this conclusion to support our thesis that for the purposes of analyzing and evaluating arguments from analogy in legal argumentation, one has to look at supporting arguments, defeaters and undercutters that can be stated as critical questions. These need to be seen as arguments of a different sort that typically surround and relate to an argument from analogy.

Consider once again the question of whether a qualification needs to be inserted by considering the kind of case in which a police officer uses an X-ray machine to allow him to see what is inside a container without opening it or taking possession of it. Would the information collected in such a case be classified as a search? Once again, in this case we need to see that the issue, although partly one of analogy, is also partly one that involves argument from classification. It was noted earlier that one tool that could be used to deal with a new kind of case of this sort might be the adding of a qualification. For example, the question might be rephrased to ask whether what the policeman is doing is a search if he is allowed to see what is inside a container without opening it and without using a technological device.

Consider once again the argument formulated in section six, taken by Brewer to be deductively valid. What happens when this argument is confronted with a case in which the police officer obtains the information using a technological device? Let's expand the reasoning in the case to show how this new factor affects the previous argument.

Expansion of the Warranting Rule Argument in the Drug-Sniffing Dog Example

If a police officer obtains information about a person or thing in a public place without intrusion on the person or taking possession of or interfering with the use of the thing, it is not a search for purposes of the Fourth Amendment.

In the case of the drug-sniffing dog, the police officer obtained information about a person or thing in a public place without intrusion on the person or taking possession of or interfering with the use of the thing.

Conclusion 1: The case of the drug-sniffing dog is not classified as a search for purposes of the Fourth Amendment.

Using the argument diagramming system Araucaria,[5] the structure of this argument can be represented as in Figure 2.1. As shown in the diagram, argument from verbal classification is the scheme that links the two premises together and leads to the conclusion. Two things need to be noted. One is that in this case, the argument is a negative type of

---

[5] Araucaria (Reed and Rowe, 2005) aids a user in constructing a diagram of the structure of an argument by inserting the text to be analyzed as a text document. Each statement is represented in a text box that appears on the screen. The user can then draw in arrows from each premise to each conclusion that it supports, producing an argument diagram connecting the premises and conclusions, as well as displaying argumentation schemes.

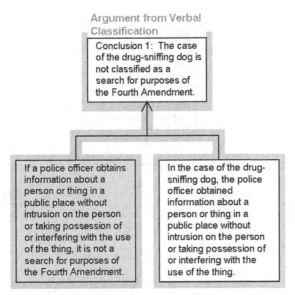

FIGURE 2.1. Araucaria diagram of the WRA in the drug-sniffing dog example.

argument from verbal classification, a type of argument not previously recognized in the literature. The other is that the argument has the form of a defeasible *modus ponens* (DMP) type of argument. It is the thesis of Verheij (2003a) however, that all of the defeasible argumentation schemes have this form (in broad outline).

Let's go on to extend the reconstruction of the argument a little further. In a new case, the police officer obtained information about a person or thing in a public place without intrusion on the person or taking possession of or interfering with the use of the thing by using a technological device to see what was inside a container possessed by the suspect without touching or opening it.

> Conclusion 2: It now becomes doubtful, in the new case, whether this should be classified as not being a search for purposes of the Fourth Amendment, as the case of the drug-sniffing dog was by Conclusion 1.

This part of the argument is diagrammed in Figure 2.2, showing how an implicit premise is added in the shaded box. This diagram shows how classification is involved, but is incorporated into an argument from analogy based on a similarity between two cases. The old case is a precedent that supports a classification. An implicit premise is assumed, to the effect that the new case is similar to the old one. A conclusion is inferred about how the new case might be classified.

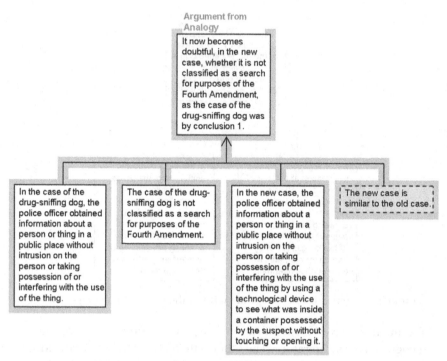

FIGURE 2.2. Argument diagram for the reconstructed development of the drug-sniffing dog example.

We can now look at the original argument in this new way, seeing it as part of a chain of reasoning in which new cases can come in, setting or overruling precedents from previous arguments. The previous argument looked like it could be deductively valid, because the general rule stated in the first (conditional) premise looked like a universal (absolute) generalization. But in the expanded version, we see how the original argument could be defeated by a new case. As Weinreb (2005, p. 25) proposes, the original generalization now needs to be qualified as follows.

> If a police officer obtains information about a person or thing in a public place without intrusion on the person or taking possession of or interfering with the use of the thing, and without using a technological device, it is not a search for purposes of the Fourth Amendment.

Given the insertion of this new qualification (antecedent of the conditional), the new conclusion is that what the police officer did when using the X-ray device is classified as a search.

This reformulation of the original argument is interesting, because it suggests that the argument is not deductively valid after all. It is based on a defeasible generalization of a kind that is subject to exceptions, one that can be defeated in the future by an exception that may arise through a new case to be decided. This view of the matter suggests that we should reformulate the major premise, as follows, to classify it as a defeasible generalization.

> Generally, but subject to exceptions, if a police officer obtains infor-
> mation about a person or thing in a public place without intrusion
> on the person or taking possession of or interfering with the use of
> the thing, it is not a search for purposes of the Fourth Amendment.

Moreover, this general rule may fail to apply in some instances. It should not be seen as an absolute generalization of the kind modeled by the universal quantifier in classical deductive logic. Thus Brewer's view of the case as fitting a deductively valid form of argument from analogy falls apart, once we take a closer look at the example of the drug-sniffing dog.

Really, the question here is how a search should be defined, where we know that the term 'search' is open-textured, even though it is increasingly being made more precise by precedents brought forward in the courts. To handle such cases, Weinreb (2005, p. 31) suggests that there has to be a sequence of argumentation moves described as a process of "reciprocal reflective adjustment" in which a given rule is refined by taking new exceptions and precedents into account. He suggests that the sequence of argumentation in such a case goes back and forth from the specific case to the statute, and is mediated by precedents and by new cases considered by the courts that intervene between the statute, at one end, and the specific case being ruled on, at the other end. As we have noted, Brewer also sees the sequence of argumentation in the drug-sniffing dog case as containing not only an argument from analogy but also a wider process of reasoning in which it is embedded. What these considerations suggest is that the analyses of the drug-sniffing dog example considered in section one, by seeing this whole broad network of reasoning as an argument from analogy, have cast too wide a net. A more realistic analysis is that there is a central argument from analogy involved, but that it is situated in a massive network of evidence containing many other kinds of arguments that either support the argument from analogy or attack it. In a typical case, well exemplified by the drug-sniffing dog case, many of the most important arguments surrounding the argument from analogy, and most closely related to it as interlocking supporting or attacking arguments, are arguments from classification and precedent.

10. CONCLUSIONS

In this chapter we have successfully developed tools enabling us to identify precise forms of argument from analogy, and to analyze both common and controversial examples of argument from analogy in legal reasoning. By this means, we have proposed a third way between the skeptics, who doubt whether argument from analogy is a form of rational argument at all, and the mystics, who think that argument from analogy lacks rational force even though it can be justified in the law as having procedural benefits. Basically, we have done this using two methods. One is to go between the horns of the dilemma posed in traditional logic by seeing argument from analogy as not having to fit one of the two traditional paradigms of rational argument, deductive or inductive reasoning. We have argued that in some cases argument from analogy can assume deductive and inductive forms, but in the broad majority of typical cases of argument from analogy as used in legal reasoning, it has a defeasible form that is neither deductive nor inductive, and is best modeled using the defeasible argumentation schemes we have presented.[6] Another important thesis of this chapter has been that the typical kinds of cases of arguments cited as argument from analogy in legal reasoning, like the case of the drug-sniffing dog, are not simply arguments from analogy but involve other argumentation schemes, especially argument from verbal classification and argument from precedent. We have shown how, in such typical cases, a network of argumentation is formed, and how the arguments in it can be visualized using an argument diagram, which typically combines argumentation fitting these various schemes into a connected mass of evidence. Thus we see that prior treatments of the subject of argument from analogy in law are oversimplified, that they tend to analyze these common cases of legal reasoning as being simply arguments from analogy, ignoring other argumentation schemes that are woven together with the argument from analogy and are an important part of it. In distinguishing these different types of arguments, and in

---

[6] Juthe (2005) supports our conclusion that arguments from analogy of the most common and typical kinds are not deductive in nature, but he takes the even stronger view that such arguments are always from particular to particular, and are therefore never deductively valid. His thesis is that in an argument from analogy, "the conclusion never follows solely in virtue of the semantics or the syntactical structure of the argument" (p. 24).

showing how they fit together with the argument from analogy in such typical legal cases, we have indicated how the literature that has tried to grapple with the logical problems posed by these cases needs to be broadened to take these other argumentation schemes into account.

Earlier we said that the interesting question raised by the case of the drug-sniffing dog is how arguments from classification and arguments from analogy can be combined with other argumentation schemes in the wider process of legal reasoning in a case that involves judging the admissibility of evidence in a trial. The process of untangling the individual stages in this broader process of reasoning requires three steps. The first step is the definition of argument from analogy as a specific argumentation scheme. The second step is that of distinguishing it from closely related schemes like argument from classification and argument from precedent. The third is the examination of the problem of how the critical questions match each of these argumentation schemes and how they can be used as devices for undercutting or rebutting an argument that fits a particular scheme in a given case. In the past it has not been unreasonable to try to portray argument from analogy as deductively valid or, alternatively, to see it as merely a mystical device that has no logical structure we can firmly pin down. But once the argumentation scheme for argument from analogy has been precisely defined, and differentiated from closely related schemes like argument from classification and argument from precedent, the air of mystery disappears. We can see that argument from analogy has a form that can be precisely defined. Now we are partway to our goal, because we can see how to identify and analyze argument from analogy, even though some questions remain about the precise form the scheme should take.

It may well turn out that the approach we have taken to developing argumentation schemes for argument from analogy, of a kind that would be adequate to modeling this argument as used in case-based reasoning of the kind so commonly found in law, is incomplete. One reason is that legal reasoning is based on the structural similarity of the rationales of two cases, when one is used as a precedent by analogy to another. This kind of similarity is inherently complex. It may not be possible to model it in the simpler type of case we have mainly been concerned with here, in which one situation or event is held to be similar to another. Another reason is that according to the theory of Verheij and Hage (1994), in order to analyze argument from analogy as used in legal reasoning, one has to take into account the principles and goals that underlie a legal rule as applied to a case, and thereby to see how these factors generate reasons to solve the case. They comment that classical logics, such as sentential and first-order predicate logic, are not well suited to dealing with this kind of reasoning. Also, classical deductive logic is not adequate to fully treating rule application disputes where reasons both for and against drawing a conclusion can be offered. By analyzing a case from the Dutch Supreme Court, Hage (2005) has brought out more limitations of attempts to treat argument from analogy in law as based on the kind of deductive inferences we are familiar with as the model of logical reasoning in classical logic.

We agree with Govier, who argued that many of the most typical cases of argument from analogy should not be classified as deductive, because they are persuasive in nature and lose their persuasive force when recast into deductively valid form. We also agree with Guarini and Govier that the so-called principle or generalization on which an argument from analogy is based is uncovered only through a period of successive refinements during a process of reflection and analysis. We see this process as best modeled in the form of a dialogue (or investigative procedure) in which the generalization is refined through a process of finding exceptions to it that can then be built into the antecedents of the generalization. As the literature on defeasible reasoning in artificial intelligence has abundantly illustrated, such generalizations can be defeated by contrary instances in the future, instances of a kind that cannot be anticipated in all cases. We see this process as one in which one side in a dialogue brings forward an argument from analogy, combined with other argumentation schemes, and the other side critically questions this argument, or subjects it to criticism using counterarguments. We see the argumentation in the typical case of a legal dispute, like that of the drug-sniffing dog, as best analyzed and evaluated as a rule-governed dialogue in which each side has a burden of proof. How strong the argument needs to be in order to be acceptable is determined on the basis of whether it meets the requirements of this burden of proof. Although we cannot include our whole approach to argument evaluation here, and only allude to it by citing some of the literature, the general approach can be summarized as one that treats defeasible argumentation schemes using a dialogue model in which the critical questions are used as the vehicle for evaluating the argument in a given case. Our analysis of argument from analogy clearly fits this general format.

We offer the following solution to the problem of analyzing how argument from analogy works as a fallacy. In many cases, arguing from analogy is quite reasonable. Indeed, as we've seen, our legal system of reasoning is fundamentally based on it. So when does it become fallacious? We theorize that this form of argument becomes fallacious when the similarity is so striking that the argument from analogy appears impervious to critical questioning. Consider again Judge Kozinski's story describing how when the movie *Total Recall* came out in 1990, he found the similarities between the movie and his unpublished novel striking. The story lines appeared to him to be identical, and he was strongly impressed by a scene in which the villain kicks over a fish tank, which seemed to him very similar to a scene in his novel. He was therefore convinced that his novel had to have been pirated. The key premise of this argument from analogy was the perceived striking similarity. He drew what appeared to him to be the inescapable conclusion that his story had been stolen.

Based on such a striking similarity, an argument from analogy can look like a very strong form of argument, like a deductively valid argument or a very strong inductive argument. However, as we've seen, in the most typical cases in legal reasoning argument from analogy is a defeasible type of reasoning. It is an inherently weak and fallible argument on its own, which can nevertheless carry some evidential weight in a larger mass of evidence of which it is a part. We have shown how argument from analogy in typical legal cases is combined with other arguments, like argument from precedent and argument from verbal classification. It is also commonly combined with other forms of argument, like argument from expert opinion and appeal to witness testimony. Argument from analogy has a legitimate place as a vehicle for presenting persuasive evidence that can shift a burden of proof, even though it is inherently defeasible and subject to critical questioning. The problem of its being a fallacy comes in when the strength of the argument is overrated, when it is not evaluated in the context of a mass of relevant evidence in a case but is seen as a strong, even overwhelming, argument that stands on its own as convincing. Later, after a period of reflection, Judge Kozinski came to realize his error. He took other evidence into account. He had never finished the novel, nor had he ever sent it to anyone to read. He realized that he had leaped to a hasty conclusion. The structure of the argumentation is shown in the Araucaria diagram in Figure 2.3. The darkened box represents a refutation.

The initial stage of the argumentation is pictured on the right, leading to the conclusion that the novel must have been stolen. At the second stage, Judge Kozinski was led to retract his earlier conclusion and form a new one based on the new evidence shown in the bottom two boxes on the left. He now concluded that, despite the similarities, it was just a

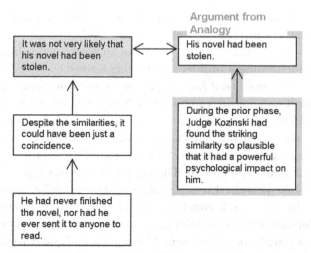

FIGURE 2.3. Araucaria diagram of the fallacious inference in the striking similarity case.

coincidence. During the prior stage, he had found the striking similarity so plausible and seductive that he could not resist the overwhelming conviction that his novel had been stolen. The occurrence of the fallacy of jumping from an analogy to a conclusion is indicated by the sequence of argumentation joining the two stages. The first stage shows the powerful psychological impact of the striking similarity. The second stage shows a retraction of the earlier conclusion, based on a more sober reassessment of the total evidence. The new evidence supports the refutation of the prior conclusion.

The fallacy of deceptive argument from analogy can be compared to some other fallacies of the traditional kind, where there is an underlying reasonable basis as a heuristic, but where the fallacy is the tendency to leap too quickly to a conclusion without considering other factors. The *post hoc* fallacy, for example, is the fallacy of leaping too quickly to a causal conclusion on the basis of a correlation. The correlation may exist, and therefore the premise of the causal argument may be true. Moreover, the correlation may suggest that the possibility of a causal conclusion should be further investigated. The problem is, however, that the correlation may be simply a coincidence, or it may be caused by some other factor that is common to both variables that are correlated. So arguing from a correlation to a causal conclusion is a legitimate heuristic, but leaping too quickly from a correlation to the causal conclusion, without considering or investigating other factors that may be involved, is a fallacy. Appearances offer evidence, and are a main source of evidence, but the inferences drawn from them can lead to a conclusion that turns out to be wrong.

There is actually an argumentation scheme corresponding to this form of reasoning called argument from appearance. Henry Prakken (2003) used the example of this argument: this object looks like an affidavit; therefore, it is an affidavit. The problem with the *post hoc* fallacy is that we tend to see the world in causal terms, so when we see a striking chain of events in which one event occurs, and then another event occurs right after it, the second event strongly appears to be causally related to the first one. Hence we tend to leap to a causal conclusion. But this very strong conviction or appearance of causality can easily be mistaken. Similarity is part of the evidence for copyright cases, and if there is a striking similarity, that could be further evidence for drawing the conclusion that there is a possible violation of copyright that needs to be investigated by looking into other relevant factors. But simply leaping to the conclusion that there has been a violation of copyright, based only on the evidence of the striking similarity between two cases, is a fallacy. The striking similarity premise suggests the conclusion that there may have been a violation of copyright, but this inference is only a very weak form of abductive reasoning that considers one possible explanation instead of investigating alternative explanations at sufficient depth to reach a conclusion based on evidence. Our limited experience tells us that copying is the only explanation for the striking similarity, and we draw on the heuristic that we believe in our perceptions.

The remaining problem that also seems to enshroud argument from analogy in an air of mystery is that of how to model the critical questions. The critical questions appear to be part of a dialogue through which the scheme is matched with devices that can be used to pinpoint weak parts of the argument that should properly be subject to doubt. But how can this process, through which the question is used to raise doubts in relation to the scheme, be modeled? The answer we have proposed is taken from recent advances in the field of artificial intelligence. The general research proposal is to see the critical questions as modeled by counterarguments that can be expressed as sets of propositions (Bex et al., 2003; Reed and Walton, 2005; Walton and Gordon, 2005). First of all, critical questions that merely ask whether or not a premise holds can be classified as redundant, because these kinds of questions can be asked automatically by the automated system designed to evaluate the argument from analogy. The next step is to classify the remaining critical questions as being counterarguments of various kinds. Some of them can be classified as undercutters, others as rebuttals. The problem is then how to model undercutters and rebuttals using the current technology of artificial intelligence. We do not comment further on these investigations here, except to point to directions of advance in the recent literature.

Our conclusion is to recommend version one of the argumentation scheme for argument from analogy, reserving application of version two to cases where relevance needs to be decided. The relevant similarity premise makes version two a precise tool that could be useful for comparing cases in evaluating arguments from analogy. We repeat, however, that we see the major premise of version one as fulfilling quite an important role in the scheme for argument from analogy. We see the major premise as a generalization stating that one case has been found to be similar enough to another case to warrant a conclusion drawn by argument from analogy proceeding from the source case to the target case. For such a premise to be established, or to be well supported by the evidence, there needs to be a significant match between the source case and the target case. Thus, for example, there can be a similarity between a red ball and a red apple in one respect, namely, the respect that both are red. However, this similarity would not be sufficient for inferring that because the apple is good to eat, the ball would also be good to eat.

The final remaining problem is how similar two cases have to be for us to be able to draw an inference by a reasonably strong argument from the source case to the target case. We do not attempt to give a general answer to this question here. All we have been able to do is to show that in many common cases, because of our skill at matching patterns by bringing to bear knowledge that enables us to see one case as similar to another, we can make and judge such inferences. Our approach has been to take the first step of explaining how an argument from analogy can be identified, and how the critical questions matching the argument can be modeled as rebutters or undercutters of a kind that pinpoint weaknesses and gaps in the given argument. In effect, we've taken the negative route of showing how a given argument from analogy, once accepted or seen as being plausible in a given case, can be questioned or criticized by locating weak points in it and exploring the backing evidence and supporting arguments on which it is based. We think that this is an important first step in beginning the ultimate project of learning how to evaluate arguments from analogy as strong or weak.

3

Knowledge-Related, Practical, and Other Schemes

In this chapter, we outline a number of schemes that do not fall into any single classification but are useful to know about before considering the schemes discussed in the next chapters. We begin with a group of schemes based on assumptions about knowledge and go on to consider a group related to actions, goals, and means. These latter include schemes representing practical reasoning and arguing from the consequences of an action that is being considered. Finally, we consider some other schemes that are not easily classified, including arguments from composition and division, fear appeal arguments, appeals to pity, and argument from alternatives and opposites. Some work has been done on some of these schemes, while others remain largely unstudied.

1. ARGUMENTS FROM KNOWLEDGE

In arguments from expert opinion, position to know, and witness testimony, the acceptability of the conclusion is drawn from a knowledge base that the source is assumed to have access to. The relation between the latter and the plausibility of the conclusion is based on a kind of default reasoning: it is assumed that the proposition in question is contained in a certain subject domain that is known by the source. For this reason, it is an assumption in this kind of argumentation that many or most of the true propositions of the domain are known. The warrant of this kind of argumentation is therefore closely similar to that of the argument from ignorance. The falsity of a proposition is inferred from its absence from the domain of knowledge. The reasoning runs as follows: the proposition is not known to be true; therefore, it is assumed that it is false. This form is closely related to the source-based arguments, because it is based on a negative scheme: the expert source does not know that the proposition is true; therefore, (plausibly) it is false. The greater presence of true rather than false propositions in the expert domain is a good reason to believe that, in case it is true, the alleged proposition will be present in his database. Thus from its absence, its presumed falsity follows. Of course, the strength of such an argument will depend on the exhaustiveness of the database.

The problem of what is meant by falsity is, however, an extremely complex problem. The *argumentum ad ignorantiam*, so closely related to negation as default – or intuitionistic negation, as it is known in logic – is only one of the negative schemes related to a domain of knowledge. Another is the following scheme: the proposition may be falsified because its opposite, contradictory, or contrary is known to be true. Argument from ignorance, however, is relevant in the discussion about source-based topics, because the conclusion is reached through the relationship between the proposition and the whole relevant database. Like the positive arguments from source, its level of completeness determines the strength of the conclusion.

In addition to this characteristic, in the evaluation of arguments from knowledge it is necessary to consider the link between the expert and the proposition that is his assertion. The possibility of lying, misinterpretation, or misquotation can affect the plausibility of the conclusion. For this reason, all these topics follow the same general format: the conclusion follows from a premise regarding the domain of knowledge and another one concerning the assertion made by the source. The difference between them lies in the connection between the database and the supported proposition. The latter, in fact, defines the field it belongs to: it may be part of practical notions, related to experience, or part of technical and cultural knowledge, or it may be simply a fact, an event. For this reason, the arguments are differentiated in position to know, expert opinion, and witness testimony, respectively. The proposition, moreover, restricts the domain to a specific field that influences the choice of the source used to support it. In relation to the specific area of interest, the most complete available database is chosen.

In order to show the difference between the argument from position to know or witness testimony and the argument from expert opinion, it is useful to distinguish between access to facts, on the one hand, and experience or expertise, on the other. The definition of 'expert' lies in the notion of specialized, extraordinary skills or mastery in a particular field. This presupposes, first of all, study or practice and second, acknowledgement by peer experts (Walton, 1997, p. 121). Both an expert and a person in a position to know, or a witness, have a privileged access to a particular kind of information, but the former also has access to the whole domain of knowledge the proposition belongs to. For this reason, expertise or experience and testimony must be distinguished under the aspect of the interpretation of the known event. The latter can report the facts, but does not have the knowledge necessary to connect it with the domain of related information. Such a source cannot be expected, in other words, to give an authoritative opinion.

The distinction between expertise and experience may be clearly drawn by explaining the difference between cultural and practical knowledge. Being in a position to know means having a privileged access to facts or having a wide database of previous similar events, while the privileged kind of information the expert has access to represents the possible interpretations of these facts or events. The person in a position to know can express a valid opinion about the requested information, because he is in a position to interpret facts. Thus the expert opinion may be the best among the possible available explanations of a given set of facts. The difference between experience and expertise may be clarified through the analysis of the following example.

> If we want to know what it feels like to be in battle, we are rightly inclined to pay more attention to the report of a soldier who has been in the front lines than to the account of a man, however imaginative, who has never been exposed to the shooting. For, in such a case, we feel that the personal experience of the person giving testimony is essential to the reliability of his account. (Walton, 1997, p. 83)

The soldier is in a position to know about the feelings of one personally exposed to battle conditions and so can give a personal opinion about being in a battle. However, if we want to have an analysis of the psychological impact of the battle on a person, we should address this question to an expert in that particular field.

The argument from *position to know* takes the following form (Walton, 2002, p. 46):

> *Major Premise*: Source *a* is in a position to know about things in a certain subject domain *S* containing proposition *A*.
> *Minor Premise*: *a* asserts that *A* is true (false).
> *Conclusion*: *A* is true (false).

The first critical question is directed against the major premise: if the source is not in a position to know, the whole argument collapses. The remaining two critical questions focus on the problem of the assertion: it may be unreliable because the source is dishonest, or because his words have been distorted.

CQ1: Is *a* in a position to know whether *A* is true (false)?
CQ2: Is *a* an honest (trustworthy, reliable) source?
CQ3: Did *a* assert that *A* is true (false)?

The argument from witness testimony can be a strong argument when there is no possibility of direct access to the facts. The main problem regarding this argumentation scheme is the reliability of the source. For this reason, in a trial, the witness takes an oath before testifying. The truth-telling premise is an important difference from the previous scheme. The two topics are distinct as well in virtue of another important feature. The argument from position to know may often be used in deliberation or in persuasion dialogue, whereas argument from witness testimony is mainly used in legal dialogue, in courts. Moreover, in such cases, the witness must strictly report the facts that he is directly acquainted with.

A typical example is the following argument.

Witness *A* states that Peter shot George.
Therefore, Peter shot George. (Verheij, 2000)

In this case, no interpretation is given, and although the argument is defeasible, especially in the absence of contrary evidence, it may be considered evidence supporting the conclusion that Peter is guilty. The following scheme represents the general form of the *argument from witness testimony* (Prakken, Reed, and Walton, 2003, p. 35):

Position-to-Know Premise: Witness *W* is in position to know whether *A* is true or not.
Truth-Telling Premise: Witness *W* is telling the truth (as *W* knows it).
Statement Premise: Witness *W* states that *A* is true (false).
Conclusion: *A* may be plausibly taken to be true (false).

Witness testimony is a strong argument when there is no direct access to the facts. For this reason, however, there are no direct empirical means to evaluate or verify a speaker's words except for comparison to other available evidence and evaluation of its consistency, both internal and external. The reliability of the source being the most important condition of the scheme, the critical questions are directed against this major premise. Consistency of the statement with the rest of the testimony, with the known facts, and with other testimony, along with source bias, are the most important factors used to evaluate the trustworthiness of the witness and, consequently, the plausibility of his statement.

CQ1: Is what the witness said internally consistent?

CQ2: Is what the witness said consistent with the known facts of the case (based on evidence apart from what the witness testified to)?

CQ3: Is what the witness said consistent with what other witnesses have (independently) testified to?

CQ4: Is there some kind of bias that can be attributed to the account given by the witness?

CQ5: How plausible is the statement *A* asserted by the witness?

The difference between arguments from witness testimony and position-to-know arguments, on the one hand, and arguments from expert opinion, on the other hand, concern the nature of the available database and the information given. The following example clarifies this distinction:

A *Newsweek* article, reporting on the debate on corporal punishment (paddling) in the classroom cites split opinions in state legislatures and among various advocacy and professional groups who have taken a position on the issue (James N. Baker, with Daniel Shapiro, Pat Wingent and Nadine Joseph, "Paddling: Still a Sore Point", *Newsweek*, June 22, 1967, p. 61). Don F. Wilson, president of the Ohio Education Association, is quoted in support of the claim made by paddling advocates that corporal punishment is necessary for classroom discipline: "A lot of child-development experts might change their minds if they were in a classroom." Immediately following this quotation, University of Kansas child-development Professor Donald M. Baer is quoted as concurring: "Corporal punishment can be effective in changing behavior." Immediately following this citation, the article states: "Most experts, and many parents, vehemently disagree." After citing a number of other studies by psychologists and university research centers on problems related to the issue, the article concludes with the statement that "experts say there may be no consensus to end the practice" of paddling as long as adults feel that what was right for them may also be right for their children. (Walton 1997, pp.138–139)

The experts are requested to provide opinions on a problem, based on interpretations of the facts. This is the form of the *argument from expert opinion* (Walton, 2002, p. 50):

Major Premise: Source *E* is an expert in subject domain *S* containing proposition *A*.

Minor Premise: *E* asserts that proposition *A* is true (false).

Conclusion: *A* is true (false).

The main critical points of the argument are the question of expertise and the question of the relevance of the source's expertise in relation to the expressed opinion or assertion. The first question has as its scope the assumed expertise of the source. For this reason, its focus is on the evaluation of the effective skills of the source through the analysis of his job, his qualifications, the testimony of peer experts on his competence, as well as his experience, skills, and publications. The field question is directed against the correspondence between the domain of expertise and the proposition: the domain itself may be examined from the point of view of type (knowledge or technical skill), its relation to other fields, and its development. The assertion is the topic of the third and fourth questions. The opinion question regards the problem of quotation and interpretation of the source's words, while the trustworthiness attack undermines the argument on the basis of source bias or dishonesty. The last two questions are directed against the plausibility of the link between premises and conclusion. The consistency and backup-evidence questions express the possible objection that "experts may be wrong." Their opinion may not be accepted by other experts, or it may not be based on strong evidence. Both these attacks weaken the acceptability of the conclusion, but they do not compromise the argument's validity.

Argument from expert opinion has already been outlined in Chapter 1. The following is a more extensive and detailed account of the critical questions and subquestions (Walton, 1997, pp. 211–225):

1. *Expertise Question:* How credible is E as an expert source?
   1.1. What is E's name, job or official capacity, location, and employer?
   1.2. What degrees, professional qualifications, or certification by licensing agencies does E hold?
   1.3. Can testimony of peer experts in the same field be given to support E's competence?
   1.4. What is E's record of experience or other indications of practiced skill in S?
   1.5. What is E's record of peer-reviewed publications or contributions to knowledge in S?
2. *Field Question:* Is E an expert in the field that A is in?
   2.1. Is the field of expertise cited in the appeal a genuine area of knowledge or an area of technical skill that supports a claim to knowledge?

2.2. If *E* is an expert in a field closely related to the field cited in the appeal, how close is the relationship between the expertise in the two fields?

2.3. Is the issue one where expert knowledge in *any* field is directly relevant to deciding the issue?

2.4. Is the field of expertise cited an area where there are changes in techniques or rapid developments in new knowledge? If so, is the expert up to date in these developments?

3. *Opinion Question*: What did *E* assert that implies *A*?

   3.1. Was *E* quoted as asserting *A*? Was a reference to the source of the quote given, and can it be verified that *E* actually said *A*?

   3.2. If *E* did not say *A* exactly, then what did *E* assert, and how was *A* inferred?

   3.3. If the inference to *A* was based on more than one premise, could one premise have come from *E* and the other from a different expert? If so, is there evidence of disagreement between what the two experts (separately) asserted?

   3.4. Is what *E* asserted clear? If not, was the process of interpretation of what *A* said by the respondent who used *E*'s opinion justified? Are other interpretations plausible? Could important qualifications have been left out?

4. *Trustworthiness Question*: Is *E* personally reliable as a source?

   4.1. Is *E* biased?

   4.2. Is *E* honest?

   4.3. Is *E* conscientious?

5. *Consistency Question*: Is *A* consistent with what other experts assert?

   5.1. Does *A* have general acceptance in *S*?

   5.2. If not, can *E* explain why not, and give reasons why there is good evidence for *A*?

6. *Backup Evidence Question*: Is *E*'s assertion based on evidence?

   6.1. What is the internal evidence the expert herself used to arrive at this opinion as her conclusion?

   6.2. If there is external evidence (e.g., physical evidence reported independently of the expert), can the expert deal with this adequately?

   6.3. Can it be shown that the opinion given is one that is scientifically verifiable?

In any given case, an argument from expert opinion needs to be evaluated in light of how finely it has been questioned and examined in the dialogue in the case.

2. PRACTICAL REASONING

In reasoning from alternatives and consequences, the agent's goal is to evaluate an action through its consequences, in order to decide whether or not to carry it out. On the other hand, he can reason backward, from the action to its antecedent necessary or sufficient conditions. He can assess, in other words, the means taken to bring it about. For instance, in the following example, the speaker's decision to invest money is conditioned by the availability of information about the economic market. The basis for making such a decision is the consultation of an expert.

> Suppose I am deliberating with my spouse on what to do with our pension investment fund – whether to buy stocks, bonds, or some other type of investments. We consult with a financial adviser, an expert source of information who can tell us what is happening in the stock market, and so forth, at the present time. (Walton, 1997, p. 113)

This kind of inference is defined as practical reasoning, and may take two forms. In the necessary-condition scheme, the action cannot be carried out without bringing about at least one of the possible alternatives. For instance, if I want to have a car, I have to buy, steal, or build it, or ask somebody to give it to me as a present. I cannot obtain it without realizing at least one of these or comparable conditions. My action is almost always bound to other actions. On the other hand, as in the case just presented, the condition may be sufficient but not necessary. It may be possible, in other words, to carry out the action without satisfying the condition.

This kind of argumentation scheme appears to be closely related to causal argumentation schemes. Aristotle called this *phronesis*, or practical wisdom. Nowadays it is widely recognized as central to AI. It is based on the notion of a rational agent, an entity that has goals and the capability of carrying out actions based on its goals and its knowledge of given circumstances (Wooldridge, 2000). Practical reasoning, according to the analyses of Clarke (1985), Bratman (1987), Audi (1989), and Walton (1990), is a chaining together (Walton, 1990) by an agent of what are called practical inferences. A practical inference has two characteristic types of premises. One states that the rational agent has a particular goal. The other states some kind of action through which the agent could carry out the goal. In the argumentation scheme for practical reasoning presented here, the rational agent is referred to by a first-person pronoun like 'I' or 'my', indicating that it is this agent who is putting forward the argument. The set $S_0, S_1, \ldots S_n$ represents a sequence of states of affairs that can be ordered temporally from earlier to later. In adopting this notation, a shift is made from the earlier practice of using capital letters to stand for events, to follow the kind of notation used in the scheme for argument from consequences presented earlier. A state of affairs is meant to be like a statement, but one describing some event or occurrence that can be brought about by an agent. It may be a human action, or it may be a natural event.

> *Goal Premise*: Bringing about S_n is my goal.
> *Means Premise*: In order to bring about S_n, I need to bring about S_i.
> *Conclusion*: Therefore, I need to bring about S_i.

The 'need to' in the conclusion expresses what is called a "practical ought," conveying the idea that if the rational agent is prudent, it ought to bring about S_i if it is committed to both premises of the argumentation scheme for practical reasoning. This argumentation scheme has been recognized as having two variants (Walton, 1997, p. 164). There is a necessary-condition version and a sufficient-condition inference that has the same structure except that the means premise expresses a sufficient condition rather than a necessary condition. In realistic cases of practical reasoning, the necessary and sufficient schemata are combined into longer sequences that include some necessary-condition links and some sufficient-condition links. The sequence chains forward from an initial state to a final goal state.

Necessary-Condition Schema

(*N1*) My goal is to bring about A (*Goal Premise*).

(*N2*) I reasonably consider on the given information that bringing about at least one of [$B_0, B_1, \ldots, B_n$] is necessary to bring about A (*Alternatives Premise*).

(*N3*) I have selected one member B_i as an acceptable, or as the most acceptable necessary condition for A (*Selection Premise*).

(*N4*) Nothing unchangeable prevents me from bringing about B_i as far as I know (*Practicality Premise*).

(*N5*) Bringing about *A* is more acceptable to me than not bringing about B_i *(Side-Effects Premise).*

Therefore, it is required that I bring about B_i *(Conclusion).*

Sufficient-Condition Schema

(*S1*) My goal is to bring about *A (Goal Premise).*

(*S2*) I reasonably consider on the given information that each one of $[B_0, B_1, \ldots, B_n]$ is sufficient to bring about *A (Alternatives Premise).*

(*S3*) I have selected one member B_i as an acceptable, or as the most acceptable sufficient condition for *A (Selection Premise).*

(*S4*) Nothing unchangeable prevents me from bringing about B_i as far as I know *(Practicality Premise).*

(S5) Bringing about *A* is more acceptable to me than not bringing about B_i *(Side-Effects Premise).*

Practical reasoning is put forward by a rational agent, typically in a deliberation dialogue with another rational agent or group of agents. The respondent agent can then pose critical questions in the dialogue. Following the analysis of Walton (1990) once again, five critical questions are given.

> *Other-Means Question:* Are there alternative possible actions to bring about S_i that could also lead to the goal?
> *Best-Means Question:* Is S_i the best (or most favorable) of the alternatives?
> *Other-Goals Question:* Do I have goals other than S_i whose achievement is preferable and that should have priority?
> *Possibility Question:* Is it possible to bring about S_i in the given circumstances?
> *Side Effects Question:* Would bringing about S_i have known bad consequences that ought to be taken into account?

As practical reasoning is used in a given case, the putting forward of an argument having the form of practical reasoning shifts a weight of plausibility from the premises to the conclusion, indicating that the agent should provisionally go ahead with committing to S_i. However, practical reasoning is commonly defeasible, subject to further dialogue. Asking any one of these five critical questions can defeat the argumentation, at least temporarily. Once the proponent gives an adequate answer to the question, however, the burden shifts back to the respondent. He must then commit to the conclusion unless he asks another appropriate critical question. This process can continue until the dialogue reaches the closing stage. Except under closure conditions, practical reasoning is a defeasible form of argumentation. It can default, or be defeated, at any point during the argumentation stage of a dialogue, as long as that stage continues.

The practical form of the argument from expert opinion is two-person practical reasoning. The agent, in this kind of argument, has a precise goal to achieve, and he asks the expert questions about the conditions and means to carry it out. An example of this topic is the Lorenzo's Oil case (Walton, 1997, pp. 134–136). Augusto and Michaela Odone's son, Lorenzo, was diagnosed in 1984 as having a rare and incurable disease, ALD, caused by the body's inability to eliminate VLCFA, very long chains of fatty acids. Doctors put the child on a special diet without VLCFA, in order to diminish this dangerous element. The Odones, frustrated by the treatment failure brought about by decreasing the level of VLCFA, tried to find a cure for the disease themselves. In a conversation between the two parties, the Odones and the doctors, two opposite points of view arose. The doctors used the clinic and the treatments to collect clinical evidence for the treatment of ALD, while the patient's parents had the task of saving their son's life. They requested the physicians to give them the means to achieve their goal, and the doctors replied from the perspective of their own interests and objectives. This communication failure is an interesting example of *two-person practical reasoning*, which takes the following form (Walton, 1997, p. 163):

X (the agent) intends to realize A, and tells Y (the expert) this.

As Y sees the situation, B is a necessary (sufficient) condition for carrying out A, and Y tells X this.

Therefore, X should carry out B, unless he has better reasons not to.

Critical Questions

CQ1: Does X have other goals (of higher priority) that might conflict with the goal of realizing A?

CQ2: Are there alternative means available to X (other than B) for carrying out A?

CQ3: Would carrying out B have known side effects that might conflict with X's other goals?

CQ4: Is it possible for X to bring about B?

CQ5: Are other actions, in addition to B, required for X to bring about A?

The Lorenzo's Oil case shows how this kind of dialogue can be very frustrating for those who have to deal with experts. The Odones had evidence that a special oil was working in the case of their son, but the physicians called such evidence "merely anecdotal," and they routinely dismissed it. From their point of view, drugs or other treatments must be clinically tested before any decision to use them can be made. Thus the dialogue between the user of the knowledge and its suppliers, the experts, can admit of many difficulties.

3. LACK-OF-KNOWLEDGE ARGUMENTS

As mentioned earlier, arguments from source are connected to the *ad ignorantiam* scheme. The *negative form of argument from knowledge*, in fact, may be considered a subspecies of the *argumentum ad ignorantiam* (Walton, 1997, p. 95):

> All the propositions in E's knowledge base K on subject S are true
> (*Truth Condition*).
> All the propositions on subject S not in E's knowledge base K are false
> (*Completeness Condition*).
> E is asked whether A is true.
> E replies that A is not true (as far as E knows).
> Therefore, A is false.

The main distinctive feature of this scheme is the completeness condition. In order for the argument to be maximally strong, the knowledge base must be considered complete (epistemic closure). That is, the expert is assumed to have an exhaustive knowledge of the field, one that allows the inference from the absence of evidence to the falsity of the conclusion. This device is frequently used in courts to shift the burden of proof. For instance, a common argument might be:

> As far as any evidence that has been brought out at the present time,
> there is no danger (Walton, 1997, p. 97).

This topic is an instance of the general form of *ad ignorantiam*: the absence of positive proof for the truth of the proposition is considered a reason to believe in its falsity. This argument is the basis for the presumption of innocence in criminal law. In the absence of evidence, the accused is assumed to be legally innocent. The form of argument is the following (Walton, 1995, p. 150):

> *Major Premise*: If A were true, then A would be known to be true.
> *Minor Premise*: It is not the case that A is known to be true.
> *Conclusion*: Therefore, A is not true.

The following example is a clear instance of this form of argument:

> I do not know that there is a skunk in the cabin.
> Therefore, it is false that there is a skunk in the cabin. (Walton, 1995, p. 150)

This may be applied to two particular cases: reasoning from normal expectations and negative practical reasoning. In the first case, the knowledge base is given by the set of the facts that in a normal situation would be true, or are expected to be true. The absence of positive evidence for a proposition of this sort is a reason to conclude that the proposition is false. For instance:

> A: Is Leona Helmsley still in jail? She's probably out by now.
> B: Maybe she's still there, because we'd probably hear about it if she got out. (Walton, 1996, p. 85)

If an important person had been released, the normal expectation would be wide media coverage of the event. Given this kind of knowledge base, it is possible to infer from the failure of the expectations that the event has not occurred. The reasoning is from expectations, and *epistemic argument from ignorance* (Walton, 1992, p. 386):

> It has not been established that all the true propositions in D are contained in K.
> A is a special type of proposition such that if A were true, A would normally or usually be expected to be in K.
> A is in D.
> A is not in K.
> For all A in D, A is either true or false.
> Therefore, it is plausible to presume that A is false (subject to further investigations in D).

The other subspecies of the *ad ignorantiam* scheme is a form of practical argument. It is a deliberative inference, used when there is no knowledge about a particular fact or proposition the action is based on. In this scheme, the possible consequences following from the truth of the proposition are compared to the consequences following from its falsity. The practical decision is taken on the presumption that the worst possibility is the case. For instance:

> It is a rule of safety in handling firearms that if you do not know for sure that a weapon is unloaded to act in accordance with the

presumption that the weapon is (or may be) loaded (Walton, 1996, p. 86)

The comparison of the possible consequences is the rationale underlying how the presumption is placed. The lack of conclusive evidence in this case is a reason to believe that the proposition is true. This negative practical reasoning follows the following scheme (Walton, 1996, p. 86):

> I do not know whether A is true or not.
> I have to act on the presumption that A is true or not true.
> If I act on the presumption that A is true, and A is not true, consequences B will follow.
> If I act on the presumption that A is not true, and A is true, consequences C will follow.
> Consequences B (C) are more serious than consequences C (B).
> Therefore, I act on the presumption that A is not true (true).

The argument from ignorance, both in its subspecies and in particular cases, depends on the knowledge base. The reasoning stems from what is expected to happen or the possible known consequences.

4. ARGUMENTS FROM CONSEQUENCES

Argument from consequences may be considered as a generic category of causal inference, a topic treated in Chapter 5. Causal reasoning is one of the possible aspects, or meanings, of the consequential relation between two elements. The relation may be temporal or possibilistic, for instance. In some cases the correlation may be clearly identifiable as a causal, temporal, or possibilistic link, but the meaning of the consequence itself is not defined in the reasoning. The topic, in other words, does not presuppose an explanation of the inferential passage, but only that the two facts are successive and somehow related. The temporal succession may be, in other words, causally motivated, but the motivation itself may or may not be expressed or mentioned.

Two main categories of topics stem from this argumentation scheme: practical reasoning and the danger and pity appeals. The distinction is made on the basis of the direction of the argument. The first topic is deliberative, that is, it is aimed at decision making relative to certain goals. Emotional appeals, by contrast, are directed to the hearer. They may be described as means to persuade the interlocutor to act in a certain way. Both are classes of practical arguments. They have the performance of actions as the ultimate goal, and are related to inference from consequences. Practical reasoning is often a backward process from the outcome of a consequence to its premises, while danger and pity arguments proceed from the possible negative consequences that carrying out or not carrying out an action may have, respectively, on the hearer himself or on a third party.

Practical reasoning and argument from consequences represent two directions of the same kind of reasoning. The positive or negative evaluation of an event is a reason to bring it about or not to bring it about. The outcome of first type of inference, however, is the evaluation of the given possible course of action; in the second case, the course of action is discovered as the best possible means to realize the goal. While appeal to danger is a kind of negative argument from consequences, the pathetic premises of the appeal to pity are means directed to initiating a practical inference. The possible fact referred to as undesired and avoidable is relevant for the conversation only if it constitutes an effective speaker's concern. The emotional claim aims at making the problem relevant, in order to subsequently suggest a possible way to solve it through a practical inference. By examining the structure of consequential and practical topics, it is possible to understand the mechanism underlying these pathetic appeals.

Argument from consequences has a positive and a negative form (Walton, 1995, p. 155; Walton, 2000, p. 141):

Argument from Positive Consequences

If A is brought about, then good consequences will occur.
Therefore, A should be brought about.

Argument from Negative Consequences

If I (an agent) bring about (don't bring about) A, then B will occur.
B is a bad outcome (from the point of view of my goals).
Therefore, I should not (practically speaking) bring about A.

Argument from Negative Consequences (Prudential Inference)

You were considering not doing A
But, if you do not do A, some consequence B, which will be very bad for you, will occur or is likely to occur.
Therefore, you ought to reconsider, and (other things being equals) you ought (prudentially) to do A.

Critical Questions

CQ1: How strong is the likelihood that the cited consequences will (may, must) occur?

CQ2: What evidence supports the claim that the cited consequences will (may, must) occur, and is it sufficient to support the strength of the claim adequately?

CQ3: Are there other opposite consequences (bad as opposed to good, for example) that should be taken into account?

The following example is interesting to compare to fear or threat appeals. The advice given by the adviser is based on a negative evaluation of the consequences of the action being considered. But the adviser makes no threat, and his argument is not really an appeal to fear.

Candidate: I plan to tell the people in my next speech that I am not going to raise taxes.

Adviser: If you are going to say that, you had better be committed to it, because one of the consequences of your saying it is that people are going to hold you to it. If you later raise taxes, people are going to criticize and even ridicule you for it. My advice would be not to say it in the first place. (Walton, 1995, p. 182)

The adviser is only giving a warning here, but it is a warning that concerns the negative consequences of an action that the candidate in considering. The consequences could perhaps be somewhat fear-inducing or threatening to the candidate. But still, the advisor is not saying that he will see to it that the consequences come about, nor are the consequences specifically meant to target the emotion of fear. Thus here we have a case of argument from negative consequences, but it would be an error to classify it as a fear or threat appeal.

5. FEAR AND DANGER APPEALS

Arguments from fear appeal, as mentioned earlier, have a structure similar to that of arguments from negative consequences, but the difference lies in the nature of the appeal as a speech act. Between these two types of arguments we can classify the argument from danger appeal. Danger is an aspect of reality independent of the agent (Walton, 2000, p. 163). It is objective, while fears are emotional reactions of an agent to a situation. For this reason, the danger appeal argument may be considered as a special kind of argument from negative consequence based on a claim about the respondent's suffering serious harm or loss of life. We can supplement the examples given earlier with one involving a danger appeal.

There is a movie that is intended to convince people not to smoke cigarettes. It features interviews with people dying of lung cancer, mainly cowboys, in an ironic counterpoint to those Marlboro ads that pictured handsome cowboys smoking. So, for instance, you see one poor fellow riding the range with oxygen bottles strapped inside his saddlebags and the tube up his nose. (Walton 2000, p. 82)

The circumstances are objective, and thus the appeal is not just to the emotions. It is also based on the serious possible dangers of smoking. The ad appeals to irony by satirizing the famous Marlborough ad, making its macho image appear ridiculous, and thus deflating its emotional appeal to the smoker. In reality, smoking is an extremely dangerous habit that can lead to congestive lung disease. The ad is pointing out this danger in a graphic way. It is arguing to the viewer that, because of this very real danger, he or she should reconsider smoking.

Here is the scheme for danger appeal (Walton, 2000, p. 173):

If you (the respondent) bring about *A*, then *B* will occur.

B is a danger to you.

Therefore, (on balance) you should not bring about *A*.

We can compare the examples from negative consequences and danger appeal to the fear appeal argument. These arguments are based on an appeal to the subjective emotions of an agent. They represent an exploitation of an emotional and personal response to a given situation, whereas danger is an objective property of the situation external to the agent (Walton, 2000, p. 162). As mentioned earlier, the role of feeling and emotion is critical in classifying these kinds of argument. For instance, we can analyze the following example:

> An insurance salesman drops in to "inform" you of the advantages of buying a policy and, incidentally, of the possible disadvantages of not buying one. He describes the plight of Benny Morelli on the other side of town whose house burned down. Benny kept all his money in a heavy oak chest, which, naturally, was completely destroyed. Benny's wife had always worked to make ends meet, but now his oldest son had to drop out of school to help carry the financial burden. Benny's new car (which also lacked insurance) was demolished when the burning house fell on it. (Walton, 2000, pp. 53–54)

In this case, the argument fits the scheme of argument from negative consequences, and also that of danger appeal, but the situation itself is not really dangerous, and neither are the consequences that are said to proceed from the premises very likely to occur. Benny's problems were not caused solely by his failure to have an insurance policy, and drawing the conclusion that insurance is necessary from an exceptional case is not very reasonable. The appeal is to danger, but because the picture is painted in such graphic terms calculated to appeal to emotion, the argument is an appeal to fear. The extreme seriousness of the events narrated awakens feelings of fear that seem to make the conclusion plausible. The point is proved, however, not by analyzing the objective facts, but by raising fears about dangers. The salesman makes it seem to the respondent that the only way he can prevent the horrible disaster that happened to Benny is to take out an insurance policy.

The example fits the scheme for fear appeal (Walton, 2000, p. 22):

If you do not bring about A, then D will occur.
D is very bad for you.
Therefore, you ought to stop D if possible.
But the only way for you to stop D is to bring about A.
Therefore, you ought to bring about A.

The argument from negative consequences, as already shown, has as a subtype the prudential scheme. The advice given in such an argument, however, may be used to convey a threat. The agent commits himself, in other words, to bringing about a dangerous situation for the hearer. This example may clarify the distinction:

> A known gangster says to the owner of a small business: "You should pay us protection money, because this is a very dangerous neighborhood. The last guy who didn't pay had his store looted and destroyed, right after he failed to pay." (Walton, 2000, p. 123)

The scheme for the reasoning in this case is apparently very similar to that of the danger appeal, but we can observe that the gangster is making an indirect threat. Only if we interpret the speech act as an indirect threat does the passage make sense. The scheme of argument from consequences, specifically negative consequences, and threat appeal are involved (Walton, 2000, p. 123). First consider the argument from negative consequences contained in the following warning.

You were considering not doing A.
But, if you don't do A, some consequence B, which will be very bad for you, will occur or is likely to occur.
Therefore, you ought to reconsider, and (other things being equal) you ought (prudentially) to do A.

A warning of this kind is different from a threat. The making of a threat is a particular speech act that involves the speaker's commitment to bringing about the event in case the hearer does not carry out the designated action. There are three conditions that characterize the speech act of making a threat (Walton, 2000, pp. 113, 114):

> *Preparatory condition*: the hearer has reasons to believe that the speaker can bring about the event in question; without the intervention of the speaker, it is presumed by both the speaker and the hearer that the event will not occur.

> *Sincerity condition*: both the speaker and the hearer presume that the occurrence of the event will not be in the hearer's interests, and that the hearer would want to avoid its occurrence if possible, and that the hearer would take steps to do so if necessary.

> *Essential condition*: the speaker is making a commitment to see to it that the event will occur unless the hearer carries out the particular action designated by the speaker.

These conditions underlie the structure of the argument from fear appeal (Walton, 1995, p. 157):

> If you bring about A, some cited bad consequences, B, will follow.
> I am in a position to bring about B.
> I hereby assert that in fact I will see to it that B occurs if you bring about A.
> Therefore, you had better not bring about A.

For instance, in the following example the Nazis' speech act can be defined as a threat because they are in a position to bring about the unfortunate consequences for the reader, and their modus operandi of using violence to obtain their ends is well known.

> If I cancel my subscription to the German paper, then, the Nazis say, I will be subject to "unfortunate consequences", which would probably include at least severe bodily injury. (Walton, 2000, p. 85)

Here again, as in the insurance example, we have a case of an indirect threat. The argument can be put in disjunctive form as an instance of practical reasoning (Walton, 2000, p. 142):

> You (the respondent) must bring about A or B will occur.
> B is bad or undesirable, from your point of view.
> Therefore, you should (ought to, practically speaking) bring about A.

But as applied to the Nazi case, this also involves a threat. Thus we can see how the scheme for argument from negative consequences supplies the underlying basis of arguments used as scare tactics based on appeals to danger or fear.

As shown earlier, appeal to threat may also assume the form of argument from alternatives. An additional premise presents the clause that the agent is committed to bringing about the undesirable action. In another example, the speaker commits himself to bringing about the consequences of the action posited as contrary to the advised one.

> You ought to study hard; otherwise, I'll discontinue your allowance. (Walton, 2000, p. 58)

The argument of the disjunctive *ad baculum* threat is, therefore, a subspecies of the argumentation scheme just presented (Walton, 2000, p. 140):

> You (the respondent) must bring about A or I (the proponent) will undertake to see to it that B will occur.
> B is bad or undesirable, from your point of view.
> Therefore, you should (ought to, practically speaking) bring about A.

The disjunctive way of putting the argument contrasts the bad consequences with an alternative, thereby appearing to make the alternative attractive, or even the only possible course of action.

6. ARGUMENTS FROM ALTERNATIVES AND OPPOSITES

Appeals to fear and danger do not always follow the general scheme of the argument from negative consequences that may be considered as the plausible counterpart of the logical *modus tollens*. They may in some instances have the form of the disjunctive syllogism: from the disjunction of two propositions and the negation of one of them, the truth of the other follows. It may be represented as below:

p∨q
¬q
Therefore, p.

In argumentative discourse, this scheme may be used for a pragmatic argument, to persuade the hearer to act in a certain way by showing the alternative. The action is directly opposed, however, not to its contradictory or opposite, but to the consequences of the contradictory. For instance, in the following case, obedience is opposed to the consequences of disobedience. We can observe, therefore, that this kind of reasoning involves a use of argument from consequences.

> You ought to obey the Ten Commandments: otherwise, you will suffer eternal damnation. (Walton, 2000, p. 59)

This example follows the scheme of argument from alternatives, the practical counterpart of the disjunctive syllogism. The opposition is neither necessary nor linguistic, but belongs to the domain of decision making. It is an incompatibility between actions and consequences of actions, aiming at evaluating the best decision.

The analogical argument may regard the properties of two opposites – for instance, war and peace, or hatred and love. Opposites and contraries are the polarities of a respectively continuous and close paradigm. They are, in other words, elements sharing common characteristics, represented by the class shown as conflicting. The similarity is extended to the predication: from the opposition under a determinate point of view, an opposition in consequences or characteristics is inferred. The structural analogy in categorization is reflected in a structural analogy in predication. For instance, in the following case, from the opposition between war and peace, the opposition in their effects is inferred.

> If the war is responsible for the present evils, one must repair them with the aid of peace. (*Rhetoric*, Book II, Chapter 23, 1397a)

The argument fits the following schemes:

Positive Form of Argument from Opposites

The opposite of subject S has the property P.
Therefore, S has the property *not-P* (the opposite of property P).

In some instances, the argument will have the following negative form:

Negative Form of Argument from Opposites

The opposite of subject S has the property *not-P*.
Therefore, S has the property P (the opposite of property *not-P*).

Argument from opposites is a very common scheme of wide generality, treated more fully in Chapter 7.

7. PLEAS FOR HELP AND EXCUSES

Practical reasoning is the argumentation scheme that underlies the pathetic arguments from need for help, distress, and plea from excuse. These forms of argument are constituted by an emotional appeal, involving the hearer in a problem and a sequence of practical reasoning aiming at proposing a means to solve it. As in cases of fear appeals, appeals to pity lead to inferences through emotions. The problem described in the premises does not involve the speaker directly: it may involve other people, or other beings, but not the interlocutor's life or interests. For this reason, it does not always constitute a rational reason to act. The task of the emotional appeal is to make the problem relevant for the audience, to transform it into a personal concern for the other party. A special kind of empathy is therefore created between the two persons, defined by ten points (Walton, 1997b, pp. 73, 74):

1. There are two parties, x and y, who exist in a relationship to each other; x is the pitier, or subject, and y is the pitied party, or object of pity.
2. x can identify with y – that is, x and y are enough alike, or have enough in common, so that x can relate to y as another person.
3. y is in a situation or set of circumstances C, as far as x knows (or at least x thinks this to be the case).
4. x does in fact put himself mentally into the situation C of y, so that he can appreciate how y feels about it.
5. C is a bad or catastrophic situation for y (an event or set of circumstances that would be generally judged as particularly unfortunate).
6. y thinks that C is bad, his attitude toward it is one of pain or suffering, at least to some extent, and generally it would be judged as a situation where feeling pain and suffering would be normal and appropriate.
7. y does not deserve to be in this set of circumstances.
8. x does not have this same or comparable bad situation (C) in his personal circumstances.
9. x is liable to suffer from this same type of bad situation (C), meaning that it is possible, and a risk for x.
10. x knows that proposition nine is true, or thinks it is true with justification.

The speaker, in his appeal, points out the reasons why the hearer should be involved in the bad situation described. He provides the premises for development of the feeling of pity.

The second step of the argument is constituted by a sequence of practical reasoning aiming at finding a possible and not-too-costly means for the resolution of the problem. The argument may not be referred to an external interlocutor. In this case, the topic is simply a sequence of reasoning about a situation leading first to empathy and then to an action as a response, as in the following example:

> BoyScout Bob sees a frail, elderly lady trying to cross the street. She is clearly in need of help getting across the busy intersection. Bob takes her arm and helps her to cross the street. (Walton, 1997b, p. 104)

The argument in this case follows the scheme of the argument from need for help (Walton, 1997, p. 104):

> For all x and y, y ought to help x, if x is in a situation where x needs help, and y can help, and y's giving help would not be too costly for y.
> x is in a situation where some action A by y would help x.
> y can carry out A.
> y's carrying out A would not be too costly for y – that is, the negative side effects would not be too great, as y sees it.
> Therefore, y ought to carry out A.

The following critical questions (Walton, 1997b, p. 155) are focused on the practical reasoning in the argument: the action proposed as a possible way to resolve the problem should not be impossible or too costly for the speaker, and it must really help the other party. If these critical points are not respected, the argument itself may lose its plausibility.

> CQ1: Would the proposed action A really help x?
> CQ2: Is it possible for x to carry out A?
> CQ3: Would there be negative side effects of carrying out A that would be too great?

The urgency and the seriousness of the problem are extremely important factors for the defeasibility of the argument. In a critical situation, such a use of practical reasoning is very hard to refuse, or to brush aside. For instance, in the following case, the distress of the person constitutes a persuasive reason to take action to help her:

> "I was taken downstairs blindfolded, while being beaten and kicked." Those are the words of Saadet Akkaya, who was 16 years old when Turkish officials arrested her 2 years ago. During her interrogation and torture, Saadet's ordeal included being stripped and tied to a cross with ropes. "When they took away the chair, I was hanging my arms. Then they gave me electric shocks to sensitive parts of my body. At the same time, a policeman threw water over me." Sadeet was also subjected to other forms of torture, abuse and ill-treatment. After undergoing 15 days of torture, she confessed to activities about which she knew nothing. Saadet is still imprisoned in Saq- malcilar. Under this story, it said "This barbarism can be stopped. Please help." In the accompanying letter, it said, "Our need for your support is so terribly urgent because, even as I write this letter to you, some place in the world – in Communist countries, in Western societies, in the Third World – innocent victims of government abuse are imprisoned, suffering unspeak- able physical and mental agonies." In this letter it also gives details like, "the guards taunted the prisoner as they applied electrical shocks to her body," and "her cries were echoed by the screams of other victims." (Walton 1997, p. 106)

If the hearer is emotionally involved in the situation, it is very hard for him to argue against the necessity for an urgent action. This form of reasoning is a subscheme of the need for help, called argument from distress (Walton, 1997b, p. 195):

Individual x is in distress (is suffering).
If y brings about A, it will relieve or help to relieve this distress.
Therefore, y ought to bring about A.

These are the critical questions, similar to the general scheme (Walton, 1997b, p. 155):

CQ1: Is x really in distress?
CQ2: Will y's bringing about A really help or relieve this distress?
CQ3: Is it possible for y to bring about A?
CQ4: Would negative side effects of y's bringing about A be too great?

Another argumentation scheme is related to the appeal to pity: the plea for excuse. In this topic, the speaker admits to a transgression of a rule or law, but bases his defense or justification on an emotional appeal. He does not point out possible objective exceptions to the application of the rule, but instead appeals to excuses, reasons to believe that the case should be dealt with in a special manner. The appeal to empathy plays an important role in the characterization of the defense as an excuse. The hearer is invited to consider the speaker's situation and his reasons to act illegally. The fact that the hearer mentally puts himself in the other party's situation gives rise to empathy that may lead to an understanding and justification of the transgression. For instance, in the university cases, a student's dramatic personal situation is not by itself considered an exception in the case of deadlines. It can, however, be a valid excuse, even if defeasible, especially when the appeal to emotions is extremely strong, as in the following example:

> I missed your test last week because I recently discovered my parents are divorcing. This was when I had to go home to see my brother who was just in a car crash. Otherwise I would have been prepared for your test and would have taken it. I know you have a "no make-up" policy. But I really couldn't help what happened to my parents and brother. It is only fair to permit me to make up a missed quiz. One ought to do what is fair. So you ought to allow me to take the test, (Walton, 1997, p. 22)

Professors are very familiar with such excuses, and have to make rulings on them all the time. Each case needs to be decided on its own merits, and student pleaders may of course cite cases that they take to represent similar exceptions to the rule. This is the scheme of the argument from plea for excuse, with its relative critical questions (Walton, 1997, pp. 154, 156):

> Normally, rule R requires or forbids a type of action or inaction T, which carries with it a sanction (penalty) S.
> I (the pleader) have committed T.
> But I can cite special circumstances that constitute an excuse, E.
> Therefore, in this instance, I ought to be exempted from S.

Some excuses are clearly strong, while others, although weak, may be presented in a highly dramatic manner and exploit strong emotions. Although there is no magic formula for deciding all such cases, such arguments are defeasible, and it may help to ask critical questions.

Critical Questions

CQ1: Does E fall under one of the recognized categories of excuses for this type of case, and, if so, can this inclusion be justified in this case?

CQ2: If E does not fall under a recognized category, then what about this case is special that justifies the claim to exemption?

CQ3: If E does not fall under a recognized category, would it set a precedent, and if so, would this pose a problem in future cases?

The need for rules in political and legal institutions brings with it a need to examine arguments used to justify rule infractions. Such arguments are extremely common, and a creative pleader can use great ingenuity to try to get special treatment. Still, the key weakness of an argument of this sort can often be located merely by asking critical questions. Students will often persist with such emotional appeals, even after being warned that they can be fallacious, possibly because they have worked so well for them in the past.

8. COMPOSITION AND DIVISION ARGUMENTS

The standard treatment of the fallacy of composition and division is based on the distinction between collective and distributive terms, as well as on the differentiation between wholes and parts. Collective terms refer to a whole, or a collection, while distributive terms apply only to individuals or parts (Engel, 1980, p. 105; Copi and Jackson, 1992, p. 116). The same terms may be taken to have different properties according to the meanings they assume in specific contexts. Adjectives may be either absolute, like 'red' or 'square', or relative, like 'large' or 'heavy'. The meaning of a relative term is determined by the class of reference. For example, the adjective 'heavy', when attributed to the components of a car, is different in meaning when it is predicated of the car itself. Because the classes of comparison are different, the predicates themselves are different in meaning. As a consequence, the relation between predicates may need to be examined. The fact that the components of the car are heavy, as components of a car, may not imply that they are heavy when considered in relation to the weight of the car as a whole. In the case of, say, books, the problem is different. The pages of a volume might be heavy as pages, but the book, as a book, might be light, because it might be short. In this case, there are other variables influencing the determination of the predicate. The two meanings of the same adjective are not related by compositional properties.

The classical notion of the distributivity of the adjectives may be integrated into a wider theory, related to the concepts of part and whole and to the connections between them. In Walton (1989, p. 106), the treatment of the topics of composition and division is related to the notion of aggregate.

Aggregates are physical entities in space-time, susceptible to change and action. The parts of an aggregate, in addition, do not necessarily make it up. For instance (Walton, 1989, p. 108), if the links are components that make up a chain, their atoms are not relevant under this aspect. For this reason, there are two main rules for the aggregates:

For any aggregate there is some property that does not compose it.
For any aggregate there is some property that does not divide it.

A car, for instance, made of its parts, is built by appropriately connecting its components: for this reason, the properties relevant for the relation belong, for instance, to the category of substance, or quality, but not to dimensions or shape. On the other hand, if we consider a book and its pages, the kind of relation is "to be constituted by ordered superimposition of a variable number of elements." In this case, the shape and dimensions are relevant categories of comparison, but not the weight.

The treatment of the scheme using the categories of distributive and collective terms may be deepened by an analysis of the reasoning behind the distribution itself. The focus is shifted to the compositionality and the relation between the elements and the aggregate. Distributivity of a property might be replaced by the notion of compositional hereditariness. The general scheme of the *argument from composition* is the following (Walton, 1989, p. 130):

All the parts of X have property Y.
Therefore, X has property Y.

The critical aspect of these topics is the nature of the analogy between the aggregate and its elements and the relevant properties for this relation:

CQ1: Is property Y compositionally hereditary with regard to aggregate X? That is, when every part that composes X has property Y, does X (the whole) have property Y?

The *argument from division* presents an analogous form, but the direction of the analogy, instead of being from the elements to the aggregate, is from the aggregate to the parts. For instance, in the following example, the predicate 'heavy' is shifted from the aggregate to its parts. The property, however, is not divisionally hereditary: there is not, in this case, a relation between the two meanings of heavy.

This machine is heavy.
Therefore, all the parts of this machine are heavy. (Walton, 1989, p. 130)

In this case, the following scheme is applied (Walton, 1989, p. 130), but the critical question fails, determining the argument's defeat.

X has property *Y*.
Therefore, all the parts of *X* have property *Y*.

Critical Question

CQ1: Is property *Y* divisionally hereditary with regard to aggregate *X*? That is, when *X* (the whole) has property *Y*, does every part that composes *X* have property *Y*?

9. SLIPPERY SLOPE ARGUMENTS

All slippery slope arguments are based on an underlying form of reasoning called argument from gradualism, in which a respondent is persuaded of the conclusion through a series of small steps. The conclusion is presented as being related to the sequence of intermediary steps proceeding from the accepted premises. For this reason, the commitment to the conclusion is a result of a sequence of previous commitments, as in the following example:

A government knows that it needs to get an 18 percent value-added tax (VAT), sometimes called a goods and services tax (GST), in order to deal with its budget deficit. However, the public would never vote for or approve such a large tax, in one single step. Therefore, the government adopts the strategy of introducing a 3 percent VAT and then increasing it every few years, when politically appropriate, until the 18 percent level is reached. (Walton, 1995, p. 95)

Scheme for Argument from Gradualism

Proposition *A* is true (acceptable to the respondent).
There is an intervening sequence of propositions, $B_1, B_2, \ldots B_{n-1}$, B_n, *C*, such that the following conditionals are true: If *A*, then B_1; if B_1, then $B_2, \ldots$; if B_{n-1}, then B_n; if B_n, then *C*.
The conditional "If *A* then *C*" is not, by itself, acceptable to the respondent, nor are shorter sequences (than the one specified in the second premise) from *A* to *C* acceptable to the respondent.
Therefore, the proposition *C* is true (acceptable to the respondent).
(Walton, 1999, p. 93)

This scheme is represented by the ancient example of the heap. The sequence of similarities may lead to the acceptance of absurd conclusions.

If you take one grain away from a heap, it makes no significant difference, you still have a heap. Each time you repeat this step, it makes no difference, because one grain is too small to make a difference between something being a heap or not. But repeated long enough, the conclusion of this reasoning will become absurd, for it will become obvious that what is left can no longer be described as a heap. (Walton, 1996a, p. 237)

The example of the heap, and comparable examples like the bald man argument, are instances of a pattern of reasoning that works by setting up a series of steps of inference that are chained together in a certain way. There is always a grey area that is based on a vague term, like 'heap'. As the chain of reasoning moves forward, once it goes into this grey area, which cannot be demarcated precisely, it cannot be reversed.

This failure of reversal due to a vague term is characteristic of the scheme for the *sorites* slippery slope (Walton, 1992a, p. 56).

Scheme for the *Sorites* Slippery Slope Argument

Initial Base Premise: It is clearly beyond contention that a_k has P.
General Inductive Premise: If a_k has P, then a_{k-1} has P.
Reapplication Sequence: A sequence of *modus ponens* subarguments linking premises and conclusions from the clear area through the grey area.
Conclusion: a_i may have P, for all we know (or can prove).

Arguments of this kind are common in medical ethics, where it is hard or even impossible to define terms like 'person', as opposed to 'fetus', in a precise manner. They are also very common in law, where terms like 'free speech' cannot be defined without leaving a grey area (Schauer, 1995).

The series of analogical steps may be connected with the scheme from verbal classification in a verbal classification slippery slope argument. The term classified in the first premise is connected by a series of comparisons to similar cases. In the form presented here, a_1 is posed as similar to $a_2 \ldots a_n$. The reasoning from analogy is reiterated in each comparison, classifying each similar term as F. An important implicit premise is called the *hypothetical syllogism* rule: if all these analogical inferences are correct, then if a_1 has property F, then a_n has property F. The last inference is based on the last premise and has a *modus tollens* form: a_n does not have property F; therefore, a_1 does not have property F. The scheme might be compared to the strategy of *reductio ad absurdum*. The chain of reasoning leads to a conclusion that is absurd, that is, unacceptable to the proponent of the claim. From the rejection of the conclusion, by *modus tollens*, the implausibility of the premise is proved. The structure of the scheme is represented here, following Walton (1995, p. 160):

> Individual a_1 has property F.
> If a_1 has F, then a_2 has F.
> Property F is vague, and so generally, if a_i has F, then you can't deny that the next closely neighboring individual a_n has F.
> But quite clearly it is false that a_n has F.
> Therefore, you can't truly say that a_1 has F.

Here is an example:

Ted: Surely the baby in the womb must be defined as a person, with rights, during these latter stages, because a surgeon can do intrauterine surgery to correct the baby's heart defect, in some case of this sort. The baby is the doctor's patient, therefore it must be a person. Moreover, in many such cases, the baby, if delivered by Caesarian section, could be supported by intensive care, without the mother's support.

Marcia: Well, yes, but I think the baby is a person just before it is born, but not before that.

Ted: But where do you draw the line? If the baby is a person in these latter stages near birth, then it is also a person in the earlier stages, where it cannot survive on its own, but where it has all the same features like a heart, lungs, limbs, and so forth. I don't see any point where you can draw the line, other than by having to admit that it could be a person from the moment of conception (an absurd proposition to admit, for anyone advocating your viewpoint). I mean if it is a person on one day, then the day before it couldn't have been that much different, so that it couldn't be a person that day. (Walton, 1995, p. 106)

As already noted, such arguments are very common in law and medical ethics, and have been the basis of many philosophical disputes in these fields.

Now we return to argument from precedent, the scheme displayed in the previous section, and show how a special type of slippery slope argument is based on it. Here the general form of the argument from precedent slippery slope is quoted from Walton (1996, p. 98).

Scheme for Argument from Precedent Slippery Slope

Claim to Exceptional Status Premise: Case C_0 is claimed to be an exception to the rule R (an excusable or exceptional case).
Related Cases Premise: Case C_0 is similar to case C_1, that is, if C_0 is held to be an exception, then C_1 must be held to be an exception too (in order to be

consistent in treating equal cases alike). A sequence of similar pairs of cases $\{C_i, C_j\}$ binds us by case-to-case consistency to the series $C_0, C_1, \ldots C_n$.

Intolerable Outcome Premise: Treating case C_0 as an exception to the rule R would be intolerable (for the various kinds of reasons that could be relevant).

Conclusion: Case C_0 cannot be judged to be an exception to the rule (an excusable or exceptional case).

In the following example, the prohibition upon professing kinds of religion different from the legitimate one is defended by showing the unacceptable consequence of allowing a precedent.

> One participant argued that with all the different minority groups, once you accept one kind of religion as legitimate, you are going to have to accept many other kinds of religious groups as having a legitimate right to have prayers or religious services in the classroom. This participant said: "It's a Pandora's box. You know that Satanism is a religion too!" (Walton, 1995, p. 100)

In this case, the implausibility of the reasoning may be analyzed through the device of using critical questions to pinpoint weak points in the argument. The doubtfulness of inferring the conclusion from the premise is shown by the second question. It rebuts the argument by showing that one of its necessary conditions fails to apply. Argument from classification is the key. The classification of Satanism as a religion is questioned.

Critical Questions for Argument from Precedent Slippery Slope

CQ1: Would C_0 set a precedent?

CQ2: What is the evidence showing why each of the cited intervening precedents would occur?

CQ3: Is C_n as intolerable as it is portrayed?

Argument from precedent slippery slope is a counterargument to argument from precedent. Both are related to argument from verbal classification. It is a complex subspecies of argument from precedent. As slope arguments, they all share a general pattern. The *reductio ad absurdum* pattern moves through a related sequence of predications in this schematic pattern.

10. ATTACKING VERBAL CLASSIFICATION AND SLIPPERY SLOPE ARGUMENTS

In Chapter 8 it will be shown how attacks, rebuttals, and refutations are defined and analyzed. But to close Chapter 3, it is interesting to see how such counterarguments work in practice. The slippery slope argument is itself a form of attack on some previous argument or proposed policy for taking action. But it can be attacked by other counterarguments derived from argument from analogy, argument from classification, and other schemes we have studied. One of these is the argument from verbal classification, fitting the scheme called defeasible argument from a verbal classification in Chapter 2, section 6. Such arguments from verbal classification are often attacked by counterarguments claiming that the verbal classification is too vague, or is arbitrary. Here is a typical type of example.

> Marcia and Ted are debating on the issue of abortion. Ted, who is pro-life, argues: "There can be no abortion when the fetus becomes a person." Marcia replies: "That's hopelessly vague! There is no way to exactly define when the fetus has become a 'person'. You don't have a leg to stand on there! (Walton, 1995c, p. 102)

In this argument, Ted's standpoint is defended by using the term "person." This term is not defined, and the meaning that one party would likely accept would not be accepted by the other party. By inference, the mere use of the term triggers an accusation of homicide. Thus the use of the term cannot be supported by a verbal classification that is commonly shared. A certain usage is presumed by the speaker as common knowledge, but is clearly not acceptable to the respondent. This communicative move fails as an argument, because the assumption is not shared. The move has not made sense and, if not defended, collapses.

This refutation tactic is a meta-dialogical move, because the criticism is addressed to the assumptions and to the use of language in the speaker's sequence of reasoning. The opening dialogue, that is, the possible defense of the argument, is situated on another level. The discussion moves to a level where agreements about the common meaning of terms in the argument need to be discussed and evaluated.

This is the form of the argument from vagueness of a verbal classification (Walton, 1996, pp. 102, 103):

Argument from Vagueness of a Verbal Classification

If an argument A occurs in a context of dialogue that requires a certain level of precision, but some property F that occurs in A is defined in a way that is too vague to meet the requirements of that level of precision, then A ought to be rejected as deficient.

A occurs in a context of dialogue that requires a certain level of precision that is appropriate to that context.

Some property *F* that occurs in *A* is defined in a way that is too vague to meet the requirements of the level of precision appropriate to the context.

Therefore, *A* ought be rejected as deficient.

Critical Questions for Argument from Vagueness of a Verbal Classification

CQ1: Does the context of dialogue in which *A* occurs demand some particular level of precision in the key terms used?

CQ2: Is some property *F* that occurs in *A* too vague to meet the proper level of standard of precision?

CQ3: Why is this degree of vagueness a problem in relation to the dialogue in which *A* was advanced?

The problem of defining the class of which the subject is predicated may present another problem. If, as seen earlier, if it is considered vague, and is not accepted for this reason, the definition may also be considered arbitrary. In the scheme just presented, the criteria are not established, and the meaning of the word is taken for granted. Such reasoning should be rejected on the basis that it leads to a conclusion that is not dialogically relevant, because it stems from commitments that are not commonly shared.

In the following example, the speaker's argument holds in virtue of the fact that the proposed definition of "person" is adequate to the standards required by the context. Here this means that it is precise, uncontroversial, and supported by sufficient evidence. The attack is based on the meta-refusal of this assumption.

Ted: The fetus should be considered a person through the third trimester.

Marcia: You mean to say that the day before the third trimester, the fetus is not a person. And then the first day of the third trimester, it is a person. That is an arbitrary way of drawing the line. (Walton, 1995c, p. 104)

In the following scheme for argument from arbitrariness of verbal classification (Walton, 1996, p. 104), the role of context of dialogue is critical. The classification is judged uncontroversial if it respects the conditions the dialectical situation imposes on the common commitment sets. The level of precision and the agreement on determinate standards is a consequence of the topic and kind of discussion.

Scheme for Argument from Arbitrariness of a Verbal Classification

If an argument *A* occurs in a context of dialogue that requires a certain level of nonarbitrary definition for a key property *F* that occurs in *A*, and *F* is defined in an arbitrary way in *A*, then *A* ought to be rejected as deficient.

A occurs in a context of dialogue that requires a nonarbitrary definition for a key property *F* that occurs in *A*.

Some property *F* that occurs in *A* is defined in a way that is arbitrary.

Therefore, *A* ought to be rejected as deficient.

The most interesting of the following critical questions regard the problem of the context of dialogue: the relevance of the moves following the analyzed topic stems from the requirements of the dialectical situation.

Critical Questions for Scheme from Arbitrariness of a Verbal Classification

CQ1: Does the context of dialogue in which *A* occurs require a nonarbitrary definition of *F*?

CQ2: Is some property *F* that occurs in *A* defined in an arbitrary way?

CQ3: Why is arbitrariness of definition a problem in the context of dialogue in which *A* was advanced?

These refutation schemes are dialectical. Arguments directed against the vagueness or arbitrariness of a classification shift the dialogue onto a different level that concerns the legitimacy of the sequence of moves in a first-level dialogue. They are therefore classified as meta-dialectical argumentation schemes. It will be shown in Chapter 7 how the study of argument refutation needs to be based on meta-dialogical factors like burden of proof.

4

Arguments from Generally Accepted Opinions, Commitment, and Character

The schemes studied in this chapter typically represent arguments of a kind that are weak and subject to default. They work best as uses of plausible reasoning in situations of uncertainty and lack of knowledge. However, they are so fragile and prone to error that they have traditionally been treated in logic as fallacies under the headings *argumentum ad populum* and *argumentum ad hominem*. Even so, despite their inherent fragility and proneness to exploit prejudice, they can be reasonable arguments in many instances, and sometimes they may be the best kind of evidence we have to make a rational decision. First we address the kind of argument Aristotle called endoxic, meaning that it is based on an opinion accepted by the majority and/or by the wise (the experts). In logic, this form of argument is called appeal to popular opinion, but it might be less negative to label it argument from informed public opinion. Then we address ethotic argumentation, recognized by Aristotle as especially powerful. *Ethos* is the moral character of the speaker. It can be used to support an argument, but in the *argumentum ad hominem*, or personal attack on an arguer's character, it is used to discredit his argument. Many argumentation schemes fitting this general type have now been identified and studied, and much of the chapter is taken up with reviewing and discussing these schemes.

The connection between these two apparently different classes of argumentation schemes can be found in the analysis of the meaning of "plausibility." Plausibility is a common basis of acceptability: it represents a rational interlocutor's acceptance of a speaker's position as a commitment in a dialogue. In dialogue games (Hamblin, 1970), the notion of truth has been replaced by the pragmatic concept of acceptance. Accepting a proposition on the part of a participant in a dialogue means inserting it into his own commitment store. This store is the set of propositions that the participant in the game is committed to, meaning either that he has to defend it if challenged or that he agrees to it tentatively for the sake of continuing the argument. Commitments do not always need to be consistent, but it is a main rule of dialogue, and one of the principal conditions of rationality (Perelman and Olbrechts-Tyteca, 1969, p. 206), that a participant can be criticized for being committed to contradictory standpoints. Persistence in demonstrable inconsistency can lead to the temporary exclusion of the subject from the community of rational speakers, and can provide an argument for his lack of credibility. In such a case of provable contradiction, an arguer needs to reject one of his commitments, or somehow resolve the apparent inconsistency. Failure to do so may lead to the loss of the dialogue game. Thus an allegation of inconsistency is a serious form of argumentation against an opponent, and a very common and powerful one, notably in political and legal argumentation.

1. ARGUMENTS FROM POPULAR OPINION

As far as we know, the argument from popularity was first considered a fallacy of the kind appropriately treated in a logic textbook in the Port Royal logic, in 1662. Antoine Arnaud (1964, p. 287) warned of this fallacy:

> Men follow the ridiculous procedure of believing a thing true according to the number of witnesses to its truth. A contemporary author has wisely pointed out that in difficult matters that are left to the province of reason, it is more likely that an individual will discover the truth than that many will. The following is not a valid inference: The majority hold this opinion; therefore, it is the truest.

From this passage, we can understand the viewpoint of the modern logic textbook writers. The focus is on the truth of a sentence, and the fallacies are described from the point of view of illusions of truth. This modern perspective is different from the ancient one. Aristotle distinguished dialectic from logic through the differentiation between necessary and acceptable consequences. The argumentation process does not aim at necessary arguments based only on true premises and conclusions, but at plausible propositions (Walton, 1999, p. 148). Such argument premises can be based on *endoxa*, on what is known or widely accepted by everybody, or by the majority, including the wise. From this perspective, arguments that appeal to popular opinion are not always considered fallacious. Classifying one's own thesis, or premise supporting it, as customarily or generally accepted, may give a reason to support it, and thus may rightly make an opponent hesitate to refuse to grant it. People are, in fact, usually careful about rejecting what is *eiothos*, what is commonly held, and often justifiably so. Seneca articulated this scheme, giving it an argument form (Walton, 1999, p. 147): "Everybody believes that *A*; therefore, *A* is true." He recognized that this kind of argument is only provisional and defeasible, and that it can be misused, but also that it can be reasonable in some instances. In the Middle Ages, Boethius defined this topic as an argument from judgment and characterized it by the following maxim (Boethius, 1978, p. 54): "what seems true to everyone or the many or the wise should not be gain-said." Here the locus is referred not to what is usual, but to the *endoxon*. In Thomas Aquinas, the analysis of this argumentation scheme is related to the process of deliberation: when exact knowledge about an event is lacking, it is necessary to consider the choice among the known alternatives. So conceived, the argument from popular opinion is treated as a practical form of reasoning. In the ancient treatment of argumentation generally, dialectical argumentation, following Aristotle, was based on *eikos*, or acceptability, referring to the relation between an opinion and an argument, rather than that between an argument and the truth of the premises or conclusion. The notion of conclusive evidence at some point replaced, in the history of the discipline, the concept of argument evaluation as a weighing of a balance of considerations under realistic conditions of uncertainty and lack of knowledge.

The standard treatment of the fallacy from popular opinion highlights the rhetorical power and emotional character of this scheme. The basic forms underlying the treatment of the *ad populum* fallacy in the logic textbooks are represented by the following. This generic pattern of argument is called the pop scheme (Walton, 1999, p. 224).

> S1: Everybody (in a particular reference group) accepts that *A*.
> Therefore, *A* is true (or you should accept *A*).
> S2: Everybody (in a particular reference group) rejects *A*.
> Therefore, *A* is false (or you should reject *A*).

These two basic forms are identified in Freeman (1995, p. 70) with the bandwagon appeal and the abandon-ship fallacy, respectively. One characteristic of this scheme is particularly important. The acceptance of the proposition *A* in the scheme refers to "everybody." When the scheme from popular opinion is used, the person to whom it is addressed is always identified with a particular group of people. Thus the force of the argument depends on the set of commitments and values shared by the community that the hearer emotionally and culturally belongs to. Moreover, the conclusion should not be expressed as a logical necessity or a categorical claim of truth, but only as a pragmatic (dialogical) attitude toward the proposition, that is, acceptance or rejection.

In practice, the appeal to popular opinion may refer to a large majority in the cited reference group rather than to the whole group (see the variant noted just below). In general, the argument from popular opinion may be undermined under three aspects: the actual agreement of the majority with the proposition; the weakness of the argument itself when used to prove the truth of a proposition; and the link between popular opinion and true opinion. The first critical question attacks the premise, while the second is directed against the force of the conclusion. The latter, in fact, is only plausible: if contrary evidence is presented, the whole argument is likely to be refused. The last question criticizes the connection between premise and conclusion, having as its target the implicit premise of the scheme: the fact that the majority is likely to be right. While in the first two critical points of the topic the burden of proof is on the opponent who tries to undermine the reasoning, in the last case the obligation to provide satisfactory evidence is on the proponent of the discussed position. The respondent questions the sense of the consequence itself. If the majority is not presupposed likely to be right, there is no reason to conclude that its opinion is acceptable. These are the critical questions for the pop scheme *ad populum* argument (Walton, 1989, p. 89):

CQ1: Does a large majority of the cited reference group accept *A* as true?
CQ2: Is there other relevant evidence available that would support the assumption that *A* is not true?
CQ3: What reason is there for thinking that the view of this large majority is likely to be right?

The pop scheme is just a basic form that represents the essential structure of the appeal to popular opinion type of argument. By adding other factors, several possible variants may be created.

2. VARIANTS OF THE BASIC FORM

From these observations, we can reconstruct the *ad populum* scheme, or pop scheme, in a different guise as a presumptive form of argument.

If a large majority in a particular reference group G accepts A as true (false), then there exists a defeasible presumption in favor of (against) A.

A large majority accepts A as true (false).

Therefore, there exists a presumption in favor of (against) A.

This form of the argument may cite polls or other statistical findings that are supposed to measure public opinion. The following example is a common case of this kind of argument in matters of public deliberation on an issue.

> VANCOUVER – The Vancouver aquarium will stop capturing killer whales but will continue to show them, officials said yesterday. Curator Dr. Murray Newman said a new policy proposes rescue programs and breeding programs among whales already in captivity as ways to stock aquariums. He also said the aquarium will not return its current stock of killer whales and dolphins to the seas. "The decision not to collect whales is a reflection of human sensitivity," Newman told a news conference in the aquarium's underground boardroom, where several times the huge whales swam by and looked in. He said a year-long study showed the public preferred the aquarium not to collect whales. "At the same time we feel we have a mandate to tell people about them and keep them before the public eye." There have been three orca deaths at the Vancouver aquarium in three years. (Walton, 1996, pp. 83–84)

In this case, the fact that the public is against the collection of whales increases the force of one of the arguments on the issue. A balance of considerations between the two sides is shifted in favor of ending the existing practice. It is interesting to notice that the reference group is the public, the relevant community when making decisions about the issue, one of public concern. The general scheme can be analyzed following the criteria of the scheme. The reference group determines the type of argument that is associated with the *ad populum* topic. This kind of group is relevant because it motivates the aspect under which the opinion of the community should be important for the hearer, and considers the possible types of relations between the person and the group or between the group and the proposition constituting the reasons of the appeal itself. The subtypes can be considered as resting on a typology of groups and relative rhetorical tactics.

The first two subtypes concern the appeal to knowledge. In the first case, the group is identified with people who are in position to know, and therefore the strength of the plausible consequence follows from the uniformity of the reputable opinion.

2.1. Position-to-Know *Ad Populum* Argument

Everybody in this group *G* accepts *A*.
This group is in a special position to know that *A* is true.
Therefore, *A* is (plausibly) true.

In the following example, all the people who are considered to be best judges agree on the excellence of the lake as a place to swim. This general opinion provides a good reason to believe in the plausibility of the conclusion, because they are presumably in a position to know.

> Nearly everyone who lives in Cedar Rapids thinks that the lake is a good place to swim in the summer.
>
> Therefore, the lake in Cedar Rapids is probably (plausibly) a good place to swim in the summer. (Walton, 1996, p. 83)

The second case of *ad populum* subarguments from knowledge is the appeal to expert opinion. In this scheme, the reference group is constituted by experts.

2.2. Expert Opinion *Ad Populum* Argument

Everybody in this group *G* accepts *A*.
G is a group of experts in a domain of knowledge.
Therefore, *A* is true.

The following example refers to a discussion of Galileo's proof that objects of unequal weight fall at the same speed. Galileo's argument rejects the *ad populum* conclusion, based on the authority of the Aristotelian theory. The fact that many experts, following Aristotle, believed in a formulation of the gravitational law different from Galileo's was a good reason to believe that their theory was correct. Of course, it was not.

> So many people accept that heavier things fall faster to the ground [therefore it is plausible that] heavier things fall faster to the ground. (Dauer, 1989, p. 80–81, quoted in Walton, 1999, p. 43)

The experts can be wrong, but if they are in the majority, it can be very hard to argue against them.

The third subtype is a form of argument related to the topics from knowledge. The depth and presumed thoughtfulness of the deliberation is presented as a reason to believe that the people involved have carefully and thoughtfully weighed all the possibilities. In addition, the uniformity of their opinion is a sign that their position is preferable.

2.3. Deliberation *Ad Populum* Argument

Group *G* has deliberated intelligently and extensively on whether to accept proposition *A*.
Everybody in *G* accepts *A*.
Therefore, *A* is (plausibly) true.

In the following example, the majority decision is opposed to the opponent's opinion: the fact that everybody has thoughtfully reflected on the verdict is a good reason to believe that it is not very plausible that they are all mistaken. But here again the appeal can be fallacious, because it puts pressure on the ones who disagree with the majority, and who may be right.

> Suppose after long deliberation ten or eleven members of the jury favour one verdict, and one or two members favour another. At that point, the majority will almost certainly appeal to the weight of its numbers in an effort to persuade the dissenters to agree with them. The one or two in the minority will probably be subjected to such arguments as: "Look, everyone else on the jury agrees that the defendant is guilty; you are the only one who doubts it. Since there are so many more of us who have come to the guilty conclusion, doesn't that show you that your own conclusion must be mistaken? Be reasonable, and accept the view of the overwhelming majority. After all, eleven heads are better than one." (Waller, 1988, 132, cited in Walton, 1999, p. 36)

Cases like this one show how an appeal to majority opinion can be used to put pressure on someone, who may then start to doubt his or her own opinion.

The fourth subtype is from excuse. The fact that everybody in a group accepts a policy is a reason to believe that it is acceptable to the reference group and, therefore, that the speaker's action is morally justified. The moral character of the group itself is relevant in this topic: the righteousness of the members is a reason to believe that the practice has not been accepted with harmful intentions.

2.4. Moral Justification (Excuse Subtype) *Ad Populum* Argument

Everybody in a (morally good) group *G* does *x* (or accepts proposition *A* as a policy).

Doing *x* (or accepting *A*) shows that *x* (or policy *A*) is an acceptable norm of conduct for *G*.

I (the speaker) am a member of *G*.

Therefore, my doing *x* (or accepting *A*) is morally justified as an acceptable action (or policy).

For instance, the allegation, in the following case, is that a certain type of treatment is normal procedure among physicians. Such an argument can provide a very good defense for the accused. The presumed good intentions of the physician provide an implicit reason to believe that the medical procedure should not to be considered malpractice.

> A physician is accused of malpractice on the grounds that the medical treatment he applied to a patient turned out badly. The physician defends himself by arguing that this type of treatment is the normal and accepted procedure for the condition of this patient. In other words, he argues that all the other physicians are using the same type of treatment. (Walton, 1999, p. 209)

Any physician who departs from the generally accepted standards of care, from what is accepted by the majority of physicians as a standard treatment, can be in a vulnerable situation.

The fifth subtype is from moral justification. In this argument, the group is identified with positive moral qualities, and thus the argument is from *ethos*. The plausibility of the argument is based on general acceptance of the proposition by a group taken to have a morally right view. It is interesting to point out that this argument proceeds from a kind of rhetoric of belonging. The desire to belong to a group (in this case, a group formed by morally good people) compels the respondent to adhere to the conduct of its members.

2.5. Moral Justification *Ad Populum* Argument

Everybody who is good, or who represents a group *G* with good qualities, accepts policy *P*.

Your goal is (or should be) to be a good person, or a member of a group with good qualities.

Therefore, you should accept *P*.

For instance, the golden rule is widely approved. On this basis it can be argued that it is acceptable as a moral principle in certain circumstances. In the following example, the argument would be quite strong if the circumstances and qualifications were specified. The excessive strength of the conclusion, however, makes the argument (as quoted) fallacious.

> The Golden Rule is basic to every system of ethics ever devised, and everyone accepts it in one form or other. It is, therefore, an undeniably sound moral principle. (Copi, 1986, p. 64, quoted in Walton, 1999, p. 37)

The notion of belonging to a group is the main feature that characterizes, in different ways, the appeals to common folks, vanity, and snobbery. In some cases it is the speaker who purports to belong to the audience's community, while in other cases the hearer is presumed to agree with group opinions or to follow the group's actions in order to be accepted by its members.

The common folks appeal may be described under two aspects. On the one hand, the common background shared with the audience may be considered a reason to believe in the speaker's honesty: he is a member of the community, and a member is unlikely to deceive his friends. On the other hand, the line of consequence can be compared to the ethical argument: the fact that the speaker shares the interlocutor's values is a good reason to believe that he has moral integrity and that everyone can judge his ethical conduct.

2.6. Common Folks *Ad Populum* Argument

I (the speaker) am an ordinary person, that is, I share a common background with the members of this audience (group *G*).
You (the respondent) are a member of this audience (group *G*).
Therefore, you should accept what I say.

In the following example, the speaker uses the common folks appeal, identifying himself with the moral values of the community in order to show his ethical correctness.

> I am sure you will recognize that I am more competent than my opponent. When I was in high school I had to get up at four-thirty every morning to deliver papers. In college I was barely able to make C's and had to do janitorial work in order to make ends meet to put myself through school. Therefore, I would make a better congressman. (Walton, 1999, p. 213)

The direction of the common folks appeal differs somewhat from that of arguments having the form of the rhetoric of belonging. The latter are based on the hearer's wish to belong to a particular (often prestigious) group. The rhetoric of belonging also has a negative aspect. There is a wish not be excluded.

2.7. Rhetoric of Belonging *Ad Populum* Argument

Everybody in this group *G* accepts *A*.
Being a member of this group *G* is highly valued by you (the respondent).
If you do not accept *A*, you will be excluded from this group *G*.
Therefore, you should accept *A*.

In the following example, the values of honesty and dedication are presented as terms of acceptance in the American movement: people that do not share them are excluded from the (ideal) community.

> I think we can all agree that the overwhelming majority of the leadership of the American movement is composed of decent, honest, dedicated people who have made a great contribution involving great personal sacrifice, helping to build a decent American labour movement. We happen to believe that leadership in the American movement is a sacred trust. We happen to believe that this is no place for people who want to use the labour movement to make a fast buck. (Bailey, 1983, p. 134, quoted in Walton, 1999, p. 217)

The rhetoric of belonging argument has a positive scheme that may assume two particularly interesting forms, the snob and vanity appeals. They focus on the elite or on the admired community. In the first case, the group is shown as a target to reach through imitation, while in the vanity argument the hearer is seen as already belonging to an admired class.

2.8. Snob Appeal *Ad Populum* Argument

Everybody in this group *G* accepts *A* (or has some property *P*, or possesses some object *O*).
This group *G* is elite, that is, everyone who belongs to it has prestige.
Prestige is an important goal for you (the respondent)
If you accept *A* (acquire property *P* or buy object *O*), then you will become a member of the group *G*.
Therefore, you ought to accept *A* (etc.).

In the advertisement below, the possession of a Rolls Royce is suggested as a means for one to be distinguished from ordinary people.

> A Rolls Royce is not for everyone. If you qualify as one of the select few, this distinguished classic may be seen and driven at British Motor Cars Ltd. (By appointment only, please) (Walton, 1999, p. 91)

As mentioned earlier, the appeal to vanity merges flattery and group identification: the interlocutor is urged to assent in order to show his distinctive qualities or culture.

2.9. Appeal to Vanity *Ad Populum* Argument

Everybody in this group of admired (popular) people G accepts A (possesses A, etc.).
If you carry out action x, then you will belong to this group G.
Therefore, you should carry out action x.

For example, in the following commercial the hearer is flattered, and it is suggested that people having his or her wonderful quality usually own the advertised item.

> You were probably born with a bigger share of intelligence than most of your fellow men and women... and taught how to use it. And you appreciate the difference. You aren't ashamed of having brains. You enjoy using them. And that's why *The Hundred Greatest Classics* belong in your home. (Walton, 1999, pp. 215–216)

From the typologies briefly explained here we can observe that it is partly the reference group that defines the type of appeal.

The argument from popularity also has a practical form: the reference group's opinions are considered in relation to what decision to take when deliberating a course of action. This is the form of the scheme from popular practice.

> If a large majority of a particular reference group G does A, or acts as though A is the right (or an acceptable) thing to do, then A is a prudent course of action.
> A large majority of a particular reference group G acts as though A is the right thing to do.
> Therefore, A is a prudent course of action.

The argument from popular practice, when used in deliberation, may be quite strong if there is a lack of sufficient information about the decision.

> In a sailboat race, there were a lot of markers that had to be passed, and it was very easy for the participants to become disoriented and get lost. The competitors made elaborate charts before the race, and during the race spent a lot of time using a compass to try to figure out the route. The captain of one sailboat was asked what strategy he used. His reply: "Well, we try to prepare carefully by making good charts. But if you are really getting

lost, you often follow the other fellows who seem to be very successful in getting ahead." (Walton, 1996, p. 84)

The reference group in this case is constituted by the successful participants in the sailboat race. The argument proceeds from a motivation of position to know: the group is the source for the decision making because it is expected to have a better knowledge of the route.

3. ARGUMENT FROM COMMITMENT

Under the heading of argument from commitment can be grouped many kinds of arguments relating to different aspects of the dialogical exchange. Some are related to the prohibition on holding inconsistent commitments, while in others it is the *ethos* (ethical character) of the speaker that is related to an arguer's commitment in various ways. Many of the rationales of these connections can be found in the very meaning of the concept of character. Character is the habit or disposition of a person to act in certain relatively consistent ways over a lifetime, and it is judged relative to the role the person plays in a dialogue in a given case (Walton, 1998, p. 178). Therefore, *ethos*, the ancient name for character, relates to obligations and commitments that a participant in a dialogue incurs concerning sincerity, honesty, trustworthiness, and other aspects of character relating to reputation and collaborative action in a joint enterprise. According to Kupperman (1991, p. 17), character represents the link between an agent's externally evident choices and actions and his internal feelings, thoughts, and motives. Present and past actions in dialogues determine the set of clear or implicit commitments a person has, the set of propositions and choices he has previously accepted and that his future decisions must be consistent with. Character depends, therefore, on choices, as Aristotle claims (Walton, p. 179): it is a status, a pattern of action connected to practical wisdom. It is a quality presumed to be stable over time, and for this reason it must be distinguished from commitments relative to a particular situation. Arguments from *ethos* involve judgments about an agent's presumed goals, and rest on the assumption that the agent's past actions, decisions, and behaviors in different situations are known according to some evidence. Only from the person's history of commitments it is possible to understand his present internal disposition to act and what are taken to be his virtues or vices. On the other hand, argument from commitment is often judged relative to decisions taken over a short period of time, under conditions of uncertainty. Thus they are defeasible, and the particular circumstances of a case, as known, need to be taken into account.

Argumentation based on the presumed trustworthiness of an agent has recently become a topic of some importance in multi-agent computing. In multi-agent systems technology, software devices called agents, used to gather information on the worldwide web and carry out services, need to carry out actions based on this information in an autonomous way. For example, a personal money management agent may need not only to collect the latest stock and mutual fund information, but also act on this information, as told to do so under certain conditions, in order to manage a client's portfolio. But this demands a certain degree of autonomy, as not all conditions requiring action can be formulated exactly. In negotiation, for example, such an autonomous agent may need to conceal information or even to engage in deceptive argumentation with another agent. Rational agents need to be proactive as well as reactive and to have social ability (Wooldridge, 2000, p. 3). Such agents must communicate with each other in order to pass along information, and must make commitments to collaborative actions. Thus, in designing communication protocols, argumentation needs to be based on assumptions about the trustworthiness and sincerity of a speech partner that are important for collaborative participation in a dialogue. Thus it is important for an agent to be able to judge whether another agent is a reliable source of information. Judging the *ethos* of another agent is therefore an important skill. A reputation rating comes from direct observations of an agent's past conduct, referrals from other agents, and from ratings by these agents (Yu and Singh, 2000, p. 4). The trust rating measures the trust one agent should have in another as a source of reliable information, and is updated as a dialogue proceeds. If one agent lies, or gives false information, and another agent finds this out, it can downgrade the first agent's trust rating. Other agents can be told about this new rating (p. 6).

Thus character and commitment are closely intertwined in argumentation. While the particular commitments focus on the acceptability of the particular proposition in relation to the rules of the game, perceived ethical qualities of character affect the credibility of the person as a source and as a collaborator in argumentation. Such judgments in argumentation are often based on the perceived connection between an agent's words and deeds, as will be shown later. A liar, or a hypocrite whose deeds are inconsistent with what he says, cannot be considered trustworthy, and therefore his commitments are the basis for judging that his argument

cannot be deemed acceptable. The fundamental presupposition of communication, trust (Rigotti, 1997), fails to hold in these cases, leading to
certain kinds of arguments based on trust and commitment. From these
linkages it becomes possible to understand the close relation between
arguments from popularity, arguments from expert opinion, and arguments from character. All proceed from basic principles of argument
acceptability based on an agent's credibility, supported by the agent's
presumed knowledge, trustworthiness, and personal qualities of character. We begin with argument from commitment.

The scheme for *argument from commitment* is the following (Walton,
1996, p. 56):

> *Major Premise*: If arguer *a* has committed herself to proposition *A* at some
> point in a dialogue, then it may be inferred that she is also committed to
> proposition *B*, should the question of whether *B* is true become an issue
> later in the dialogue.
> *Minor Premise*: Arguer *a* has committed herself to proposition *A* at some point
> in a dialogue.
> *Conclusion*: At some later point in the dialogue, where the issue of *B* arises,
> arguer *a* may be said to be committed to proposition *B*.

The following dialogue is an example of this kind of argumentation. Ed's
political faith, often shown in the past, is a good reason to believe that he
will be consistent with his convictions by choosing the same side in a new
situation involving matters concerning public policy. Being a communist
may not be one of Ed's explicit commitments, as far as we know, but
it may be a commitment that we can infer from Ed's previous dialogue
moves and actions. Suppose that Ed often goes around shouting "Power
to the people!" and expresses his admiration for the views of Marx and
Lenin. Bob knows this, and so infers that in the current labor dispute,
Ed will very likely take the union side. It would be surprising if he did
not. In this dialogue, Bob brings this commitment to light and makes it
explicit through a dialogue with Ed.

Bob: Ed, you are a communist, aren't you?

Ed: Of course. You know that

Bob: Well, then you should be on the side of the union in this recent labor
dispute. (Walton, 1996c, p. 55)

If Ed were to refuse to accept the assumption that he is a communist,
Bob would have to defend his point and provide evidence to defend his
view. But Ed freely accepts that commitment, and thus Bob is justified

in drawing the conclusion that Ed will be on the union side in this case. Of course, the argument is defeasible. Ed could deny that he is on the union side, in this case, because their demands have gone too far. Bob's move is dialogical, and open to critical questioning. These are the critical questions corresponding to argument from commitment (Walton, 1996, p. 56):

> CQ1: Is *a* really committed to A, and if so, what evidence supports the claim that she is so committed?
>
> CQ2: If the evidence for commitment is indirect or weak, could there also be contrary evidence, or at least room for the rebuttal that this case is an exception?
>
> CQ3: Is the proposition A, as cited in the premise, identical to the proposition A as cited in the conclusion? If not, what exactly is the nature of the relationship between the two propositions?

Bob's use of argument from commitment could be judged on its merits, once we know more about the evidence in the case. Ed may have changed his mind, or the present labor case may be an exception for Ed, as already noted.

The general scheme of argument from commitment is the basis for several other schemes, like argument from sunk cost and circumstantial ad hominem arguments. The argument from sunk cost may be considered a practical form of argument from commitment in which a decision about a possible course of action follows from past commitments, often called pre-commitments. In the example cited in Chapter 1, Susan was dealing with the choice between completing her Ph.D. program and going to law school. Her conclusion was reached not only by means of a comparison of consequences, in the argument from waste. Her decision to attend law school meant retracting her commitment to her present course of action.

> A PhD student, Susan, has spent more than five years trying to finish her thesis, but there are some problems. Her adviser keeps leaving town, and delays are continued. She contemplates going to law school, where you can get a degree in a definite period. But then she thinks: "Well, I have put so much work into this thing. It would be a pity to give up now." (Walton 1996, p. 81, cited in Chapter 1).

The context is one of deliberations and actions, but at the same time it is based on argument from commitment. *Argument from sunk costs* is represented by the following scheme (Walton, 2002, p. 489). The symbol t_1 represents the time of the proponent's commitment to a certain action (pre-commitment), while t_2 represented the time of proponent's confrontation with the decision whether to carry out the pre-commitment or not.

> There is a choice at $t2$ between A and *not-A*.
> At $t2$ I am pre-committed to A because of what I did or committed myself to at $t1$.
> Therefore, I should choose A.

What needs to be emphasized here is that argument from commitment is the basis of many other argumentation schemes, like the sunk costs argument, where it is combined with practical reasoning. The problem is that argument from commitment is such a basic and ubiquitous form of argument that it often goes unrecognized as being the basis of an argument. Its unrecognized importance as an argumentation scheme has led to many confusions. It has often been equated with the ad hominem argument, for example. This equation is very confusing indeed, because it is the basis of the circumstantial type of ad hominem argument that is also based on an allegation of inconsistency. To sort this confusion out, we need to begin by examining argument from inconsistency.

4. ARGUMENTS FROM INCONSISTENCY

Argument from inconsistent commitment proceeds from a rule of rejection: in a case of inconsistent commitments, at least one of the commitments has to be rejected in order to preserve the consistency of the set. The analysis of this subject reintroduces the perennial problem of implicit commitments. In applying argument from commitment to cases, the inconsistency often follows from the proponent's past statements or actions that the opponent brings to light through questioning. In this argumentation, a previously hidden inconsistency may be brought to the surface. In other cases, however, the inconsistency can easily be detected in an arguer's commitment set. The following is the scheme for *argument from inconsistent commitment* (Walton, 1998, p. 220):

> a is committed to proposition A (generally, or in virtue of what she has said in the past).
> a is committed to proposition $\neg A$, which is the conclusion of the argument α that a presently advocates.
> Therefore, a's argument α should not be accepted.

In the following sequence of dialogue, Ed does not support the union cause, a socialist political party. He is faced with his old commitments, that is, his past support of the communist cause. The two positions are presented as contradictory, and his opponent concludes that one commitment should be rejected.

> You admit you are a communist, and you have always supported communism very strongly in the past. Yet, in this case you are not supporting the union cause? Come on, Ed! You can't have it both ways. (Walton, 1998, p. 108)

As mentioned earlier, the opposition between commitments is not always cut-and-dried. Critical questioning can often reveal subtleties that explain or resolve the apparent inconsistency.

The first three critical questions in the following set attack the incompatibility of the two commitments. The attack is meta-discursive: the respondent comments on and challenges the possibility of the proponent's move; he refuses the latter's presuppositions. The burden of proof is on the proponent: if it is satisfied, the argument may be acceptable. In the fourth question, the opponent attacks the link between premise and conclusion. He may argue for an alternative to the retraction of the commitment, by showing that the seriousness of the conflict does not lead to an unacceptable incoherence. The last two questions are directed against the force of the conclusion: it must be evaluated on the basis of the gravity of the inconsistency, and claims stronger than the simple questioning of the speaker's credibility should be carefully assessed.

These are the critical questions (Walton, 1998, p. 220):

CQ1: What are the propositions alleged to be practically inconsistent, and are they practically inconsistent?

CQ2: If the identified propositions are not practically (pragmatically) inconsistent, as things stand, are there at least some grounds for a claim of practical inconsistency that can be evaluated from the textual evidence of the discourse?

CQ3: Even if there is not an explicit practical inconsistency, what is the connection between the pair of propositions alleged to be inconsistent?

CQ4: If there is a practical inconsistency that can be identified as the focus of the attack, how serious a flaw is it? Could the apparent conflict be resolved or explained without destroying the consistency of the commitment in the dialogue?

CQ5: Does it follow from a's inconsistent commitment that his argument should not be accepted?

CQ6: Is the conclusion the weaker claim that a's credibility is open to question or the stronger claim that the conclusion of his argument is false?

The pragmatic inconsistency introduces a problem, different from one posed by logical inconsistency, of interpretation of actions that the commitments supposedly imply. The opposition is not between commitments following from statements, but between present positions adopted and the implications of past commitments or actions. The arguer's words and deeds are interpreted, and from this interpretation the commitments underlying the decisions to pursue a course of action are reconstructed. The following scheme represents the general form of the *argument from pragmatic inconsistency* (Walton, 1998, p. 218):

> *a* advocates argument *α*, which has proposition *A* as its conclusion.
> *a* has carried out an action, or set of actions, that imply that *a* is personally committed to ¬*A* (the opposite or negation of *A*).
> Therefore, *a*'s argument *α* should not be accepted.

The following argument is a good example of the opposition between the interpretation of facts and stated commitments. The love of nature is opposed to the choice of supporting its destruction by using wooden material in the construction of the house. The fact that the house is made of wood is *presented* as contrary to the environmentalist claims: that does not mean that it *is* contrary. The second critical question of the scheme can thus be directed against the argument in this passage.

> As I sat eating my whole-wheat cinnamon bun, my eyes wandered around the large room. The walls were peppered with posters, petitions and other pleas for everyone to love the goddess, keep the karma and save the trees. As I grudgingly tried to remain sympathetic to all this peace and love, I noticed that this spiritual haven for all those who fight to make the world a better, greener place, was built almost entirely out of wood. The tables and chairs were made of rough-hewn logs. The walls were paneled in cedar. The roof was supported by massive exposed beams that could only have come form old-grown trees. Even the counters and shelving were made of wood, wood, wood. And, as the crowning touch, a huge wooden staircase spiraled up to the second floor. The hypocrisy of it all suddenly struck me like a 200-year-old spruce succumbing to a chainsaw. Everyone wants to save the trees, but each one of us is as guilty as MacMillan Bloedel for their destruction. The demand for single-family homes with private driveways and yards leads developers to level woods and farmland for new suburbs. All these new homeowners insist on hardwood floors and wood-burning fireplaces. Gone are patios made with paving stones nestled into the grass. Now backyard lawns that rarely feel the tread of human feet are dominated by cedar decks, and the bigger the better. (Walton, 1998, p. 10)

The weak points of the argument can be recognized in the effective commitment of the proponent to the claimed proposition and the strength of the conclusion. The distinctive critical question, however, regards the problem of the interpretation and presentation of the facts. This example, using another description of the wooden house, might be a strong premise to support the conclusion that the individuals in question effectively love nature, because they refuse to use artificial materials.

These are the critical questions matching the scheme of argument from pragmatic inconsistency (Walton, 1998, p. 251).

CQ1: Did *a* advocate *α* in a strong way by indicating her personal commitment to *A*?
CQ2: In what words was the action described, and does that description imply that *a* is personally committed to the opposite of *A*?
CQ3: Why is the pragmatic inconsistency indicated by satisfactory answers to CQ1 and CQ2 a relevant reason for not accepting argument *α*?

The double standard argument may be considered a subspecies of argument from pragmatic inconsistency. In this argument, there is an alleged incoherence in the policies adopted with respect to two similar cases: the agent, choosing two incompatible courses of action in two similar situations, is presented as supporting or defending conflicting commitments. The double standard argument may be classified, for this reason, as a species of argument from pragmatic inconsistency. The use of one policy in one case implies that the agent is committed to a determined position with respect to that specific type of case. Since the second case is allegedly similar to the first, the agent is expected to have the same commitments toward it and therefore to act in a similar way. The failure of this expectation is related to a difference in the standpoints: with respect to the same type of situation the agent holds incompatible commitments, and he is therefore alleged to be pragmatically inconsistent.

Double standard arguments, like argument from pragmatic inconsistency, may be described as a bridge between arguments from alleged inconsistency and ad hominem attacks. While in the first case the critique of dialogical or pragmatic decisions aims at a retraction of commitment, in the latter case the same strategy is used to draw a moral judgment about the person's character (*ethos*) and from this to draw a conclusion about the nonacceptability of the person's argument.

The following is the scheme *argument from double standard* (Walton, 1998, p. 253):

The respondent has one policy with respect to *a*.
The respondent has another (different) policy with respect to *b*.
a is similar to *b* (or comparable to *b* in some relevant respect).
Therefore, the respondent is using a double standard.

In the following excerpt, the credibility of the government is attacked by showing inconsistency in its decisions:

> Liberal finance critic Gerry Phillips accused the government yesterday of a "double standard" because it has launched a major effort to stem welfare cheating, including a toll-free provincial fraud hotline, while taking a sympathetic view of tax evasion. "On the one hand, people who cheat on welfare are crooks and you have to prosecute them, but if you're just cheating on taxes, it's human nature," Mr. Phillips said of government. Tax fraud is likely 10 times larger than welfare fraud, he said, and the government should make "the same concerted effort" tracking it as has been put in place for social-assistance cheating. (Walton, 1998, p. 222)

Here are the critical questions for the argument from double standard (Walton, 1998, p. 253):

CQ1: What is the respondent's policy with respect to *a*?
CQ2: What is the respondent's policy with respect to *b*?
CQ3: How is the one policy different from the other?
CQ4: How is *a* similar (or comparable) to *b*?
CQ5: Can the differences in policies be explained, or is it significant as evidence that the respondent's policies are not consistent in some important way?

The most powerful critical questions are the third and the fourth, focused on the problem of the similarity of the cases and the differences in the relevant policies. The argument presupposes the comparability of two events: if this assumption is not accepted by the interlocutor, the argument collapses, and is defeated. Therefore, this kind of attack reverses the burden of proof on the opponent. The policies are assumed by the opponent to be incompatible: if the decision showing them as conflicting is not shared by the other party, or does not stand up to her questioning, the speaker has to defend his position. The last question is relative to the force of the conclusion: the inconsistency may be explained, or not taken to be serious, and therefore the potential attack on the respondent's credibility may be reduced or avoided.

5. ETHOTIC ARGUMENTS

The argument from *ethos*, based on the credibility of the person, has two forms, a positive and a negative one. While the positive topic provides a good reason to accept the proposed position, the negative one is closely related to the generic ad hominem argument. Ethotic topoi are the link between arguments from commitment and personal attacks. The ad hominem argument is based essentially on two topics. One is that from the bad character of an arguer it is reasonable to infer his non-credibility, and hence the lack of worth of his argument. The other is that from incoherence in an arguer's actions or words it is possible to draw a conclusion to a serious ethical failure in the agent's character, and to take that as an impugning of his credibility. The following scheme represents the Aristotelian *ethotic argument* (Walton, 1995, p. 152):

> *Major Premise*: If *x* is a person of good moral character, then what *x* says should be accepted as more plausible.
> *Minor Premise*: *a* is a person of good moral character.
> *Conclusion*: Therefore, what *a* says should be accepted as more plausible.

The following are the critical questions matching this scheme.

> CQ1: Is *a* a person of good moral character?
> CQ2: Is character relevant in the dialogue?
> CQ3: Is the weight of presumption claimed warranted strongly enough by the evidence given?

The most important qusestion in connection with many cases where a fallacy has been alleged is the second. Argument from *ethos* is strong only in the context of a lack of direct evidence in an investigation or dialogue, when other means to assess the speaker's position are not available. Relevance, therefore, may be considered a most critical point in such cases.

A typical circumstance where ethotic appeal is regarded as a strong argument is that of political debate. For the electorate, in many instances, there are no other reasons to believe in the speaker's words and promises except for his consistency and moral integrity. For many of us in democratic societies do not have direct knowledge of all the facts on which social polices are based. Thus we go by a politician's commitments, or what we think they are. A person with clear core convictions that appear to be consistent with positions he has advocated is more likely to pursue a definite policy. On the other hand, "flip-flopping" may be easily understood as a sign of incoherence in decisions and moral weakness of character, suggesting a lack of integrity. The following example presents a use of ethotic argumentation in democratic politics.

> Part of the problem is the sense that his character is elusive; that there is little he will draw a line or fight for. David Gergen, who is leaving the Clinton administration, said recently that U.S. voters want to know what Mr. Clinton's "core convictions" are as he faces the next two years. "Bill Clinton ... has few core values on which his presidency is built," said James

Lake, a former communications adviser to President Reagan and George Bush. "Ted Kennedy –I never agree with him, but he stands for something, he's consistent, you know where he stands. Ronald Reagan – people disagree with him, time after time, but you know where he stood. No one can say that they know what Bill Clinton stands for." Privately, many Democrats say the same thing and fear that, as a result, the Clinton presidency cannot recover. (Walton, 1998, p. 198)

In this example the apparently inconsistent actions of Bill Clinton are presented as a strong reason to distrust his character, and in turn to offer an argument for the people not to vote for him. If they cannot understand his policy as clear and consistent, they have no way of estimating his future decisions. Therefore, the argument runs, they cannot trust him and believe in his present avowed commitments.

The basic type of ad hominem argument derives a judgment on the speaker's words by drawing an inference about his presumed bad character. This form of attack is the *argument from negative ethos*, also defined as the *generic ad hominem* argument (Walton, 1998, p. 249). It can be called the direct ad hominem argument, and in many textbooks it is called the abusive type of ad hominem argument (a negative wording, suggesting that the argument is fallacious). It is a very simple form of argument, but is the basis of many more complex types of ad hominem argument.

> *a* is a person of bad character
> Therefore, *a*'s argument α should not be accepted

The following critical points are parallel to those of the positive version of the ethotic argument (Walton, 1998, p. 250):

> CQ1: Is the premise true (or well supported) that *a* is a person of bad character?
> CQ2: Is the issue of character relevant in the dialogue in which the argument was used?
> CQ3: Is the conclusion of the argument that α should be (absolutely) rejected even if other evidence to support α has been presented, or is the conclusion merely (the relative claim) that α should be assigned a reduced weight of credibility, relative to the total body of evidence available?

The main difference between the positive and the negative form of appeal to character lies in the way the argument is directed. The ad hominem attack leads to the rejection of an arguer's argument, while in the positive scheme the argument is inferred to be more plausible.

There are many subspecies of the direct type of ad hominem argument that attack different aspects of character. Argument from veracity, one of these variants, is extremely powerful in political discussions. For example, in the following ethotic argument put forward against Bill Clinton, the alleged failure of correspondence between personal convictions and commitments is used to attack his character for veracity, and thus to try to destroy his credibility.

> Today's bipartisan consensus is that Clinton is neither bad nor dangerous, just silly. Plainly put, almost no one thinks he believes a word he says. Or, more precisely, he believes everything he says at the moment he emphatically says it, and continues to believe it at full throttle right up to the moment he repudiates it. He has the weird sincerity of the intellectual sociopath, convinced that when he speaks, truth is an opinion but convenience is an imperative. (Walton, 1998, p. 199)

In this case we are not told which argument attributed to Clinton is being attacked. But assuming there is such an argument under attack, the case could be said to have the form of *argument from veracity* represented by the following argumentation scheme (Walton, 2002, p. 51).

a has bad character for veracity.
Therefore, *a*'s argument should not be accepted.

Veracity is different from integrity, but closely related to it. Veracity is regard for the truth of a matter.

In other decision-making contexts, prudence is a relevant quality of character that may be significant in deciding whether a source is credible. When other, more decisive arguments or information are not available, this kind of attack on perceived prudence may be reasonable. On the other hand, such arguments are often based on flimsy evidence and mere innuendo. Each case must be judged on its individual merits, depending on the evidence given to support a charge. Consider the following very common sort of example.

> A friend has recommended a new investment opportunity, but your significant other rejects the recommendation with the remark, "How could you possibly value the advice of that crazy?"[1]

In this example, the argument may not be fallacious if the judgment made against the person is based on hard evidence or documented past experience. The fact that the friend often makes rash decisions could be a good reason to believe that he is not very reliable in the matter of safe

---

[1] Example from http://www2.sjsu.edu/depts/itl/graphics/adhom/adhom.html, with small changes.

investments. Such decisions are very important in business, for example. On the other hand, the argument could be fallacious if no real evidence backs it up and if the argument is only meant to smear the person's credibility. Innuendo is often the basis of such character attacks.

The reasoning in such a case fits the scheme of *argument from prudence* (Walton, 2002, p. 51):

> *a* has bad character for prudent judgment.
> Therefore, *a*'s argument should not be accepted.

Prudence is a matter of both character and reputation, as well as one of practical reasoning.

The argument from perception regards the skills of understanding a situation: in the following example, Marx's work is criticized on the basis that he is held to be uninterested in investigating facts and, therefore, that he could not have a good understanding of capitalism. The attack on his knowledge of the real factual circumstances is relevant in the context of a discussion where the careful observation of the economic phenomena is the basis for an economic theory.

> What Marx could not or would not grasp, because he made no effort to understand how industry worked, was that from the very dawn of the Industrial Revolution, 1760–1790, the most efficient manufacturers, who had ample access to capital, habitually favored better conditions for their workforce; they therefore tended to support factory legislation and, what was equally important, its effective enforcement, because it eliminated what they regarded as unfair competition. So conditions improved, and because conditions improved, the workers failed to rise, as Marx predicted they would. What emerges from a reading of *Capital* is Marx's fundamental failure to understand capitalism. He failed precisely because he was unscientific: he would not investigate the facts himself, or use objectively the facts investigated by others. From start to finish, not just the *Capital* but all his work reflects a disregard for truth which at times amounts to contempt. (Walton, 1998, p. 217)

As with all direct ad hominem attacks, this argument needs to be judged on the evidence put forward to back it up. Marxists could find all kinds of examples that would suggest that Marx did have a good grasp of the facts on which he based his theories. Thus such arguments tend to be defeasible, and subject to attack and rebuttal by those opposed to them. Still, they can be a reasonable basis for making a decision under uncertainty.

Still another scheme fitting under direct ad hominem attack is the *argument from perception* (Walton, 2002, p. 51):

a has bad character for realistic perception of a situation.
Therefore, *a*'s argument should not be accepted.

This form of argument is similar to the previous one, except that the focus of the attack is on the person's perception of a situation. A failure to be aware of relevant facts is a problem for a practical reasoner, and it can be attacked as a character defect that is relevant to an arguer's credibility.

In every context of discussion, logical reasoning is one of the most important qualities for the construction of an argument. The claim of a speaker's irrationality or deficiency in reasoning may be highly unreliable when considering how to judge his argument. For example, in the following case, the person is labeled as not very enlightened in understanding social problems. What does this mean? What evidence backs it up? We may not know, but still, the move is directed against his character and may impact our judgment of his ability to produce reasonable arguments in that specific field.

> We need not consider this piece of social legislation. It was, as you know, introduced by Senator Farrell, who is just not a very socially enlightened person. (Engel, 1980, p. 130)

The argument in this example is fallacious, because other evidence that may be available is not taken into consideration, and because the appeal to *ethos* is presented as conclusive.

The attack in the case just presented is a species of ad hominem that can be called *argument from logical reasoning*. It is represented by the following scheme (Walton, 2002, p. 51):

a has bad character for logical reasoning.
Therefore, *a*'s argument should not be accepted.

This form of argument may be based on allegations that a person is mentally ill, for example, if that is used to suggest that any argument he puts forward can be discounted as illogical. For example, the philosophy of Friedrich Nietzsche might be attacked on the ground that he was mentally ill in later life. This fact, coupled with the fact that his philosophical views later became fashionable among the Nazis, could seem to many to condemn his philosophical work. Sorting out whether such an ad hominem attack based on argument from logical reasoning is reasonable could be difficult. It could involve the untangling of a lot of allegations and historical facts, as well as looking at what Nietzsche really wrote and how that should be interpreted. Still, the mere argument that he was mentally ill might seem to many to be a powerful argument against him as a philosopher to be taken seriously. Many, on hearing this allegation, might simply dismiss his views as "crazy."

One of the most common subspecies of ethotic argument is the argument from failure of moral standards in past conduct, perhaps based on convictions or actions taken against the person on grounds that he committed ethical breaches. Such examples can be based on very little evidence, or on allegations that may not be true, but the personal attack argument can have a powerful effect. In the following example, Francis Bacon's intellectual work is deemed untrustworthy on the basis of an event of his life. The problem here is that character is evaluated from a particular alleged fact, and evaluation of his thought based on other, more relevant criteria of assessment may be ignored.

> It may be argued that [Francis] Bacon's philosophy is untrustworthy because he was removed from his chancellorship for dishonesty. (Copi and Cohen, 1961, p. 54, quoted in Walton, 1996, p. 55)

A major historical controversy has swirled around the allegation made in this case. Recently, it has been argued that Bacon was unfairly attacked by powerful circles in the government of the time who were opposed to his views. In addition to being a philosopher and scientist, Bacon was active in law and politics, having been a member of the House of Commons and lord chancellor. In 1621, he was accused of bribery and corrupt dealings and admitted his guilt in a trial. He was fined, and never sat in Parliament again. A famous essay, "Lord Bacon" by Thomas Babington Macaulay (1856), attacked Bacon's character as corrupt and dishonest. However, doubts persisted about whether this attack was really justified by the facts. A thoroughly researched recent work by Mathews (1996) clears Bacon of the charges of corruption and fraud and rehabilitates his character. Mathews shows that the allegations against Bacon were uncritically passed on from one writer to another without any real investigation of their truth. Those who wrote about Bacon took their material from sources that just assumed that Bacon was a villain. This case shows how persuasive an ad hominem attack on a philosopher can be in getting everyone to dismiss his views.

The argumentation scheme in this kind of case is that of the *ad hominem argument from moral character* (Walton, 2002, p. 51):

> *a* has bad character for personal moral standards.
> Therefore, *a*'s argument should not be accepted.

Sorting out the Bacon case has involved a lot of historical research, and the investigation still may not be entirely conclusive. Examining the facts to see if the premise of the argument is true, or can be supported by the evidence, can involve a lot of investigation. Still, a reported breach of moral standards may be quite enough to discredit someone's argument, especially in politics or matters of ethical philosophy, where character is an issue. Only a small basis in fact may be enough to convict someone in the court of public opinion.

The structure of the direct ad hominem argument is relatively simple compared to that of the circumstantial type, which is always based on an allegation of some sort of inconsistency. The alleged inconsistency is then used to attack the person's character, which is then used to attack his argument. However, not all arguments from an imputed inconsistency should properly be classified as ad hominem arguments. There must always be a personal attack, that is, an attack on the arguer's ethical character, at the center of any ad hominem argument.

6. CIRCUMSTANTIAL AD HOMINEM

As said before, ethotic arguments are based on a premise expressing a moral judgment about a person, involving his or her history. Circumstantial attacks involve an inference from a person's allegedly inconsistent commitments to a moral critique of the person, and thence to a rebuttal of his argument on the basis of his lack of credibility. The circumstantial attack may assume two forms, depending on whether a logical and pragmatic inconsistency is alleged. In the *logical inconsistency circumstantial ad hominem argument*, a contrast is presented between two propositions to which the opponent is supposedly committed (Walton, 1998, p. 259):

> *a* advocates argument α, which has proposition A as its conclusion.
> *a* is committed to proposition A (generally, or in virtue of what she has said in the past).
> *a* is committed to proposition $\neg A$, which is the conclusion of the argument α that *a* presently advocates.

Therefore, *a* is a morally bad person.

Therefore, *a*'s argument α should not be accepted.

The following example is interesting because it raises a fundamental problem regarding such an alleged opposition of commitments. It is a common tactic to cite statements belonging to different periods of a person's life in order to prove his inconsistency. Such arguments rest on a quite questionable presupposition, namely, the conception of life as a coherent unity in which commitments remain constant. According to this assumption, a person is expected never to change his mind. The removal of this main assumption, in these cases, is often enough to cause the attack to collapse:

> Under the heading, "The changing 'Times'," Macdonald raked the *New York Times* over the coals in the winter issue of *Politics* for 1948 because of inconsistent views on the Hiroshima bomb:
>
> The Japanese would like the world to believe that had it not been for the atomic bomb, they could have fought indefinitely . . . Revelations by their surrender envoys provide the answer to this fallacy. They were well licked before the first atomic bomb exploded over Hiroshima. (Editorial, August 23, 1945)
>
> The Japanese had been greatly weakened but they were still determined to fight to the death . . . That is the justification for the bomb's use. (Editorial, January 28, 1947) (Boller, 1967, p. 145)

In the logical variant of the argument, the conflict is shown as a logical inconsistency between statements. In its pragmatic counterpart, the circumstantial ad hominem, an opposition is presented between a sentence and the implication of a set of actions. The commitments underlying the words and the actions are criticized as inconsistent. This is the form of the *(pragmatic) circumstantial ad hominem argument* (Walton, 1998, p. 251):

> *a* advocates argument α, which has proposition *A* as its conclusion.
>
> *a* has carried out an action, or set of actions, that imply that *a* is personally committed to ¬*A* (the opposite or negation of *A*).
>
> Therefore, *a* is a morally bad person.
>
> Therefore, *a*'s argument α should not be accepted.

Circumstantial ad hominem arguments can be justified as reasonable if the proposed practical inconsistency is really present in the commitments of the speaker and if the accusation of inconsistency is relevant to the issue. For instance, in the following example, Mr. Smith's commitment to the law on the drug tests conflicts with his refusal to apply the rule to himself. The incoherence shown is relevant, because laws are assumed to be equal for everyone. The incoherence in Smith's proposing a law casts doubts on his personal fairness and on the equitability of the rule.

> Rodney Smith, of the President's Commission on organized crime, testified before a House subcommittee that he thought there were good reasons why drug tests should be mandatory for federal workers. A critic at the subcommittee meeting asked whether Mr. Smith would himself be now willing to give a urine sample. He replied that he would not. (Walton, 1987, p. 321)

On the other hand, there are problems concerning the interpretation of actions and commitments as opposed. The moral judgment of the person supposedly following from the alleged contradiction may also be unwarranted, and this too may undermine the argument from inconsistency. In some cases the alleged conflict of commitments may not hold up to scrutiny because the statements belong to different historical moments. In other cases, the inconsistency is a clear sign of an inconsistent set of commitments that does call for some response, as in the example of Mr. Smith.

The following critical questions match the circumstantial ad hominem argument (Walton, 1998, p. 225).

CQ1: What are the propositions alleged to be practically inconsistent, and are they practically inconsistent?

CQ2: If the identified propositions are not practically (pragmatically) inconsistent, as things stand, are there at least some grounds for a claim of practical inconsistency that can be evaluated from the textual evidence of the discourse?

CQ3: Even if there is not an explicit practical inconsistency, what is the connection between the pair of propositions alleged to be inconsistent?

CQ4: If there is a practical inconsistency that can be identified as the focus of the attack, how serious a flaw is it? Could the apparent conflict be resolved or explained without destroying the consistency of the commitment in the dialogue?

CQ5: Does it follow from *a*'s inconsistent commitment that *a* is a morally bad person?

CQ6: Is the conclusion the weaker claim that *a*'s credibility is open to question, or the stronger claim that *a*'s conclusion is false?

The argument from circumstantial attack has two subcategories, stemming from the speaker's commitment. In one type of argument, the proponent is alleged to be personally committed to a proposition. In another, the speaker may argue that he belongs to a group or to "everybody" of a certain class, all of whom are committed to a proposition: in this case, his belonging to 'everybody' or to the group in question indirectly binds him to the commitment, or so the argument alleges. This is an important difference, because most alleged practical inconsistencies arise from the fact that the person has membership in some community the attacker is speaking about. On the other hand, the wrong evaluation of such an assumption, or what is taken to be implied by it, is a frequent cause of argument fallaciousness.

Two subtypes are based on such forms of indirect commitment to a group. One is the scheme from *universal circumstantial ad hominem* (Walton, 1998, p. 253):

> *a* advocates argument α, which has proposition A as its conclusion, which says that everybody should be committed to A.
> *a* is bound by the 'everybody' in the first premise.
> *a* has carried out an action, or a set of actions, that imply that *a* is personally committed to $\neg A$.
> Therefore, *a* is a morally bad person.
> Therefore, *a*'s argument α should not be accepted.

The following case is the classic smoking example.

Parent: There is strong evidence of a link between smoking and chronic obstructive lung disease. Smoking is also associated with many other serious disorders. Smoking is unhealthy. So you should not smoke.

Child: But you smoke yourself. So much for your argument against smoking. (Walton 1996c, p. 58)

In this dialogue, the father claims that smoking is unhealthy, committing everyone to advocating and practicing abstinence from this vice. The child points out the incoherence in the father's argument: he, although belonging to 'everybody', is therefore logically bound to a commitment to not smoking. And yet he admits that he smokes. For this reason, his argument is not credible.

This example could be just an argument from inconsistency and not an ad hominem argument if the child were merely arguing that the father is inconsistent but not arguing that he is hypocritical, or otherwise has bad character. Let's assume from the child's strong rejection of the parent's argument that the child is basing that rejection on a premise of the parent's faulty character. As such, the argument is complex to assess, because its fallaciousness depends on the kind of argument the father used. If he is trying to persuade the son through a rational explanation of the physical consequences of smoking cigarettes, the attack is not fallacious. There is no inferential relation here between personal credibility and objective assessment of the facts in the claim that smoking is bad for health. On the other hand, if the son is throwing this objective evidence out the window just because he thinks the parent is hypocritical, that could be a big mistake. The parent's advice not to smoke, based on his claim that smoking is unhealthy, could be quite a good argument.

The analysis of this argument points out the importance of relevance in the topic assessment. The last of the listed critical questions (Walton 1998, p. 256) is focused on this problem, while the first two regard the relation between the universal commitment expressed by the speaker and his personal obligation to the proposition.

CQ1: Does *a*'s argument conclude that everybody should be committed to *A*?
CQ2: Is there any basis for *a*'s being an exception to the commitment?
CQ3: Does the action, as described, imply that *a* is personally committed to the opposite of *A*?
CQ4: Why does it follow (if it does) that the alleged practical inconsistency shows that *a* is a bad person?
CQ5: Is *a*'s being a bad person a good reason for concluding that *a*'s argument should not be accepted?

The smoking case is the classic example of circumstantial ad hominem argument because it is relatively simple in a certain respect. The parent argues that smoking is bad but smokes himself. Thus the inconsistency is direct. Smoking and not smoking are direct opposites. Many other cases are not so simple. In these other cases, the one proposition implies the negation of the other, but only by an indirect sequence of reasoning connecting the two.

The example of circumstantial ad hominem in the Smith case is a fairly clear case of group inconsistency. Mr. Smith commits the group of federal workers, which he belongs to, to undergo a drug test, but he refuses to take the test himself. In this case, the commitment is indirect and follows the form of *group circumstantial ad hominem* (Walton, 1998, p. 256):

a advocates argument α, which has proposition *A* as its conclusion, which says that everybody in group *G* should be committed to *A*.
a belongs to group *G*.
a has carried out an action, or a set of actions, that imply that *a* is personally committed to *A*.
Therefore, *a* is a morally bad person.
Therefore, *a*'s argument α should not be accepted.

The analysis of the following case helps to illustrate the pivotal role of the determination of the reference group. The reporters' attack is addressed to people in public service who receive the listed "benefits." But reporters do not belong to the group of "public servants": the identification of the two groups fails to recognize the relevant point of the argument, that is, the critique to the government's use of taxes.

> A critic argues that reporters are circumstantially inconsistent when they criticize the free lunches, air trips, and other "free benefits" that people in the public service are often said to receive by reporters. For, the critic alleges, these reporters themselves are often the recipients of these same benefits. (Walton, 1998, p. 195)

The critic's argument is fallacious because he unifies under the label "employees" two groups that are arguably different in the relevant aspect. In the following critical questions (Walton, 1998, p. 256), the first three questions are relative to the group belonging question: this kind of attack reverses the burden of proof of the discussion. The opponent, by addressing such a critique, refuses the presupposition regarding the relation between the speaker and the reference community.

> CQ1: How exactly does the argument α state or imply that everybody in group G should be committed to A?
> CQ2: Does a belong to group G?
> CQ3: Does a belong to other groups that would have goals affecting a's commitment to A?
> CQ4: Does a's action, as described, imply that a is committed to the opposite of A?
> CQ5: Why does it follow (if it does) that the alleged practical inconsistency shows that a is a bad person?
> CQ6: Is a's being a bad person a good reason for concluding that a's argument should not be accepted?

Group belonging is a relevant aspect of another kind of circumstantial attack: the guilt by association argument. A moral judgment of the person is rendered on the basis of his association with a morally reprehensible group. For instance, in the following piece of reasoning, the rebuttal of Smith's claim is based on his belonging to the Communist Party and his commitment to communist economic theory, which prevents him from considering objectively the problem of inflation.

> Smith's claim that unemployment is a graver problem than inflation may be countered on the grounds that Smith is a Communist. (Toulmin, Rieke, and Janik, 1979, p. 173, quoted in Walton, 1996, p. 55)

The example follows the topical scheme of *guilt by association* (Walton, 1998, p. 257):

> *a* is a member of, or is associated with, group *G*, which should be morally condemned.
> Therefore, *a* is a morally bad person.
> Therefore *a*'s argument *α* should not be accepted.

The following argument is a clear case of fallacious attack. The fact that the opponent shares communist ideas is not relevant for the assessment of the whole theory. The case above referred to a specific aspect of the speaker's claim, and the validity of the rebuttal depends on the strength of the conclusion, which is not specified. In the following ad hominem argument, the prejudice against communists determines a conclusive rejection of the opponent's thesis.

> This theory was introduced by a person of known Communist sympathies. There cannot be much to it. (Engel, 1980, p. 129)

The first three critical questions are focused on the assumption that the speaker effectively belongs to the group and that the latter is morally condemned. The last critique is directed against the inferential link and only weakens the consequence, without reversing the presumption.

> CQ1: What evidence is there that *a* is a member of *G*?
> CQ2: If *a* is not a member of *G*, but is associated with *G*, how close is this association?
> CQ3: Is *G* a group that should be morally condemned?
> CQ4: Is it possible that even though *a* is a member of *G*, a group that ought to be condemned, *a* is not a bad person?

The morality of a person may be judged on the basis of the situation of the speaker and his words. In the following conversation, a classic example is given of the situationally disqualifying ad hominem:

> *[Holland, December 1990]* A retired major general argues in front of his relatives that the Dutch government must give more substantial support to the allied efforts in the Gulf Area. "We ought to send ground forces," so he claims. His grandson retorts: "It's all very well for you to talk, grandpa! You don't have to go there." (Walton, 1998, p. 240)

The context of dialogue here is extremely important. Two conflicting points of view are involved in the conversation: the geopolitical and the civilian evaluation of the Dutch intervention. The term "soldier" is categorized respectively under the paradigm of "troops" and "person."

The discussion, presumably, was centered on the assessment of Dutch war politics under the aspect of the people who were going to die. For this reason, the attack on the grandfather's dialectical contribution, detached from the relevant perspective, is valid. The claim is moved against the incoherence between his condition of retired general and his consideration of the soldiers' lives. The accusation of inappropriateness here simply points out the irrelevance of the dialectical move. The general scheme is from *situationally disqualifying ad hominem* (Walton, 1998, p. 258):

> In dialogue *D*, *a* advocates argument α, which has proposition *A* as its conclusion.
> *a* has certain features in his personal situation that make it inappropriate for him to make a dialectical contribution to *D*.
> Therefore, *a* is a morally bad person.
> Therefore, *a*'s argument α should not be accepted.

The following are the critical questions for the argument:

> CQ1: What features of *a*'s personal situation make it inappropriate for him to contribute to *D*?
> CQ2: Do the features of *a*'s situation cited give any good reason to make one conclude that it is inappropriate for him to contribute to *D*?
> CQ3: Could *a*'s argument be worth considering on its merits, even though there is reason to think it inappropriate for *D*?

The Gulf War example could be interpreted as based on allegation of bias. The grandson may be suggesting that since the grandfather has nothing personally to lose, naturally he is in favor of troop deployment. This argument suggests that the grandfather has an interest at stake, and that this factor lessens the worth of his argument. This brings us to the connection between the ad hominem argument and arguments from alleged bias.

7. ARGUMENTS FROM BIAS

The definition of bias is related to the notion of critical doubt. The distinctive characteristic of critical discussion is critical doubt, a negative attitude of one party toward the viewpoint of the other one. Critical doubt means that an arguer has a contra attitude toward his opponent's view because he has critical questions about it, but it also means that he is willing to overcome this attitude if the opponent can persuade him by rational argumentation. A critical doubter evaluates the opponent's standpoint, and if he considers it more plausible than his own position, he will drop it in order to accept the interlocutor's view. The arguments a speaker provides in support of his own thesis are the rational means used to persuade the hearer of its acceptability. The hearer, on the other hand, can be open to rationally assessing it only if he can overcome critical doubt. He needs to temporarily suspend his positive attitude toward his own viewpoint and his negative attitude toward the opposite position, and to look at the issue from both the points of view. He needs to have empathy, and to be open to defeat in the face of the stronger argument. If he is a rational arguer, his ability to appreciate the reasons supporting the view of the other side is not hindered by his goal of winning the game of dialogue. Temporary suspension of the positive inclination toward his own view and commitments is, therefore, one of the most important characteristics of rational argumentation in persuasion dialogue. It is the necessary condition of critical doubt.

Bias is normal in a critical discussion, because it is proper to be an advocate of one's own viewpoint. But there is also a bad kind of bias that represents a failure of the proper exercise of critical doubt. This kind of bad bias can be defined in relation to the attitudes just cited: it is the "failure in argumentation of openness to new evidence or legitimate critical doubts that have arisen in the dialogue" (Walton, 1991, p. 14). The problem is not the speaker's partisanship toward his own point of view, but his reaction to the other party's position or critique. If the speaker's attitude is excessively strong in relation to the type of dialogue (in this case, persuasion dialogue), his behavior may be labeled as biased, referring to the bad or harmful type of bias. In a negotiation dialogue, on the other hand, persuasion and openness to other arguments is sometimes self-defeating. In these situations, an arguer is trying to make the best deal, that is, in dialectical terms, to win the game without regard for the truth of the matter. Such a dialogue game aims not at establishing the best objective position supported by the evidence on both sides, but at a victory for the more skillful negotiator. A participant is not expected to be all that open to evidence that supports the other point of view, and bias, although it can be a problem, is not problematic in the same way that it is in a persuasion dialogue. Bias, for this reason, may be considered as the tactic of pushing ahead aggressively when used in the wrong type of dialogue. The biased arguer sees everything in terms of winning gains for himself or his own group while not taking rational arguments against his own view into account.

The accusation of bias, as a consequence, is a serious attack on an opponent's credibility. For instance, in the following case, an expert opinion argument is undermined by the claim that the scientist in question is biased. By this means the effect of the argument is strongly reduced:

> A speaker in a panel discussion on the issue of industrial pollution claimed to be a neutral scientist, but a critic showed that this scientist was on the board of directors of a large industry that had often been accused of pollution. (Walton, 1995, p. 153)

The argument used in this case is from bias. Bias is presented as a reason to believe that the arguer is less likely to be objective, or to exercise an attitude of critical doubt. This accusation is particularly strong when addressed to an argument from expert opinion, because an expert's credibility largely derives from the fact that he is presumed to express an unbiased opinion. The topic is represented in the scheme of *argument from bias* (Walton, 1995, p. 153):

> *Major Premise*: If x is biased, then x is less likely to have taken the evidence on both sides into account in arriving at conclusion A.
> *Minor Premise*: Arguer a is biased.
> *Conclusion*: Arguer a is less likely to have taken the evidence on both sides into account in arriving at conclusion A.

The definition of bias just outlined is centered on the notion of type of dialogue: for this reason, the argument can be properly evaluated only in relation to a dialectical context of use. The first critical question (Walton, 1995, p. 153) points out this aspect of the topic. In addition, an allegation of bias needs to be supported by evidence, if challenged, in order to be plausible.

> CQ1: What type of dialogue are the speaker and hearer supposed to be engaged in?
> CQ2: What evidence has been given to prove that the speaker is biased?

We can observe that bias of the bad sort undermines the ability of an arguer to be properly involved in a persuasion dialogue by taking the arguments on both sides into account. No further attack necessarily follows from this move. The accusation of bias is itself a form of attack.

In the bias type of ad hominem argument, the accusation of bias is used as a premise for a negative judgment of the interlocutor's ethical character and, consequently, of his credibility. His credibility is undermined on the basis of his bias. In the discussion reported here, Wilma's argument is weakened by this form of ad hominem argument because of what is alleged to be her biased viewpoint:

> Bob and Wilma are discussing the problem of acid rain. Wilma argues that reports on the extent of the problem are greatly exaggerated and that the costs of action are prohibitive. Bob points out that Wilma is on the board of directors of a U.S. coal company and that therefore her argument should not be taken at face value. (Walton, 1991, p. 2)

The scheme Bob uses is *bias ad hominem* (Walton, 1998, p. 255):

Person a, the proponent of argument α, is biased.
Person a's bias is a failure to honestly take part in a type of dialogue D, which α is part of.
Therefore, a is a morally bad person.
Therefore, α should not be given as much credibility as it would have without the bias.

The underlying assumptions are the same as those of the argument from bias, except that the allegation of bias is taken to suggest duplicity, dishonesty, or some other sort of ethical failure. As with to all other *ad hominem* attacks, however, the nature of the personal judgment has to be carefully evaluated in light of the evidence presented. These reservations are reflected in the critical questions matching the bias ad hominem argument.

CQ1: What is the evidence that a is biased?
CQ2: If a is biased, is it a bad bias that is detrimental to a's honestly taking part in D, or a normal bias that is appropriate for the type of dialogue in which α was put forward?

Bias ad hominem aims at attacking only one argument proposed by the interlocutor. The poisoning the well argument is an extension of the former argument to other positions that the interlocutor takes, or even to positions he will take in the future. Such a general accusation of bias is used not just to damage his credibility relative to a specific case, but in general in a dialogue. The person is presented as an unfair player in the game of dialogue who always cheats, and therefore all his moves have to be discredited. They all lose their plausibility. Using this strategy, the speaker casts a negative presumption on his adversary's dialogical attitude, undermining all his possible moves at one stroke.

For instance, in the following piece of reasoning, every opponent's argument is strongly weakened by the prejudice expressed by the speaker:

> You can never believe or take seriously anything she says on the abortion issue because, as a woman, she will always take the feminist point of view which supports her own interests as a female. (Walton, 1998, p. 231)

This argument fits the scheme of the *poisoning the well ad hominem* (Walton, 1998, p. 231):

> For every argument a in dialogue *D,* person *a* is biased.
> Person *a*'s bias is a failure to honestly take part in a type of dialogue *D,* which α is part of.
> Therefore, *a* is a morally bad person.
> Therefore α should not be given as much credibility as it would have without the bias.

The critical questions matching the scheme for the poisoning the well ad hominem argument are the following.

Critical Questions

CQ1: What is the evidence that *a* has been biased with respect to every argument in the dialogue?

CQ2: Is the bias a normal partisan viewpoint that *a* has shown, or can it be shown to indicate that *a* is not honestly participating in the dialogue?

CQ3: In what respect is *a* a bad person, judging from the evidence of his participation in the dialogue that gives a reason for doubting his credibility?

The poisoning the well attack may be a consequence of the person's belonging to a group considered to be pressing for its own interests. Thus the argument is often similar to the guilt-by-association argument: from the condemnation of the group, or the prejudice cast on it, the personal attack follows. In poisoning-the-well alleging group bias, the group is labeled as biased, and thus its members' positions can be undermined by a bias attack.

For instance, in the following case, Charles Kingsley denounced the attitude of the Catholic Church towards truth and an interlocutors' opinions based on his membership in this group, accusing Cardinal Newman of bias. Kingsley, in this way, blocked the possibility for Newman to advance credible arguments. He poisoned the well, so to speak.

> The British novelist and clergyman Charles Kingsley, attacking the famous Catholic intellectual John Henry Cardinal Newman, argued thus: Cardinal

Newman's claims were not to be trusted because, as a Roman Catholic priest (Kingsley alleged), Newman's first loyalty was not to the truth. Newman countered that this *ad hominem* attack made it impossible for him and indeed for all Catholics to advance their arguments, since anything that they might say to defend themselves would then be undermined by others' alleging that, after all, truth was not their first concern. Kingsley, said Cardinal Newman, had poisoned the well of discourse. (Walton, 1998, p. 15)

Kingsley used the *poisoning the well alleging group bias argument* (Walton, 1995, p. 257):

> Person *a* has argued for thesis *A*.
> But *a* belongs to or is affiliated with group *G*.
> It is known that group *G* is a special-interest partisan group that takes up a biased (dogmatic, prejudiced, fanatical) quarrelling attitude in pushing exclusively for its own point of view.
> Therefore, one cannot engage in open-minded critical discussion of an issue with any members of *G*, and hence the arguments of *a* for *A* are not worth listening to or paying attention to in a critical discussion.

Needless to say, this was a difficult attack for Newman to respond to. It seemed to condemn everything he said, even any attempt to reply to the argument. Even so, such an argument should properly be regarded as open to critical questions.

Critical Questions

CQ1: Has *a* given any good reasons to support *A*?
CQ2: What kind of bias has *a* exhibited, and how strong is it?
CQ3: Is the kind of bias that *a* has exhibited a good reason for concluding that she is not honestly and collaboratively taking part in the dialogue?
CQ4: Is there evidence of a dialectical shift in the case, for example, from a persuasion dialogue to a negotiation?
CQ5: Is the bias indicated in CQ2 of the very strong type that warrants the conclusion that *a* is not open to any argumentation that goes against her position (or seems to her to go against her position)?

The poisoning the well argument can be extremely powerful, because it can be used to discredit an arguer in a persuasion dialogue in such a wholesale way that the audience simply counts him out and ceases to pay attention to anything he might say. It is an effective way of shutting an arguer down altogether, blocking him from taking any further meaningful part in a dialogue.

8. AD HOMINEM STRATEGIES TO REBUT A PERSONAL ATTACK

The schemes we have described here represent the different tactics used
to accuse a speaker of bad morality: the underlying strategy of personal
attack is the same in all these arguments. The refutation of the opponent's thesis proceeds from a claim against his credibility based on character attack, and the differences among the types and subtypes of the
various forms of ad hominem attack lie only in the variations on this
central theme. On the other hand, the two wrongs and *tu quoque* arguments, despite often being classified as ad hominem arguments, follow
strategies that we can consider independent of the main one. They are
schemes of dialectical replies to personal attacks as well as other forms
of attack. They are most often countermoves to accusations of blame
for having committed reprehensible actions and to the circumstantial ad
hominem attack. We can study them by examining the pragmatic difference between "blame" and "attack." In line with the first term, a negative
moral judgment of a person's action is presupposed by a condemnation
directed to the agent. From this kind of accusation no consequence follows: it cannot be considered an argument itself, if it is not used to imply
a threat, a judgment of the person, or some other dialectical action.
Attack, by contrast, is a dialectical move itself: it may be considered a
dialogical development of blame. It may involve the credibility of the
person or the acceptability of the opponent's position. The two wrongs
ad hominem rejects the possible implications of the hearer's blame: it
blocks, by undermining his credibility, any possible moves that can proceed from the accusation. For instance:

Student No. 1: I saw you copying the answer to the exam question from your
math book.

Student No. 2: At least it was my math book. Didn't you borrow John's term paper
and hand it in as your work? (Walton, 1998, p. 16)

Student No.1's credibility is damaged by attacking him with the same
type of argument he used to blame Student No.2. This is the form of the
two wrongs topic (Walton, 1998, p. 256).

Proponent: Respondent, you have committed some morally blameworthy action
[and the specific action is then cited].

Respondent: You are just as bad, for you also committed a morally blameworthy
action [then cited, generally a different type of action from the one cited by
the proponent but comparable with respect to being blameworthy]. Therefore,

you are a bad person, and your argument against me should not be accepted as having any worth.

In the analysis of the following critical questions, we can observe that the most dangerous is the last one. The counteraccusation must be of the same type as the proponent's blame; otherwise, it might be refused as nonsense.

> CQ1: Is there evidence to support the proponent's allegation that the respondent committed a blameworthy act?
> CQ2: If the answer to CQ1 is yes, then should the respondent's counteraccusation be rated as very credible?
> CQ3: Is the respondent's counteraccusation relevant to the proponent's original allegation in the dialogue?

In the two wrongs ad hominem, we can notice that the argument is circumstantial. In fact, it must be relevant to the particular reasons for the blame. On the other hand, the *tu quoque* ad hominem may be abusive: the relevance condition does not apply. As mentioned earlier, moreover, the latter topic is an offensive strategy, while the first one is protective. In the *tu quoque*, the respondent, instead of defending his position by attacking the premises or the conditions of the argument, chooses a countermove. In this way, not only is the proponent's ad hominem undermined by his loss of credibility, but the burden of proof is reversed. It is now the other party who has to defend his position if he wants to save the move.

The following example is a clear instance of the argument. Gov. Wilson replies to the opponents' accusation by defining the latter's "cheap political games": in this way, the opponent is accused of being unfair, and his conclusion is rebutted.

> California Gov. Peter Wilson has made the crackdown on illegal aliens a key theme in his '96 Republican presidential bid. But the issue could backfire. Newsweek has obtained documents showing that in 1989, as a senator, Wilson wrote the Immigration and Naturalization Service on behalf of a long-time political supporter, Anne Evans, whose San Diego Hotels had been raided on suspicion of employing illegals. Evans paid a $70,000 fine after being charged with 362 immigration-law violations. Meanwhile, following allegations that Wilson and his first wife hired illegals as household help, Democrats are placing ads in newspapers from Washington to Tijuana, seeking information from former Wilson employees. "These are cheap political games by people who got their butts kicked in the last gubernatorial campaign," said a Wilson spokesman. (Walton, 1998, p. 236).

Here is the general scheme of *tu quoque* with the related critical questions (Walton, 1998, p. 256); regarding the latter, note that the main critique is addressed to the strength of the counterargument conclusion.

Proponent: Respondent, you are a morally bad person (because you have bad character, are circumstantially inconsistent, biased, etc.); therefore, your argument should not be accepted.

Respondent: You are just as bad, therefore your ad hominem argument against me should not be accepted as having any worth.

The following critical questions match this form of argument.

Critical Questions

CQ1: Is the proponent's ad hominem argument a strong one (according to the criteria for whatever type it is)?

CQ2: Is the respondent's ad hominem a strong one (according to the criteria for whatever type it is)?

CQ3: If the proponent's ad hominem argument is strong, how much credibility should be given to the respondent as an honest arguer who can be trusted to make such an allegation?

The point to be stressed here is that the *tu quoque* and the two wrongs types of arguments can both involve replies in kind to various types of prior moves. However, some of the most common kinds of cases stressed by logic textbooks are those where one ad hominem attack is used to reply to a prior one. The danger is that the chain of ad hominem attacks can continue. Thus a persuasion dialogue can degenerate into an eristic dialogue or quarrel in which each party simply attacks the other as a bad person. The problem is that the original conflict of opinions on the given issue will never be resolved. The issue may not even be discussed at all.

5

Causal Argumentation Schemes

In this chapter, a new model of causation is formulated that views causal argumentation as defeasible. The new model structures many of the most common cases of causal argumentation as dialectical, meaning that the case is viewed in the context of an investigation or discussion in which two parties take part in a collaborative process of rational argumentation. The model is shown to apply very well, particularly during the initial stages of an investigation, where information is incomplete but preliminary hypotheses are formed. But it will also be argued that causal arguments need to be evaluated differently in different contexts. In scientific argumentation, there is an investigative process in which tentative hypotheses are formulated about a cause at an early stage, and then tested and refined at later stages. In legal argumentation, the method of evaluation typically is a trial or some other form of dispute resolution in which a causal claim made by one side is opposed to one made by the other side. For example, it could be a case in tort law concerning whether a toxic substance caused cancer in a population. In a criminal case, it could be a trial concerning the cause of an accident. In such cases, there are differing opposed views on what the cause of something is, and it is to such cases that the defeasible model applies best.

Part of the objective of the study is to begin the work of classifying a group of argumentation schemes that come under the heading of causation. The types of argumentation studied include argument from cause to effect, argument from effect to cause, argument from correlation to cause, argumentation from consequences, practical reasoning, the causal slippery slope argument, and argument from waste (also known as the sunk costs argument). The work is part of the wider research project of formalizing argumentation schemes for use in computing taken up in Chapter 10. How to analyze and develop a system of classification for schemes based on causal reasoning is a difficult problem, not least because of the lack of any universally accepted scientific theory of causation. This study is not meant to be the last word on how to formalize and classify causal argumentation schemes. It is a first step to clearing the ground for further work.

1. THE PROBLEM OF CAUSATION

Any attempt to formalize or classify argumentation schemes has to deal with causal schemes. For several of the most important schemes are either causal or have a causal aspect. Hastings (1963, pp. 143–147) distinguished between two basic types of causal argumentation – arguing from cause to effect and arguing from effect to cause. He also viewed such important forms of argumentation as argument from sign and argument from evidence to a hypothesis as being species of causal argumentation. In argument from cause to effect, for example, common in daily activities like weather prediction, the warrant is a probable generalization (p. 145). Sign reasoning Hastings saw as "almost always reasoning from effect to cause" (p. 144). Nowadays sign reasoning would be described as abductive, and the recent literature in AI has testified to the importance of this kind of argumentation. If Hastings is right, then causal argumentation is not only extremely common but also centrally important in scientific argumentation and everyday practical reasoning. (The central importance of causal reasoning in legal argumentation might also be noted here.) Thus any attempt to study argumentation schemes can hardly avoid dealing with the whole group of such schemes that are either explicitly or indirectly causal.

But how does one approach the subject of causality? How can the causal relation be defined? For unless it can be defined in some clear and precise way, any attempt to formalize causal argumentation schemes is hardly likely to be successful. And no exact theory of causal reasoning has ever achieved scientific acceptance. True, many philosophical theories of causation have been proposed. But the very number of such theories, and the difficulties they have been shown to exhibit, is an obstacle to the study of argumentation schemes. In the face of the many attacks on causal theories in philosophy, it seems wise to say nothing at all about the subject unless you have to. Scientists, for the most part, are content to admit that causality is not an exact scientific notion, and to warn of the dangers of the *post hoc* fallacy in attempting to argue from a statistical correlation to a causal conclusion. In this climate of opinion, it is wise to approach the subject of causal argumentation with caution.

Thus there is a kind of dilemma here. On the one hand, any formalization of argumentation schemes needs to deal with the causal relation and attempt to define it in some way, even if provisionally. On the other hand, to make such an attempt is to enter the arena of philosophical disputes about causality. In the field of legal argumentation – and this may be true in other fields as well – there has been a general acceptance of the view that causation cannot be defined objectively. A prevailing postmodern feeling is that causality is a subjective notion that has been used merely to prop up an older conception of law based on traditional values. But the reality is that we use causal argumentation all the time, not only in applied sciences like medicine and engineering, but also in law and in setting social policy. For example, the way in which arguments are judged in tort law, and in other branches of law like criminal law, makes little or no sense unless the arguments are based on some underlying notion of causation.

A more or less prevailing view in philosophy is that causation needs to be defined in terms of what are called necessary and sufficient conditions. Logical necessity is not meant, nor has it proved feasible so far to try to reduce necessity and sufficiency to notions of probability, or frequency of occurrence. The kind of necessity and sufficiency that is typically expressed can be conveyed by examples from an introductory logic textbook (Hurley, 2000, p. 505). When a person says that watering this plant caused it to grow, presumably she means that watering it was necessary for it to grow. As Hurley notes (p. 505), watering by itself is presumably not being held to be sufficient for plant growth, because proper soil and illumination are also required. Or to take another common example (p. 505), suppose that a person says that taking a cold shower caused his body to cool down. We can assume that he meant to say that the cold shower was sufficient for his body's cooling down. He could have achieved this effect by other means, like sitting in an air-conditioned room. Thus it is reasonable to assume that he was not claiming that taking the cold shower was a necessary condition for his body's cooling down. There are two main lessons that can be learned from these common examples. One is that in some cases of causal reasoning the cause is seen as a necessary condition, while in other cases the cause is seen as a sufficient condition. The other lesson is that causation is not seen as some sort of absolute or logically necessary connection between pairs of events. Causation is based on some sort of defeasible reasoning relative to a set of circumstances fixed by a ceteris paribus clause in which all else is held to be constant.

The problem now faced is whether there should be two analyses of causation, one in terms of necessary conditions and the other in terms of sufficient conditions, or whether the two notions can somehow be combined. To investigate this issue, let's begin by examining some versions of the necessary condition analysis.

According to Wright (1985), the "but-for test" is the most widely used criterion for judging causation in tort law. This test was formulated by Wright (1985, p. 1775) as follows: "an act (omission, condition, etc.)" is said to be "a cause of an injury if and only if, but for the act, the injury would not have occurred." When the but-for test is adopted, one's theory of causation is built on the concept of a necessary condition, meaning not logical necessity but some kind of contingent or relative necessity of a kind that can apply to human actions as well as to natural events. The following analysis of the causal relation (Kienpointner, 2002, p. 1), is built on this concept.

Event A is the cause of event B if and only if

1. B regularly follows A.
2. A occurs earlier than (or at the same time as) B.
3. A is changeable/could be changed.
4. If A did not occur, B would not occur (ceteris paribus).

The Kienpointer analysis, as indicated by the fourth clause, is based on the concept of a necessary condition. Thus its working criterion is the but-for test. Certainly this is one approach to defining causation that is applicable to some cases.

The main problem with this necessary condition type of criterion is that it does not take into account cases where causation is seen as stemming from sufficient conditions rather than necessary conditions for an event. In an example cited by Wright (1985, pp. 1777–1778), two persons start separate fires, each of which by itself is sufficient to burn down a house. The fires then converge and burn down the house. According to the but-for test, neither fire was a cause of the burning down of the house. To see why, pick one fire. It is not true that but for that fire, the house would not have burned down. The lesson seems to be that causation in law cannot be analyzed as just consisting of necessary conditions of the kind specified by the but-for test. Sufficient conditions also have to be taken into account. Many cases of overdetermination of this sort have been studied in the literature – see, for example, Scriven (1964).

To provide a more widely applicable analysis of causation, philosophers have devised a criterion that takes both necessary and sufficient conditions into account. This criterion attempts to combine the necessary and sufficient conceptions of causation into one holistic analysis of the notion. Scriven (1964, p. 408) proposed a criterion based on a set of necessary conditions such that the whole set, taken together, is sufficient for the occurrence of the outcome. The set of conditions is taken by Scriven (p. 408) to be "contingently sufficient" for the occurrence of the event, but the rest of the set apart from the designated necessary condition cannot be sufficient for the outcome (p. 408). Based on this sort of combined necessary and sufficient set approach, Mackie (1965, p. 245) formulated the famous INUS condition test for causation.

> If C is a cause of E (on a certain occasion) then C is an INUS condition of E, i.e. C is an insufficient but necessary part of a condition which is itself unnecessary but exclusively sufficient for E (on that occasion).

The INUS test adds to the but-for test by taking account of how a set of necessary conditions can be combined to form a sufficient condition for an outcome. Hart and Honore (1962) advocated a criterion of causation based on the INUS test by taking into account the problems arising from using the but-for test in legal cases.[1]

Thus, although it is fairly difficult, it seems that some sort of useful and coherent notion of causation can be achieved by building on the notions of necessary and sufficient condition. How these notions are to be analyzed in more precise terms, of course, is a subject that remains open to investigation. It will be argued later that logical necessity and sufficiency, on the model of deductive logic, is a good tool for the job in the overwhelming majority of cases. It will also be argued that probabilistic necessity and sufficiency, on the model of Bayesian reasoning, while very useful in many cases, is not the best tool for the job in a very common and typical range of cases of causal argumentation. To begin the investigation, some of these very common and typical types of causal argumentation are identified and analyzed.

[1] Hart and Honore, after examining many legal cases of causation, came to the conclusion that the INUS type of test does not work, in the end, unless supplemented by the notions of voluntariness and abnormality. Their analysis has prompted many to conclude that no objective analysis really captures legal causation. Going against this postmodernistic conclusion, Wright (1985) has argued that the analysis in terms of necessary and sufficient conditions can be fixed up so that it works to give an objective analysis of causal argumentation in law.

2. ARGUMENT FROM CAUSE TO EFFECT

Hastings (1963, p. 69) indicated the form of argumentation from cause to effect by giving an example, and using a Toulmin line diagram to indicate the warrant and how the warrant is used to support the inference from cause to effect. The warrant, in Hastings' diagram, is marked by a conditional statement expressing a prediction. Based on Hastings' example and his analysis of it, an argumentation scheme for argument from cause to effect was constructed by Walton (1996, p. 73). In this scheme, the variables *A, B, C,* . . . stand for "states of affairs," or statements describing events. The major premise of the argumentation scheme is presented in the form of a defeasible conditional of a kind that could be associated with a Toulmin warrant like that used by Hastings.

Argument from Cause to Effect

Generally, if *A* occurs, then *B* will (might) occur.
In this case, *A* occurs (might occur).
Therefore, in this case, *B* will (might occur). (Walton, 1996, p. 73)

The conditional in the major premise represents a kind of sufficiency relation. But it is not one that expresses the idea that if *A* occurs, then *B* must necessarily occur. The conditional is not absolute, like the material conditional of deductive logic. It is merely a generalization that is defeasible. All it warrants is that if *A* occurs in a particular context that an agent is familiar with, then the agent can reasonably expect *B* to occur as well, subject to possible exceptions as the agent comes to acquire new information. But should such information come into the dialogue, the agent must be prepared to give up the argument and retract the conclusion he previously accepted.

Perelman and Olbrechts-Tyteca (1969, p. 262) identified argumentation from cause to effect as "tending to show the effect which must result from a given event." Van Eeemeren and Kruiger (1987, p. 74) recognized the same kind of argumentation scheme based on the supposition of a causal relationship. Grennan (1997, p. 187) formulated an argumentation scheme for what he called cause-to-effect argumentation. It is based on a so-called warrant backing stating that generally, events of one kind cause events of another kind. This formulation explicitly recognizes the defeasible nature of argumentation from cause to effect.

So conceived, the argument from cause to effect does not assume complete information, certainty, or epistemic closure. If the premises are commitments of the respondent in dialogue, then the respondent must either tentatively commit to the conclusion, subject to possible future retraction, or ask an appropriate critical question. Hastings (1963, p. 70) proposed three critical questions corresponding to the argument from cause to effect, and Walton (1996, p. 74) followed by citing three critical questions matching the scheme above:

Critical Questions for Argument from Cause to Effect

C1: How strong is the causal generalization (if it is true at all)?
C2: Is the evidence cited (if there is any) strong enough to warrant the generalization as stated?
C3: Are there other factors that would or will interfere with or counteract the production of the effect in this case?

The respondent can challenge any of the premises, and if he is not committed to any one of them, the argument from cause to effect will not force him to commit to the conclusion. But if the proponent's argument has the form of the argumentation scheme for argument from cause to effect, and the respondent is committed to all the premises, then he must either commit to the conclusion at his next move or ask one of the appropriate critical questions. Thus the effect of the argumentation scheme in a dialogue is to make the respondent either accept the conclusion or ask a critical question.

3. ARGUMENT FROM EFFECT TO CAUSE

Hastings (1963, p. 65) saw argumentation from cause to effect as a predictive form of reasoning that reasons from the past to the future, based on a probabilistic generalization. The argumentation scheme for this type of argument can be formulated without too much difficulty. Argument from effect to cause is a different matter. While Hastings saw it as a distinct underlying form of argumentation, he saw it as taking different forms that need to be classified separately. In particular, Hastings (p. 143) saw argument from sign and argument from evidence to a hypothesis as being common forms of argument that are (often, at least) based on causal relations, and thus on an underlying sequence of argumentation from effect to cause.

Hastings writes (p. 144) that "sign reasoning is almost always reasoning from cause to effect" and depends on a "causal generalization." Nowadays argument from sign would be closely associated with abductive inference, so closely that it is hard to separate the two. A common example can be used to illustrate argument from sign.

The Bear Argument

Here are some tracks that look like they were made by a bear.
Therefore, a bear passed this way.

The tracks are taken to be a sign that a bear passed this way. But nowadays such an argument would be readily identified as an instance of inference to the best explanation, often called abductive inference. The given data are the bear tracks, and the best explanation of the data is that a bear passed this way, producing the tracks. You could say that a causal explanation underlies the argument. Presumably, the inference is reasonable on the assumption that a bear's passing that way caused the tracks. Reasoning backward from the given data of the tracks, we infer a causal explanation for how the tracks came about.

Hastings (p. 145) classifies argument from circumstantial evidence to a hypothesis as based on a chain of causal reasoning "which will explain the evidence." He seems to stop short of claiming that all argumentation from evidence to a hypothesis is based on an underlying chain of causal reasoning, but he approves the claim that most of it is. Thus while Hastings may not have thought that argument from evidence to a hypothesis is always the same as, or based on, argument from effect to cause, he did seem to be suggesting that in many instances argument from evidence to a hypothesis is based on some kind of underlying causal form of abductive inference. Thus this cluster of arguments leaves us with the big puzzle of how to formalize and classify the various important forms of argumentation involved. Not the least of these problems is that of how to formalize the notion of an abductive inference.

It looks like the argumentation scheme reasoning from effect to cause should be similar to the scheme for argument from cause to effect, only reversed.

Argument from Effect to Cause

Generally, if A occurs, then B will (might) occur.
In this case, B did in fact occur.
Therefore, in this case, A also presumably occurred.

This form of argument looks very much like the invalid form of reasoning called affirming the consequent. But is that really its form, or should some other analysis of it, in terms of best explanation, be given? In a nutshell, this is the problem of how to analyze abductive inference. The major premise in the postulated form of argument from effect to cause is a prediction. But is argument from effect to cause better seen as based on a retroduction (or retrodiction, if that is a word), from the observed data to a hypothesis about the presumed cause of the data? If so, postulating the argumentation scheme for argument from effect to cause as we have here could be misleading. It might be better to express it explicitly as a species of abductive argument.

An argumentation scheme for abductive argument based on two variables has been presented by Walton (2001). The variable F stands for a given set of facts or data in a case. They don't necessarily have to be true statements. They are called "facts" only because their truth is not in question in a dialogue. They can be questioned, but they are not in question at the present point in the dialogue. The variable E stands for an explanation. The notion of an explanation can also be defined dialectically. It is a response offered to a particular type of question that asks about a gap in some sort of account that is known to both questioner and respondent. The satisfactoriness of an explanation of this sort depends on the type of dialogue and how far the dialogue has progressed. Based on this dialectical notion of explanation, the following argumentation scheme for abductive inference is presented by Walton (2001, p. 162).

Abductive Argumentation Scheme

F is a finding or given set of facts.
E is a satisfactory explanation of F.
No alternative explanation E' given so far is as satisfactory as E.
Therefore, E is plausible, as a hypothesis.

The term 'hypothesis' in the conclusion indicates that the conclusion of an abductive inference is best seen as a tentative assumption in a dialogue. So conceived, abductive reasoning is seen, at least in the typical case, as defeasible. It draws a conclusion in a dialogue as a presumption that both parties need to accept provisionally so that the dialogue can move ahead. The abductive argument, once put forward and based on premises acceptable to the respondent, has the power to force the respondent to either accept its conclusion or ask one of the appropriate critical questions. The following critical questions are listed by Walton (2001, p. 162).

CQ1: How satisfactory is E as an explanation of F, apart from the alternative explanations available so far in the dialogue?
CQ2: How much better an explanation is E than the alternative explanations available so far in the dialogue?
CQ3: How far has the dialogue progressed? If the dialogue is an inquiry, how thorough has the investigation of the case been?

CQ4: Would it be better to continue the dialogue further, instead of drawing a conclusion at this point?

An abductive argument carries with it an ability to gently force the respondent to alter his commitment set, depending on how firmly he is committed to the premises. He can question the argument at any point in the dialogue. But when the dialogue has reached closure, the argument can have the effect of altering the burden of proof when the mass of evidence on both sides is weighed up. In many typical instances, however, abductive arguments need to be judged as relatively strong or weak at a midpoint of a dialogue. That is why the claim was made earlier that abductive arguments are typically defeasible in nature.

For these reasons, the argument from effect to cause represents only a superficial and somewhat misleading analysis of this type of argumentation. The argumentation is based not on a prediction, but on an abductive retroduction from the given effect back to a presumed or inferred cause. A better representation of the form of argument can be achieved by seeing it as the causal form of abductive inference. The given observed event is taken as the data for the abduction. A cause is then postulated as the best explanation of the event, chosen from among several competing causal hypotheses. Thus the argumentation is from effect to cause, but it represents an abductive inference from the given data to a hypothesis that is the best explanation of that data. On this basis, we now propose the following argumentation scheme for abductive causal inference.

Abductive Argumentation Scheme for Argument from Effect to Cause

F is a finding or given set of facts in the form of some event that has occurred.
E is a satisfactory causal explanation of F.
No alternative causal explanation E' given so far is as satisfactory as E.
Therefore, E is plausible, as a hypothesis for the cause of E'.

In a case like the bear example, the observed bear tracks are the given data, presumed to be the effect of some cause. In context – for example, if the bear tracks were found on a trail in a forest where bears are known to reside – the conclusion could be drawn that a bear recently passed that way. The argumentation is plausible provided that the event of a bear's passing that way is the best explanation of the given tracks. The argument can be classified as an instance of argument from sign. It can also be classified as a type of argument from effect to cause. But the deepest and most accurate analysis of its logical form is achieved by classifying it as an instance of the abductive argumentation scheme from effect to cause.

4. ARGUMENT FROM CORRELATION TO CAUSE

Another centrally important causal argumentation scheme is the argument from correlation to cause. This form of argumentation has been the focus of much study in the field of informal logic because the traditional *post hoc* fallacy has so often been acknowledged as important in the logic textbooks. *Post hoc* (*post hoc, ergo propter hoc*) is the fallacy of arguing from a correlation between two events to the conclusion that one event is the cause of the other. Of course, like so many of the traditional fallacies, this form of argument is by no means always fallacious. Part of the most important kind of evidence one can have to argue for the existence of a causal relationship between two events is a correlation between them. However, correlation is a purely statistical notion. Positive correlation means that there are a number of instances in which occurrences of one event are accompanied by occurrences of the other. Negative correlation means that there are a number of instances in which occurrences of one event are accompanied by nonoccurrence of the other. Correlation is what Pearl calls an associational concept, as contrasted with a causal concept. According to Pearl (2001, p. 3), causal and associational concepts do not mix, because "associations characterize static conditions, while causal analysis deals with changing conditions." Thus you can't argue from a statistical association between two events that one is the cause of the other. For example (Pearl, 2001, p. 3), there is nothing in a statistical correlation (association) between a symptom and a disease "to tell us that curing the former would or would not cure the latter." Of course, if there is such a correlation, it might suggest exploring the hypothesis of a possible causal connection. But just because there is a correlation, it does not follow that the existence of a causal connection can be inferred.

Arguing from observed correlations between a pair of events to the existence of a causal connection is always tenuous, for several reasons. One reason is that the correlation could be pure coincidence. Another is that some third factor could be causing both events, yielding a correlation between them but without any direct causal connection between them. Thus one could say that correlation is evidence of causation, but not conclusive evidence. Despite the fallibility of argument from correlation to causation, it is an important and very common type of argumentation.

The argument from correlation to cause has been studied as a distinctive form of argumentation in the literature. A very simple argumentation scheme representing the structure of this kind of argumentation has been postulated by Walton (1996, p. 142).

Scheme for Argument from Correlation to Cause

There is a positive correlation between A and B.
Therefore, A causes B.

This form of argument is treated by Walton as defeasible and presumptive, meaning that commitment to the premise only gives a reason for commitment to the conclusion that defaults when one of the appropriate critical questions is asked.

Matching the argument from correlation to cause is the following set of seven critical questions (Walton, 1996, pp. 142–143).

Critical Questions for Argument from Correlation to Cause

C1: Is there a positive correlation between A and B?
C2: Are there a significant number of instances of the positive correlation between A and B?
C3: Is there good evidence that the causal relationship goes from A to B and not just from B to A?
C4: Can it be ruled out that the correlation between A and B is accounted for by some third factor (a common cause) that causes both A and B?
C5: If there are intervening variables, can it be shown that the causal relationship between A and B is indirect (mediated through other causes)?
C6: If the correlation fails to hold outside a certain range of cases, then can the limits of the range be clearly indicated?
C7: Can it be shown that the increase or change in B is not solely due to the way B is defined, or to the way entities are classified as belonging to the class of Bs, or to changing standards, over time, in the way Bs are defined or classified?

If the proponent in a dialogue puts forward an argument that has the form of argument from correlation to cause, and the respondent is committed to the premises of the argument, then the respondent must also commit to the conclusion unless he can ask a critical question from this set. For example, there may be a correlation between wealth and ownership of stocks. But is such a correlation good evidence for a causal connection? It may be hard to say which factor causes the other. It is even possible to have circular causal relationships in which A causes B and B causes A. Maruyama (1968, p. 305) has given a simple example of what he calls a deviation-amplifying mutual causal process, as follows. A small crack in a rock collects some water. The water freezes and makes the crack larger. A larger crack collects more water, which makes the crack still larger. A sufficient amount of water then makes it possible for some small organisms to live in the crack. Accumulation of organic matter then makes it possible for a tree to start growing there. The roots of the tree then make the crack still larger.

And so asking the third critical question can raise doubts about inferring a causal connection exclusively in the one direction or the other. The seventh critical question is relevant because changing definitions, or standards for classification, can strongly affect causal reasoning. Many statistical findings appear to suggest a causal connection only because of bias introduced through the language used to define the events.

Thus arguments from correlation to cause are, in principle, legitimate, provided they are seen as defeasible and subject to critical questioning. Such arguments are highly fallible, however. They can easily become fallacious if critical questions are ignored or, even worse, if the argument is specifically designed to inhibit or hide the appropriate critical questions that need to be asked. We often need to argue from correlation to cause in medical investigations of the causes of diseases or in order to set social policies. In some cases, however, such arguments are based on "mutant statistics" that ignore or suppress one of the appropriate critical questions. Best (2001) has studied many examples of cases where mutant statistics have been used to argue for questionable causal conclusions by zealots who cite statistics to promote their special interests and causes. People tend to be impressed by numbers, and virtually every interest group has an arsenal of favorite statistics that are routinely trotted out to support the view advocated by the group. Sometimes these statistics are merely exaggerations obtained by using inflated definitions of key terms, as Best has shown. But along with questioning the numerical data that is used to show a correlation, it can also be very important to question how the supposed causal conclusion can be derived from the correlation data. The argumentation scheme for argument from correlation to cause is an important critical tool that can be used for this purpose.

5. CASES IN POINT

The kinds of cases to which causal argumentation schemes are most usefully applied are those where prediction is involved, or where other types of argumentation are involved that are defeasible. In such cases, there are two sides, and there is legitimate evidence on either side. Legal cases involving alleged causation are very good cases to illustrate the worth and importance of these argumentation schemes. An AI system for the reconstruction of causal evidence in legal cases of causation constructed by Prakken and Renooij (2001) is based on facts and rules. They use this system to analyze the causal argumentation in a real case (Prakken and Renooij, 2001, pp. 132–133), summarized here.

The Accident Case

A driver and passenger were returning home from a birthday party late one night when the driver lost control of the car on the highway. In the resulting crash, the passenger was injured. The passenger sued the driver, claiming that he had lost control when there was no car or other obstacle in sight. In reply, the driver claimed that the passenger had suddenly pulled the hand brake, and that that had caused the accident. This was a Dutch Supreme Court case (HR 23 October 1992, NJ 1992, 813), a civil case in which the judge is supposed to decide on the basis of the facts adduced by both parties. These facts were as follows. The police found that the accident took place just beyond an S-curve. Tire marks caused by locked tires (skid marks) were found just past the curve, and just beyond that point tire marks caused by a sliding vehicle (yaw marks) were found. When the car was found, the hand brake was in the pulled position. The driver said three times after the accident that the passenger had pulled the hand brake. The passenger was found to have consumed some alcohol. An expert witness said that pulling the hand brake can cause the wheels to lock.

What is especially interesting about the analysis of this case worked out by Prakken and Renooij is their use of a set of causal rules and a set of presumed facts listed on each side, representing the evidence brought forward as argumentation by that side. For example, on one side, the following rules are listed along with others. The arrow represents a kind of causal warrant asserting that one event is a cause of the other. Rule 1, for example, says that if skidding occurs, then an accident will or might occur. Thus, in logical terms, a rule represents a conditional of a defeasible kind.

Rule 1: skidding ⇒ accident
Rule 2: skidding ⇒ tire marks present
Rule 3: obstacles ⇒ skidding
Rule 4: loss of control ⇒ skidding
Rule 5: wheels locked ⇒ skidding
Rule 6: speeding in curve ⇒ loss of control

Each rule is defeasible. For example, the first rule, "Skidding causes accidents," is not a universal generalization. It says that as a general rule, subject to exceptions, the fact that a vehicle is skidding could, all else being equal, lead to an accident. It does not say that skidding must lead to an accident by necessity, or even by probability. If the vehicle is on a perfectly flat surface, for example, and there are no other large objects in the area, skidding would not lead to an accident. But under some circumstances, skidding would lead to an accident. For example, in normal conditions where the vehicle is going along at a good speed on the highway, and there is a ditch beside the road and other vehicles around, skidding could very definitely cause an accident. None of these six causal rules is absolute. Each warrants only the inference that if the antecedent holds, then there is a reason to think that under the right conditions the consequent holds as well.

This list of rules is not complete, but is given here only to explain how the system works. Also, as noted earlier, there is a different list of rules on the other side. The system works on the basis that given the alleged facts of the case asserted by one side, the rules can then be applied sequentially to the facts. The result is a lengthy chain of argumentation leading to the conclusion to be proved by that side. For example, the list of rules and facts on the side just indicated ultimately leads to the conclusion that the driver's loss of control caused the car to crash. On the other side, a different set of facts and rules leads to the conclusion that the passenger's pulling the hand brake caused the car to crash. Prakken and Renooij construct a small argument diagram representing the chain of argumentation on each side, and then show how a weight of (inconclusive) evidence is transferred through the chain to the ultimate conclusion purported to be proved. The details are not so important here as the fact that the case shows that causal rules are a useful way to transfer evidence in a case even if the rules and the inferences based on them are defeasible.

The causal argumentation in this case is not just defeasible. It could also be described as abductive. It is argumentation from the given fact of the crash, and other facts, to the presumed cause of the crash. Neither line of argumentation is conclusive. But each side can make a plausible case based on the facts and causal rules it cites. The conclusion of one side is opposed to that of the other. There could be more than one cause of the crash, but the assumption is that they are arguing about which is the cause that is taken to indicate legal responsibility. Thus the argumentation of one side is opposed to that of the other. Both can't be right. But still, each line of argumentation gives a reason to take the conclusion as having some weight of evidence behind it.

American tort law is full of controversial cases about alleged harms caused by factors like foods, chemicals, drugs, pollution, and so forth, where scientific experts have claimed in court that the harm was caused by the allegedly toxic substance (Alberts, 2001). The latest type of claim of this sort is that so-called junk food is causing a vast array of illnesses and medical problems. Huber (1991) chronicled many of the early cases of this sort, coining the expression "junk science" to cover many of the more spurious claims. Here we get into an area where causal claims vary from the clearly legitimate, to claims that are extremely difficult to prove or disprove conclusively by the evidence, to those that are clearly cases of fallacious *post hoc* argumentation. But many cases belong in the middle range, where it is possible to bring forward plausible argumentation on both sides. For example, claims about environmental causes of illnesses are notoriously hard to prove or to refute. These are hard cases, but they illustrate the defeasible nature of causal argumentation very well.

6. CAUSAL ARGUMENTATION AT STAGES OF AN INVESTIGATION

Many of the problems with causal argumentation come from the strong urge to try to portray its forms of argument in a context-free way, comparable to the deductive systems of logic we are familiar with. A much better and more realistic approach is to see causal argumentation as relative to a background context of investigation or discussion in which evidence is being collected and assessed. Causal arguments, as noted earlier in the example of legal argumentation, are typically defeasible and abductive. They are based on a set of facts (or presumed facts, often called data) that have been collected in an investigation so far. As new facts come in, a causal argument or assertion may default. Failure to be open to new information in such arguments is associated with fallacies, such as hasty generalization, *post hoc*, and suppression of evidence.

Scientific argumentation typically goes through a number of stages. First there is a discovery stage at which hypotheses to explain some data are advanced. Some of the hypotheses are then tested, at an experimental or further collection of data stage. Then there is a stage at which hypotheses are formulated in a more precise way, using mathematical and logical methods of formalization. This sequence of stages is not temporal, but represents a model of how the argumentation in a scientific investigation should ideally proceed. It is a matter of scientific method.

In argumentation theory, the sequence of stages in a scientific investigation can be represented normatively as a type of dialogue. Every type of dialogue has four stages, following the normative model of the critical discussion as type of dialogue (van Eemeren and Grootendorst, 1984). There is a so-called confrontation stage at which some conflict of opinions is stated, or some problem stated. There is an opening stage at which the participants try to resolve the conflict or solve the problem by using rational argumentation. There is an argumentation stage at which the strongest arguments on both sides of the issue are put forward, and objections to them are raised and discussed. Finally, there is a closing stage at which the outcome is decided by examining the evidence that has come in during the argumentation stage. In scientific argumentation, the aim may be to prove something or to explain something, but the context of the investigation is always one in which there is some unsettled issue, some question to be answered, or some problem to be solved. For example, the problem may be to find the cause of a disease or illness. At the initial stage, some hypotheses are put forward, but they may be merely guesses or conjectures at that stage. Peirce (1965) associated this early stage of hypothesis forming or "guessing" with the use of abductive reasoning.

Copi and Cohen (1998, pp. 552–556) identified seven stages of scientific investigation: identifying the problem, selecting preliminary hypotheses, collecting additional facts, formulating the explanatory hypothesis, deducing further consequences, testing the consequences, and applying the theory. It is not hard to see how, during the different stages of a scientific investigation, different types and standards of argumentation are useful and appropriate. For example, during the stage of deducing further consequences from a hypothesis, deductive reasoning will obviously be used. During the process of collecting and testing additional facts, inductive reasoning will be used. And during the early stage of selecting preliminary hypotheses, defeasible, abductive reasoning, of a kind that is neither deductive nor inductive, will be used.

One paradigm often used to represent the sequence of argumentation in scientific causal reasoning is the generate-test-debug pattern (Simmons, 1992, p. 161). This process starts with the formation of an initial hypothesis to solve a problem and then moves to testing of the hypothesis. The hypothesis may then pass or fail the test. If it passes, the hypothesis is accepted. If it fails, the debugger modifies it and resubmits it for testing. But if the process appears to be moving away from a solution, the debugger may produce a new hypothesis. The argumentation in such a sequence tends to be defeasible, for the simple reason that the hypothesis can fail. The process of scientific investigation should always be regarded as open to failure of an accepted hypothesis as new evidence comes in. This property is often called falsifiability. The debugger may also find other problems like inconsistent hypotheses or hypotheses containing cycles in causal dependencies (Simmons, 1992, p. 162).

The typical kinds of problems associated with causal argumentation and fallacies occur at the early stages of an investigation, or even prior to a proper scientific investigation. For example, a herbal remedy for an illness may become popular at health stores. People may begin to credit it as a cause for improvement in some condition. If it becomes popular enough, people may widely accept the argument that the herbal remedy is causing the improvements they feel. But is their conclusion at this stage merely based on a *post hoc* argument? To find out, a scientific investigation may be carried out by medical researchers conducting a statistical study using a control group. If the effects of the substance are found to be significant, further studies may be carried out to try to find which substances in the herb cause the effects that have been observed. Some chemical may then be identified in the herbal remedy, and a chemical hypothesis may be proposed concerning the linkage between the taking of the herb and the observed effects on human health. As the process of scientific investigation proceeds, a theory is formed. As the theory becomes more precisely formulated, and linked to underlying chemical and physiological mechanisms of kinds well known in science, deductive reasoning becomes more applicable to the argumentation. During the collecting and testing of the data, inductive reasoning is most applicable, in the form of statistical methodology. But at the early stages of hypothesis formation, the typical kind of defeasible reasoning cited earlier will be predominant.

Legal argumentation is quite different from scientific argumentation. Although it has the same four stages as a type of dialogue, the argumentation tends to be much more the defeasible sort, rather than the deductive or inductive sort. In legal argumentation, scientific evidence about causation is very important. But judges, lawyers, and juries generally are not themselves scientists. Nor are they judging or arguing in that capacity. Causal argumentation is often based on scientific reasoning – in tort cases, for example, and in the infamous "junk science" cases so often associated with fallacies. But this evidence is presented in the form of expert testimony by scientists. The judges, lawyers, and juries who receive evidence in this form cannot evaluate it by testing it scientifically themselves. As shown in the case just described, there are generally two opposed causal theories or explanations offered by the plaintiff and the defendant. Who has the stronger argument based on all the evidence in the trial is decided on the basis of what is legally considered an appropriate standard or burden of proof for that type of case. How argumentation is evaluated in a legal case is different from how it is evaluated in a scientific investigation. That being the case, causal argumentation needs to be judged differently in different contexts of investigation. Defeasible causal argumentation of the kind associated with presumptive argumentation schemes will predominate in legal argumentation, whereas it will typically predominate in scientific argumentation only at the early stages of an investigation.

7. CAUSAL ASSERTIONS AS DEFEASIBLE

An example from Pearl (2000, p. 1) shows an interesting paradox about assertive causal expressions of a common sort in natural language. Pearl asks the reader to consider two plausible premises.

(1) My neighbor's roof gets wet whenever mine does.
(2) If I hose my roof, it will get wet.

Pearl notes (p. 1) that these two premises imply the following conclusion.

(3) My neighbor's roof gets wet whenever I hose mine.

But statement (3) is implausible. What went wrong here? According to Pearl (p. 2), the paradox disappears once we recast statement (1) in a form that makes the exceptions built into the statement more explicit.

(1*) My neighbor's roof gets wet whenever mine does, except when it is covered with plastic, or when my roof is hosed, etc.

At the root of the paradox is the factor that "most assertive causal expressions in natural language are subject to exceptions, and those exceptions may cause major difficulties if processed by standard rules of deterministic logic" (Pearl, 2000, p. 1). Causal generalizations like (1) and causal conditionals like (2) are defeasible, meaning that they default, or are defeated, when new information comes into a case. Thus they can't be treated as if they were universal generalizations or material conditionals of the kind familiar in deductive logic. If they are, we get paradoxes like the one just discussed.

Most causal arguments of the kind used in law, like the ordinary examples of causal assertions just cited, are meant to be defeasible. The conclusion of the argument represents only a hypothesis that gives some sort of causal explanation of a given set of facts, but competing causal explanations are usually possible. Indeed, it is for this reason that cases in tort law about the effects of medical treatments, for example, can be disputed in lengthy arguments with plenty of evidence on both sides. Take any trial about toxic substances and cancer as an example. So-called toxic tort cases, a product of our industrialized society, involve personal injury and related harms caused by exposure to toxic substances (Alberts, 2001, p. 34). Did the toxic substance cause the plaintiff to get cancer? There are typically competing causal explanations and theories put forward by both sides, and the case needs to be judged on a basis of burden of proof. But because the injury is often a syndrome or terminal disease, because symptoms may not be unique to the disease, and because the disease can remain latent for a long time, proving or disproving causation of a toxic tort conclusively is a "challenging prospect" (Alberts, 2001, p. 34). Thus a "necessary" or "sufficient" condition of the kind appropriate for many cases of causal argumentation in law is best taken to mean "relatively necessary or sufficient relative to the given information in a case, subject to questioning as the investigation of the case proceeds."

Pearl (2000) has analyzed causal reasoning based on a Bayesian probabilistic model in which conditional probabilities are associated with the causal relationship. This model applies very well to the kind of causal reasoning employed during the later stages of a scientific investigation when causal hypotheses are tested. But there has been much discussion in artificial intelligence about the usefulness of numerical approaches as a way to model human reasoning under conditions of uncertainty (Console and Torasso, 1990, p. 86). Many in the AI community have doubts about whether humans actually do use probabilistic reasoning under conditions where information is incomplete. The problem is that at the earlier stages of a scientific investigation, or in legal argumentation, the causal hypothesis is more of a guess than a statistical assertion of some sort. Instead of a Bayesian probabilistic conditional, the reasoning appears to be based on a conditional of the form $A \Rightarrow B$, meaning that if A is accepted as true, then normally B should be accepted as true. But this kind of conditional is defeasible, meaning that it is only tentatively acceptable as an investigation or discussion proceeds. It should be regarded as open to exceptions, and as subject to defeat should new information come in. It can carry probative weight in favor of accepting a hypothesis provisionally on a balance of considerations, but it should also be open to critical questioning as the investigation goes on, until it reaches the closing stage. A is said to be sufficient for B, in this qualified sense, only relative to the facts or circumstances known at that point in the case, as they are being discovered and investigated.

Defeasible conditionals cannot be properly evaluated using semantic models of deductive logic, or using Bayesian models of probabilistic reasoning. They have to be seen as open-ended hypotheses or guesses in the context of an ongoing dialogue between a proponent and a respondent. This contextual sensitivity means that each defeasible argument needs to be evaluated differently in a different context of use. For example, a causal argument may need to be evaluated using different standards, methods, and requirements of burden of proof when used in a scientific investigation than when used in legal argumentation in a trial. The causal argument may be the same in both contexts. It may be a claim about the toxic effects of a chemical on a human body, for example. But how it should be evaluated in a scientific investigation may be different from how it should be evaluated in a legal context. At the discovery stage of a scientific investigation, however, the same defeasible model that would be appropriate for use in a case of legal argumentation may be applicable.

In all dialogues, however, defeasible causal arguments can be evaluated following a pattern, or methodical process. At the first step, the proponent brings forward the causal argument using an argumentation scheme. Let's suppose the argument is an instance of the scheme for argument from cause to effect. It will have the following form.

Major Premise: $S_i \Rightarrow S_j$
Minor Premise: S_i
Conclusion: S_j

This form of argument looks like *modus ponens*, or perhaps like a specimen of Bayesian reasoning in which the major premise is a conditional probability. But it is neither. The major premise is a defeasible conditional. Once the proponent puts the argument forward in a dialogue, the respondent must reply to it. He must challenge one of the premises, ask an appropriate critical question, or accept the conclusion. In some types of dialogue of a fairly rigorous kind, the respondent must give one of these three replies at the very next move after the proponent puts the argument forward. One of the ways he can challenge the major premise, for example, is to cite an exception to the rule expressed in the major premise. But what happens if new information comes in during the course of the dialogue revealing an exception to the rule in the major premise? Should the respondent have the right to bring up this new information later in the dialogue? Clearly the answer is that if the causal argument is defeasible, the respondent should have the right to such a move at any point before the termination of the dialogue.

This analysis brings out what is unique to defeasible arguments. Even during a fairly rigorous type of dialogue, a defeasible argument remains open to challenge, and to potential defeat, even long after the argument has been brought forward by the proponent and accepted by the respondent. The reason is that in the case of a defeasible argument, such acceptance should be regarded as tentative. At any later appropriate point in the dialogue, the respondent can cite new information that defeats the argument. This defeasibility property is preserved until dialogue closure, or what is often called epistemic closure, is achieved. This means that the argument is open to challenge by finding an exception to the rule until the investigation or discussion has been completed.

8. TOWARD A SYSTEM OF ANALYSIS AND CLASSIFICATION

The first problem in trying to classify or formalize causal argumentation schemes is to settle on some definition of causation, even if it is provisional and open to further analysis and sophistication. To do this, in line with our earlier findings, a modified version of Kienpointner's definition of the causal relation is proposed. In line with the argumentation schemes presented earlier, S_i and S_j are states of affairs (conditions, events) that can be chained into causal sequences. But if you prefer, you can also read S_i and S_j as being events that can happen or not, or conditions that obtain or not.

Definition of the Causal Relation as Defeasible Reasoning

State of affairs S_i is the cause or a cause of state of affairs S_j if and only if

1. S_j regularly follows S_i.
2. S_i occurs earlier than (or at the same time as) S_j.
3. S_i is changeable/could be changed.
4. S_i is a necessary or sufficient or INUS condition of S_j.
5. Pragmatic criteria, like voluntariness or abnormality, may single out a cause.

Several notable deviations from Kienpointner's definition require special comment. One is that the *definiens* has been changed from "the cause" to "the cause or a cause," in recognition of the possibility of multiple causes in some cases. For example, one of Prakken and Renooij's causal rules took the form skidding ⇒ accident. This rule presumably did not entitle us to infer that skidding is the only cause of accidents, or even to draw the conclusion that skidding was the only cause of the accident in this case. Presumably, all it entitles us to say is that skidding is a cause of accidents. On the other hand, the ultimate conclusion on each side in the case does seem to be a claim about the cause of the crash. The conclusion on one side is that the driver's loss of control of the vehicle was the cause of the crash, while the conclusion on the other side is that the passenger's pulling the hand brake was the cause of the crash. Thus, in any causal attribution, one should try to distinguish carefully whether a conclusion is being drawn about the exclusive cause of an event or about a cause that might be one of several causes.

Another point requiring comment is that clause four can be seen as representing three different kinds of conditionals or rules (inference warrants) of the following sorts.

Necessary Condition Rule: If S_i would not occur, then S_j would not occur (ceteris paribus).

Sufficient Condition Rule: If S_i would occur, then S_j would occur (ceteris paribus).

INUS Condition Rule: If S_i would occur within a set of conditions each also necessary for the occurrence of S_j, then S_j would occur (ceteris paribus).

In any given case, the proponent of a causal argument can select one of these clauses as representing the kind of rule he has in mind as representing the sort of claim he is making. This choice takes the ambiguity of causal arguments into account. Sometimes the claim made is that one event is a necessary condition of another event. Sometimes the claim made is that one event is a sufficient condition of another event. Sometimes the claim made is that one event is an inus condition of another event. It should be possible to make all or any of these kinds of claims, and so the analysis of causal argumentation schemes should permit all three possibilities.

Clause five is dependent on the type of dialogue involved. Obviously, in a scientific investigation, voluntariness of a human action would not be a factor in judging causation. But especially in legal argumentation, within sets of necessary and sufficient conditions, certain special antecedent conditions of an event or action will be singled out as the cause (Hart and Honore, 1962). For example, a voluntary human action might be one of many antecedent conditions of an event that is singled out as the cause of the event. Another pragmatic factor is that an event that is abnormal might be singled out as the cause.[2] This factor of abnormality indicates that all three of the causal rules listed above are based on the proponent's and the respondent's assumptions about what is normally expected to occur in a given case. The rules are not absolute. They are defeasible. They say that if one thing happens, then we can normally expect another thing to happen, but subject to exceptions and subject to default when information comes out about the special circumstances of the case. Remember that each rule has a ceteris paribus clause appended to it.

Given all these complications, it is possible to define the causal relation in outline, even though the outline has to be somewhat flexible and adaptable to the special contextual requirements of a given case. The definition given here, because it represents causal rules and inferences as defeasible, at least fits the needs of the various causal argumentation schemes as they have been formulated.

Classifying causal schemes using this definition is not easy, and has to be tentative at best. But several hypotheses are plausible. Some of the schemes clearly fit the model of causal reasoning. Argument from cause to effect definitely fits, as does the causal type of slippery slope argument. Argument from consequences, although still a little dubious, depending on exactly what can count as consequences, also seems to fit in as a species of causal reasoning. Argument from cause to effect also seems to fit, but again it is a bit hard here to close off further discussion, as the whole question of how to define abductive inference in relation to argument from sign is so controversial and unsettled. Still, it seems hard to deny that argument from effect to cause is based on causal reasoning of the type defined earlier. Argument from correlation to cause is, at least in part, a kind of causal reasoning, because the conclusion of the causal inference is a causal claim.

9. DIALECTICAL AND BAYESIAN MODELS OF CAUSAL ARGUMENTATION

When philosophers like Hume moved away from the older conception of causation as a necessary connection, the only alternative they appeared to have was to move to a probabilistic model. This move was a step in a good direction, but it still failed to come to grips with the most

---

[2] This is a good place to indicate agreement with the claim made by Kienpointner (2002, p. 1) and others cited by him that the actions of human agents cannot be reduced to causal sequences of events.

important aspects of causal argumentation as a practical, working notion. Here, a third model has been adopted that sees causal argumentation as defeasible reasoning used in a dialogue. This new dialectical model captures causation, at the stage of reasoning where it is most practically useful and important, namely, at the discovery stage of an investigation when there is more than one hypothesis about the cause, or a cause, of an event or action. In scientific argumentation, the discovery stage is the earlier phase of an inquiry when various causal explanations in the form of hypotheses are more or less plausible, and the task is to choose the best one for experimentation, further investigation, and theory building. In legal argumentation, there are typically two causal explanations for the given body of facts, but here the facts themselves are also contested, and one contention is the opposite of the other. Each side has a connected network of evidence in the form of a set of facts and causal rules leading to its conclusion. In both kinds of cases, there is uncertainty about the cause, and various hypotheses about it. In the simplest case, there are two opposed hypotheses; one is the negation of the other. It is in these cases of reasoning under uncertainty that the defeasible concept of causation works best and is most useful.

In such cases, the probabilistic model is less useful than the dialectical model, but probability is still centrally involved. Once the causal network of argumentation in a case is put into an argument diagram, Bayesian methods can be extremely useful in some cases to compute the strength of the argumentation (Pearl, 2000). But in other cases, putting numbers onto a causal graph and then using Bayesian methods to calculate an outcome is not useful. Pearl (2002, p. 1) has conceded this point by telling us that he is only "half-Bayesian." In his opinion (p. 1), causal relations are built up, at least in part, from causal generalizations that represent "everyday knowledge" like the statement "Symptoms do not cause disease." It is difficult to assign numbers to such statements with statistical precision, and it is misleading to then calculate outcomes based on Bayesian assumptions used to model the logical reasoning. How, then, can defeasible causal argumentation be evaluated?

The method proposed here is dialectical. First of all, a decision has to be made about dialectical closure relative to the argumentation in the given case. Once closure is reached, the method is to collect the whole network of argumentation in a given case into an argument diagram at the end of the case and weigh up the mass of evidence on each side. All the arguments on both sides can be arrayed at the closing stage, and then a judgment can be made about which mass of evidence has more weight in deciding the issue. If closure has not been reached at the point where a causal argument is being assessed, then one has to examine the argumentation scheme, and its place in the dialogue. Such a scheme will be defeasible, so it will be open to new information as the dialogue or investigation proceeds. This method is not so amenable to exact numbering of probabilities in a Bayesian calculation, because it is not possible to prove that each single item of evidence is independent of the others, and because there is a lot of uncertainty about how to numerically measure the plausibility or probability of many of the single items. The best one may be able to conclude is that the existence of a causal connection is plausible, but open to doubt. The important thing is to be open, and to resist the temptation to leap too quickly to any dogmatic or premature causal conclusion.

In the argument from correlation to cause, the premise is based on the finding of a statistical correlation between two events or states of affairs. It is natural to leap to a causal conclusion when the correlation becomes statistically strong enough. To make an uncritical leap of inference of this sort is, of course, to commit the famous *post hoc* fallacy. But before leaping to accept the causal conclusion, it is best to ask one or more of the critical questions matching the argumentation scheme. The defeasible model of causation suggests hesitating and making qualifications before leaping to an absolute causal conclusion in such a case. But one can see that the form of argument is partly probabilistic in nature. For the premise is based on a statistical correlation that can be observed, and whose frequency can be measured. The inference from the premise to the conclusion is harder to measure statistically, however. And indeed, a better model to capture its logic is that of dialectical argumentation. The premise, if true, shifts a weight of plausibility onto the conclusion as a tentative hypothesis that could possibly explain the data, but that should be evaluated as a competitor in relation to other opposed hypotheses depending on how well it can answer critical questions in a dialogue compared to the argumentation on the opposed side. At the early stages of a causal investigation, it may be better to talk of the plausibility of a hypothesis, rather than its probability, assuming that probability fits the Bayesian model.

6

Schemes and Enthymemes

One of the most valuable uses of schemes is to enable an argument analyst to fill in implicit assumptions needed to make sense of a given argument she is trying to analyze. Arguments that have missing (unstated) premises or conclusions are traditionally called enthymemes in logic. One problem with enthymemes is that reasonable people can have differences of opinion on what the implicit assumptions are supposed to be. Filling in the missing parts of an enthymeme may depend on interpreting the natural language text in which the argument was put forward, to try to fairly judge what the speaker meant to say. The danger of attributing such missing assumptions to an arguer is that of unwittingly committing the straw man fallacy. This fallacy is committed when an arguer misrepresents her opponent's position to make it look more extreme or unreasonable than it really is, in order to attack it more easily. In some cases, more than one interpretation of a given argument is possible. Thus the problem is to find out what kind of evidence is needed to support or question the claim that some proposition really can be inserted into an apparently incomplete argument presented in a text of discourse, without unfairly distorting what the speaker meant to say. It will be shown in this chapter, by studying key examples, how argumentation schemes constitute an important part of this evidence.

The goal of this chapter generally is to explore the role of argumentation schemes in enthymeme reconstruction. This will be done by studying selected cases of incomplete arguments in natural language discourse to see what the requirements are for filling in the unstated premises and conclusions in some systematic and useful way. It is shown that some of these cases are best handled using deductive tools, while others respond best to an analysis based on defeasible argumentation schemes. The approach is also shown to work reasonably well for weak arguments, a class of arguments that has always been difficult to analyze with respect to enthymemes.

1. INTRODUCTION

In many logic textbooks, enthymemes are treated using deductive (e.g., syllogistic) logic in order to reconstruct the given argument. The ten case studies analyzed and discussed here show why this treatment, while it is useful in some cases, is inadequate to treat the broad range of typical cases of enthymemes in natural language discourse in the best way. It is argued that these arguments cannot best be reconstructed using only deductive forms of reasoning, or inductive forms of reasoning of the modern kind associated with statistical inference. These cases are shown to require a less strict standard of reasoning that is defeasible in nature. This third kind of argumentation has been much studied in artificial intelligence (AI), where it is called plausible reasoning and is often associated with abduction (Josephson and Josephson, 1994). It is shown through these case studies that the kind of structure needed to reconstruct the missing parts of an argument is the argumentation scheme (Hastings, 1963; Perelman and Olbrechts-Tyteca, 1969; Kienpointner, 1987, 1992; Walton, 1996; Garssen, 2001). Moreover, it is shown that the argumentation schemes most useful for analysis of many enthymemes are based on defeasible generalizations of the kind that are subject to exceptions.[1] This defeasibilistic view of enthymemes is not as new as it may sound to many readers. It can be shown to be very close to what may have been Aristotle's original doctrine of the enthymeme, according to the view of some commentators (Burnyeat, 1994). According to this original view, an enthymeme is not an argument with a missing premise; rather, it is a plausible argument based on a defeasible generalization, as opposed to a deductive argument based on a universal generalization of the type represented by the universal quantifier of deductive logic.

For purposes of finding missing premises in incomplete arguments expressed in texts of natural language discourse, it would be a logician's dream to have an automated enthymeme machine. The machine would

---

[1] A defeasible generalization, in contrast to an absolute universal generalization, is one that is subject to exceptions and that is defeated (defaults) in a case where one of the exceptions occurs. Defeasible generalizations often contain expressions, like the word 'generally,' that indicate that the generalization has exceptions. In some instances, exceptions are explicitly stated in the generalization. For example, one might say, "Birds fly, except for penguins."

be a software entity that could be applied to incomplete arguments in any chunk of discourse comprised by a natural language text, like a newspaper editorial. Assuming that there is some way of identifying arguments with existing premises and (if stated) a conclusion, the function of an enthymeme machine would be to pick out the unstated premises and conclusions in these existing arguments. The task is one of identifying the commitments that could be ascribed to an arguer as a basis for posing critical questions about the argument, based on the given text of discourse. But construction of an automated enthymeme machine, if reasonable reliability and domain independence are required, becomes extremely difficult if the machine must be presumed to be capable of natural language understanding. There is a way to get around this problem, however, by beginning with a technology that supports the user's ability to mark up an argument by identifying premises, conclusion, and argumentation schemes. The Araucaria software (Reed and Rowe, 2001) is a system that, given the user's markup of an argument in a given text of discourse, aids in determining implicit premises.[2] Thus the project of building an enthymeme machine becomes immediately approachable in a small way. Supplementing familiar argument forms of deductive logic with argumentation schemes, in the way proposed in this chapter, provides a logical and philosophical basis for this new approach. The approach is shown to require two main components. One is the set of argumentation schemes and its apparatus. The other is the dialectical framework representing the different types of dialogue and features of dialogue. Most of the concern in this chapter is with the first component. But at the end, a general discussion of the dialectical component is included.

2. PRELIMINARY DISCUSSION OF THE PROBLEM

The term 'enthymeme' has been taken, since the earliest commentators on Aristotle, to refer to an argument with premises (or a conclusion) that are not explicitly stated. That definition may be historically wrong and misleading, but it is the one that has been presented as the official meaning of 'enthymeme' in logic textbooks for over two thousand years.[3]

---

[2] Araucaria is based on an argument markup language (AML) defined in an XML document type definition (DTD). It can be obtained from the Araucaria homepage: ⟨http://www.computing.dundee.ac.uk/staff/creed/araucaria⟩.

[3] Burnyeat (1994) shows that Alexander of Aphrodisias may be the origin of the traditional view that the Aristotelian enthymeme is a syllogism with an unstated premise, and

In the official account, these missing statements are generally taken to be assumptions that are needed to make the argument valid. But accurately attributing assumptions to an arguer is a difficult process. It depends on interpreting what the arguer meant to say, as far as anyone can tell from the text of discourse attributed to her. A natural language text of discourse can be difficult to interpret. It can be vague or ambiguous. An arguer may be confused, and not know herself what she means. Or in other cases, she may try to hide her meaning by using deceptive tactics and fallacies. Another problem with enthymemes (Burke, 1985; Gough and Tindale, 1985; Hitchcock, 1985) is that inserting assumptions into a text of discourse in order to make one of the arguments valid may not represent what the arguer meant to say. Maybe the argument she intended to put forward is invalid. There is even the danger of the straw man fallacy. This fallacious tactic consists of exaggerating or distorting an interpretation of an argument in order to make it look more extreme than it is, thereby making it easier to attack or refute (Scriven, 1976, pp. 85–86). Given these problems, many would despair of finding any objective method for dealing with enthymemes, and would declare that the matter is "subjective." The idea of building a mechanistic or automated enthymeme machine appears to be hopeless.

To help devise a tool that could be used to deal with the problem, Ennis (1982, pp. 63–66) drew a distinction between needed and used assumptions in enthymemes. The needed assumptions are "propositions that are needed to support the conclusion, to make the argument a good one, to make a position rational, etc." (Ennis, 1982, p. 63). The used assumptions are the missing statements that are presumably meant to be included in the argument by its proponent. Ennis (1982, p. 64) takes the difference to be that used assumptions are "unstated reasons," while needed assumptions are not. This distinction suggests that building an enthymeme machine for finding needed assumptions could be a good way of moving toward the harder project of building such a machine for

that this traditional view does not really represent what Aristotle meant by *enthymema*. What Aristotle really referred to, according to another interpretation, were eikotic or plausibilistic arguments that are syllogistic-like but based on generalizations that are not universal but hold only for the most part. If this alternative interpretation of Aristotle is correct, strictly speaking, we should discontinue using the term 'enthymeme' to refer to arguments with missing premises (or conclusions). Instead, we should use the expression 'incomplete argument'. We would prefer this latter expression, but tradition, especially one so well entrenched as this one, is hard to change. If Burnyeat's analysis is correct, the term 'enthymeme' should properly be used in its original Aristotelian sense to refer to the defeasible (presumptive) argumentation schemes of the kind cited by Walton (1996).

finding used assumptions.[4] It may turn out, then, in real cases, that the pragmatic component of the enthymeme machine is not so easily separable from the inferential component. A pragmatic tool often used to try to deal with enthymemes is the principle of charity, which offers a way of choosing between competing interpretations of an argument. This principle is usually taken to rule that one should choose the interpretation that makes the author of the argument appear more "sensible" rather than less sensible (Gough and Tindale, 1985, p. 102). Another way of expressing the principle of charity is as the following general maxim of interpretation: "When interpreting a text, make the best possible sense of it" (Johnson, 2000, p. 127). But how could this criterion be made more precise as applied to incomplete arguments? One obvious way is to rephrase the principle so that the criterion is the strength or weakness of the various interpretations as arguments. According to this criterion, the principle of charity rules that one should pick the interpretation that makes the argument strongest. But the standard objection to this version of the principle of charity is that it seems to require filling in missing assumptions until the "best possible" argument is produced (Gough and Tindale, 1985, pp. 102–103). The problem with this version of the principle is that the argument may be weak, and by making it stronger the interpreter may be distorting it. The principle of charity is too crude as a tool to help with determining missing premises (or conclusions), unless it can be made more precise in the right way. Thus pragmatic tools, although they can be of some help, do not seem to have been developed in quite the right way to work with cases of enthymemes.

Here, a number of cases are studied that bring out several aspects of incomplete arguments that have not been sufficiently appreciated. What will be suggested by these cases is that while there is a kind of formal or inferential criterion involved, it is not always that of deductive validity. What is also shown is that dialectical factors are involved as well, and that these dialectical factors pertain to the context of dialogue in which an argument was used. They have to do with the supposed purpose of an argument, in a given conversational setting or type of dialogue. One such purpose might be to seek the transfer of information between a questioner and a respondent. Formal dialectical systems representing information-seeking dialogue have been presented by Hintikka (1979, 1992, 1993, 1995). Another purpose might be to discuss an issue in order

---

[4] At least, it would suggest that from a monological point of view. Once the dialectical point of view is considered, more resources are available, as demonstrated by Gilbert's (1991) "Enthymeme Buster" algorithm.

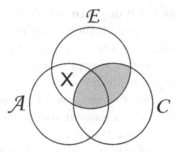

FIGURE 6.1. Venn diagram for the existential reconstruction.

to bring out the strongest arguments on both sides. For example, suppose an arguer has the goal of persuading the reader to come to accept
a particular proposition that he did not previously accept by presenting
arguments. This context of dialogue is that of a critical discussion. In
such a case, the arguer will try to use premises that the audience accepts,
or can be brought to accept, and that can be used to get the audience
to come to accept the arguer's conclusion (Van Eemeren and Grootendorst, 1992). If this approach is right, then there will be not only a
structural criterion that has to do with the form of the argument, but also
a contextual criterion that has to do with how the argument was being
used as part of some conversational exchange. In the discussion that follows, most attention is on the inferential component of the enthymeme
machine. But in the last section, the general discussion of the problem
of enthymemes includes consideration of the pragmatic component.

3. A DEDUCTIVE CASE

This case is interesting because there are two ways of filling in the missing
premise. Both ways produce a syllogism with a true premise. But one way
produces a valid syllogism, while the other produces an invalid syllogism.
The argument in question is: "No enthymemes are complete; therefore,
some arguments are not complete." Let's call this the syllogistic case,
because it turns out to have the form of a syllogism. The first candidate for
the missing premise is "Some arguments are enthymemes." This analysis
yields the following syllogism.

No enthymemes are complete.
Some enthymemes are arguments.
Therefore, some arguments are not complete.

This syllogism is valid, as shown by the Venn diagram in Figure 6.1.

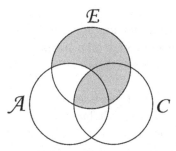

FIGURE 6.2. Venn diagram for the universal reconstruction.

But it is also possible to reconstruct the argument syllogistically as follows.

No enthymemes are complete.
All enthymemes are arguments.
Therefore, some arguments are not complete.

The missing assumption is the second premise, on this account. This syllogism can be tested for validity using the Venn diagram in Figure 6.2.

On both analyses, the missing premise is a true statement. But only on the first analysis is the syllogism unconditionally valid. On the second analysis, the argument is valid if you assume that enthymemes exist. If you adopt existential import, and assume that "All enthymemes are arguments" implies "Some enthymemes are arguments," the argument comes out as valid. But otherwise, it is not valid. The two premises are universal statements (without existential import), whereas the conclusion is a particular statement (with existential import). So although the argument is valid on the traditional Aristotelian interpretation, it is not valid on the modern Boolean interpretation.

This case seems like an easy one to resolve. The first analysis is arguably the right one, because it makes the argument come out valid without restrictions. Also, the first one seems more natural. Yet deciding what is natural may be a subjective matter that is subject to dispute. So how can we say that the first one is right just because it makes the argument come out valid without adding assumptions about existential import that may be problematic? And should a missing part always be selected in such a way that it makes the argument come out valid? If not, how can the first analysis be defended as the preferable one? After all, it may be argued, maybe the proponent of the argument really had the (invalid) second argument in mind. Some would invoke the principle of charity here, or some similar principle, arguing that the second analysis is better because it makes the argument come out stronger. But why, given two choices, should the one that makes the argument come out stronger be chosen? This question calls for some general principle that can be used to make judgments of which analysis to choose, if more than one is possible in a given case.

These issues apart, this first case does have some clear lessons. It shows that – in some cases, at least – deductive logic can be a useful tool in dealing with incomplete arguments. But as we will now go on to show by examining a range of other cases, deductive logic is not always the structural tool of choice for this purpose.

4. LIMITATIONS OF DEDUCTIVE ANALYSIS

The following example looks initially like it should be deductive, but other interpretations need to be considered. This example, which we will call the frogs case, is from an exercise in a textbook (Hurley, 2000, p. 292).

> Any drastic decline in animal life is cause for alarm, and the current decline in frogs and toads is drastic.

In this case, what is needed to make the argument explicit is the conclusion, "The current decline in frogs and toads is cause for alarm." But there is also a missing premise, which is needed to make the argument deductively valid. This premise needs to state that all frogs and toads are (forms of) animal life. In this case, the argument is plausibly cast as being deductively valid once the missing parts have been filled in.

What could be used to fill in the missing premise that frogs and toads are animals is the argumentation scheme for argument from verbal classification. The precise form of the argument from verbal classification given by Walton (1996, pp. 53–55) uses variables for individuals and properties as follows:

> Individual a has property F.
> For all x, if x has property F, then x can be classified as having property G.
> Therefore, a has property G.

The frogs case shows that the argumentation scheme for argument from verbal classification needs to be expanded to take into account cases of arguments based on subspecies relationships. In this case, it is not an individual frog or toad that is being classified, but frogs and toads generally. Still, one can see that the missing premise that needs to be filled in, in this case, is based on argumentation from verbal classification. The generalization in this case, "All frogs and toads are (forms of) animal life," is a strict or absolute one, of the kind modeled by the universal quantifier in deductive logic. These strict generalizations can be contrasted with the defeasible generalizations, the importance of which is shown in the next case.

The following example is quoted from the Sherlock Holmes story "The Adventure of Silver Blaze" (Doyle, 1932, p. 27). We will call it the Silver Blaze case.

> A dog was kept in the stable, and yet, though someone had been in and fetched out a horse, he had not barked enough to rouse the two lads in the loft. Obviously the midnight visitor was someone whom the dog knew well.

The missing premise in this case seems to be the generalization "Dogs generally bark when a person enters an area (like a stable) unless the dog knows the person well."[5] This generalization does not seem to be a strict (absolute) universally quantified statement. There are all kinds of possible exceptions. Some dogs will bark at any person who enters an area. Some dogs won't bark at any person who enters an area, or will bark at hardly any person. Some dogs are unpredictable. Or the dog in question could have been drugged. Despite such possible exceptions, the generalization does seem to hold as a reasonable warrant for an inference. But in this case, the argument does not seem to be deductively valid. The argument does carry some weight as evidence for the conclusion. It seems to be a conjecture, based on plausible reasoning. The defeasible generalization in this case can be contrasted with the strict (absolute) generalization "All frogs and toads are (forms of) animal life" in the frogs case. The word 'generally' is an indicator of this defeasibility pointing to the existence of potential exceptions. The strict generalization is falsified by a single counterexample. But when a contrary instance confronts a defeasible generalization in a given case, the generalization still holds (at a general level), even though it has defaulted in this particular case. For example, a dog that is too sick to bark falsifies the generalization "Dogs bark when a person enters an area," but it is an exception to the generalization "Dogs generally bark when a person enters an area."

[5] The missing premise in this case depends on what is often called common knowledge. This notion and its importance for enthymemes will be discussed later in this chapter.

The argument in this case could be nicely analyzed as an abductive inference, as follows. The known facts are that the intruder entered the stable and that the dog didn't bark. But this situation seems unusual, and calls for an explanation. For don't dogs normally bark when a person enters an area where the dog is kept and takes something away? Then why didn't it happen in this case? The best explanation would seem to be that the dog knew the person who entered the stable. For generally, if a dog knows the person, it won't bark. Thus the best explanation of the given data, including what happened and what didn't happen, is that the dog knew the person who entered the stable and took the horse. The line of argument in this case, so analyzed, involves the *argumentum ad ignorantiam*, the so-called argument from ignorance, or lack-of-evidence argument. Sherlock Holmes called the case "the curious incident of the dog in the night-time." To the reply "The dog did nothing in the night-time," Holmes answered, "That was the curious incident."[6] Although Holmes was known to describe the reasoning used in his detective work as deductive, in this case it is a clear example of the use of plausible, or abductive reasoning. The argumentation scheme for the argument from ignorance, according to Walton (1996, Chapter 4), is associated with the inferential rule of the Closed World Assumption in AI. Reiter (1980, p. 69) calls the Closed World Assumption the rule that if all the positive information in a database is listed, the negative information is represented by default. For example, Reiter (1980, p. 69) considers a question-answering system associated with an airline flight schedule. The user asks the question, "Does Air Canada flight 113 connect Vancouver with New York?" If there is such a flight listed in the database, the system responds "Yes." But if there is no such flight listed in the database, such a system will typically respond "no." In other words (Reiter, 1980, p. 69), "Failure to find a proof has sanctioned an inference." The system has assumed that if there were such a flight, it would be listed in the database. In other words, the system has adopted the Closed World Assumption. Or to put it another way, the system has used the argument from ignorance as the basis for its response.

Is the argument from ignorance in this case fallacious or reasonable? The use of the term 'obviously' in the conclusion indicates a kind of confidence that is typical of Holmes's 'deductions' but that is not reasonably justified by the argument. Holmes seems to be leaping a bit too quickly to a conclusion that could be wrong. The use of the word 'obviously' is

---

[6] Doyle (1932, p. 23).

textual evidence for evaluating the argument from ignorance as fallacious (with apologies to Sherlock Holmes fans).[7] On the other hand,
such an argument does carry some weight as supporting the conclusion,
giving a reason to accept it as one small argument within the mass of
relevant evidence in the case.

The Silver Blaze case is quite a nice example of an incomplete argument with a defeasible generalization as the missing assumption. It has
some interesting lessons with regard to the argument from ignorance,
or lack-of-evidence argument, as it is sometimes called. The frogs case is
more complex. It contains a missing premise and a missing conclusion.
The missing premise is a strict universal generalization, and the argumentation scheme that underlies it seems to be deductive rather than
presumptive or plausibilistic in nature.

5. USE OF ARGUMENTATION SCHEMES IN ANALYSIS

The previous section has shown that schemes can play a role even in cases
where some superficial deductive structure is apparently applicable. In
other cases, schemes play a more dominant role. Hurley (2000, p. 289)
defines an enthymeme as "an argument that is expressible as a categorical
syllogism but that is missing a premise or a conclusion," and offers the
following example. Let's call it the corporate income tax case.

The corporate income tax should be abolished; it encourages waste and
high prices.

The missing premise is said to be the statement "Whatever encourages
waste and high prices should be abolished" (p. 289). To make the argument into a categorical syllogism, this statement has to be taken to express
a universal generalization, like "All things (or perhaps practices) that
encourage waste and high prices are things (practices) that should be
abolished." One might wonder in this case whether the missing statement
should be taken to express a strictly universal generalization. Perhaps it
means something more like "In general, if a practice encourages waste
and high prices, then that is a reason to abolish it." This version of the
statement is defeasible, because it is compatible with there being reasons

---

[7] Some interesting questions are raised here about the use of discourse markers as evidence
to determine argument structure. See Snoeck Henkemans (2001). It seems unlikely that
such indicators alone are sufficiently frequent to provide a basis for giving a conclusive
determination of the argument structure.

for not abolishing the practice. It could be called a defeasible generalization or nonstrict generalization. So analyzed, the argument in the corporate income tax case falls into the category of enthymeme in the original Aristotelian sense of the term cited by Burnyeat.

Another observation about this case is that the argument seems to depend on two additional missing premises. One is a statement that could be expressed as follows: a practice that encourages waste and high prices is, all other things being equal, a bad practice. The other is the statement "If something is a bad practice, it ought to be abolished." A structure that is helpful to an argument analyst as a guide to filling in these missing premises is the argumentation scheme for the argument from negative consequences (Walton, 1996, p. 76).

> *Premise*: If action *A* is brought about, bad consequences will occur.
> *Conclusion*: Therefore, *A* should not be brought about.

This argumentation scheme can be used to give a reason to support the claim that an action should not be carried out, the reason being that bad consequences will occur. Later it will be shown that there is another argumentation scheme for what is called argument from classification. Using argument from classification, you could classify waste and high prices as being, generally speaking, bad things. Then, using argument from classification and argument from negative consequences, you could identify two generalizations that could function as unstated premises in the argument in the corporate income tax case.

> *Argument from Negative Consequences Premise*: Any practice that has bad consequences should (other things being equal) be discontinued.
> *Argument from Classification Premise*: Waste and high prices are (generally) bad things.

This way of reconstructing the argument is quite attractive, because the argumentation schemes can be used to identify the generalizations that naturally fit as the missing premises. Although we can dispute about what the missing premises really are, and about exactly what form they should take, the analysis using argumentation schemes is a good fit. It is even less problematic than the analysis of the syllogistic case. This analysis employed deductive logic, while the analysis of the corporate income tax case uses argumentation schemes.

The question of how the argumentation in the corporate income tax case should be diagrammed is interesting, because it raises the issue of

FIGURE 6.3. Diagramming an argumentation scheme.

how an argumentation scheme should be represented on an argument diagram. Consider the following representation.

1. The corporate income tax should be abolished.
2. The corporate income tax has bad consequences.
3. The corporate income tax encourages waste and high prices.
4. Waste and high prices are bad consequences.

In Figure 6.3, the 2-1 part in the shaded area is an instantiation of the argument from consequences scheme in which premise (2) is defeasible. The shaded area thus represents the structure of the argumentation scheme for argument from consequences as diagrammed by Araucaria.

The argument in the following case, like that in corporate income tax case, is better analyzed as being a defeasible inference based on a presumptive argumentation scheme. Many enthymemes have to do with practical reasoning. Consider the following example, which we will call the self-hypnosis case, from Pinto, Blair, and Parr (1993, p. 143).

> Everyone should learn self-hypnosis because it's one of the best ways to reduce stress.

This argument rests on the implicit assumption that reducing stress is a worthwhile goal for everyone. The stated premise is that self-hypnosis is a means to reduce stress – one of the best ways. The conclusion, as stated, is that everyone should learn self-hypnosis. In this case, the argument is not deductively valid. But once the nonexplicit assumption is inserted as a premise, the argument does have a recognizable form.

The argumentation scheme for practical reasoning (Walton, 1996, p. 11) is based on a form of argument that has two premises. The first premise states that an agent has a goal. The second states that the agent reasonably judges that carrying out a particular action is a means to achieve this goal. The conclusion is the statement that the agent arrives at the conclusion that he or she should carry out this particular action. Matching the argumentation scheme for practical reasoning are five appropriate critical questions (Walton, 1996, p. 11).

CQ1: Is it realistically possible to achieve the goal?
CQ2: Are there positive or negative consequences of either of the courses of action that should be taken into account?
CQ3: Are there other means of achieving the goal that should be considered?
CQ4: Which is the best of the various means available?
CQ5: Are there other goals (possibly even conflicting with the goal at issue) that should be considered?

The argument having the form of argumentation from consequences can carry some weight as a plausible argument in a given case. But if an appropriate critical question is asked, that weight is temporarily suspended until the question has been successfully answered. Thus arguments of this form are defeasible.

In the self-hypnosis case, the missing premise is the statement "Reducing stress is a worthwhile goal for everyone." This statement looks like a strict universal generalization, but is it? Statements about goals are generally defeasible (including the one in this very sentence). It can be argued, in this case, that reducing stress is not a worthwhile goal for absolutely everyone, but is a worthwhile goal for most of us, these days, who live such stressful lives. At any rate, this case suggests that goal-directed practical reasoning is a form of argumentation underlying many cases of enthymemes.

6. USE OF SCHEMES IN ANALYZING WEAK ARGUMENTS

The following example is taken from a letter to *Chatelaine* magazine, May, 1982. We will call it the abortion case. It represents the kind of case in which, once the implicit assumption is identified, it is highly questionable whether it is justified.

When a murderer is found guilty, he is punished regardless of his reasons for killing. Similarly, anyone partaking in an abortion is guilty of having deprived an individual of her or his right to life.

The implicit conclusion is the statement that anyone partaking in an abortion should be punished. Why? It seems that this statement is supported by the drawing of an analogy between the case of one person murdering another person and the case of someone partaking in an abortion. Since the two cases are alleged to be similar, it is alleged that what is true of one should also be true of the other. The argumentation scheme for argument from analogy is given by Walton (1996, pp. 77–80). The argument is based on the assumption that because a murderer is punished, then, by analogy, an abortion partaker should also be punished. In this case, the argumentation scheme for argument from analogy can be used to show that the argument is based on an implicit generalization that the two kinds of cases, murder and abortion, are similar. Such a generalization is defeasible, for any two such real cases will fail to be similar in some respects.

A good question that may be asked about incomplete arguments is whether they always have to come out as valid (or as structurally correct, by some standard) once the missing parts are filled in. Another question is whether the missing premises or conclusion are statements that have to be true, or at least plausible. The following case, an interesting one to consider, suggests an answer to this question. This case, which we will call the attendance case, is in the form of a dialogue, as given by Farrell (2000, p. 98):

> *Student*: You have no right to flunk me.
> *Professor*: Why?
> *Student*: I came to every class.

The implicit assumption of the student's argument seems to be the following premise: all students who come to every class should pass the course. Another premise that is explicitly stated is that this student (the speaker in the dialogue) came to every class. The implied conclusion is that this student should pass the course. Presumably, then, this conclusion is used as part of another argument with the conclusion that the professor has no right to fail this student. In this case, the missing premise, "All students who come to every class should pass the course," is (presumably) false. For presumably, attendance by itself is not sufficient for a passing grade.

In this type of case, there is a missing premise that is assumed by the argument, but the most natural candidate for the missing premise seems to be a statement that is false, or at least highly questionable. If this kind of reconstruction of such arguments is right, then it follows that, in some cases, filling in the missing premises (or conclusions) results in a bad argument. It results in an argument with a premise that is false, or at least implausible. It would seem to follow, then, that not all missing premises (or conclusions) have to be true or plausible propositions.

Some incomplete arguments, when completed, come out as bad arguments. Cases like this one are interesting, because they seem to suggest that filling in missing assumptions in incomplete arguments has a critical component. If the student were asked whether she really meant to state the missing premise just cited, she might react defensively, and might not want to admit it, even though her argument does not seem to make much sense otherwise. So, in this case, it might be too strong to assert that her original incomplete argument is identical to the completed version. A better approach might be as follows. A critic might pose critical questions in a dialogue by asking the student if that is what she really meant to argue. The critical discussion could then continue from there. At any rate, the issue of whether filling in incomplete arguments presupposes some sort of context of a critical discussion is considered in the section on dialectical aspects of enthymemes. One can normally find another context of dialogue that supports a reasonable interpretation. For example, suppose that an instructor had mentioned to his student at the beginning of the school term that there is a strong statistical correlation between attending every lecture and getting an A grade. In this context, a student might construct an argument like: "A high proportion of students that attend every class pass the course; I attended every class; therefore, I should pass the course."

The attendance case is curious. Normal practice in reconstructing arguments would be to try to base the argument on a missing assumption that is true or at least plausible, as already indicated. But in this case, the natural candidate for the missing premise is a statement that is false, or at least highly questionable. Is it a counterexample to this normal practice? It seems that it is not. As stated earlier, given a choice of missing assumptions that would make a given argument structurally correct, the rule is to select the more plausible one over the less plausible one. In the attendance case, the reason, "I came to class," is connected in the dialogue to the conclusion, "You had no right to flunk me," in a way that indicates that the assumption the inference rests on is the generalization "All students who come to every class should pass the course." But there could be a choice between a more plausible and a less plausible missing premise. A more plausible (or less implausible) choice would be, "Generally, all other factors being equal, if a student comes to every class then he or she should pass the course." But this statement does not look very plausible either. It looks like the only way to link the two premises of the argument together inferentially, and to come up with a valid (or

structurally correct) argument, is to choose a false or implausible generalization as the missing premise. Thus the attendance case throws some
light on how to build the enthymeme machine. It suggests that the best
choice for the missing premise does not always have to be a statement that
is true, or even very plausible. It can be a statement that is questionable.
The function of the enthymeme machine, in such a case, would not be
to determine finally that such and such statement is the missing premise,
closing off further discussion. Instead, it would be to prompt the asking
of a critical question, such as "This statement or that statement is what
is required in order to complete your argument, so which one do you
accept, or do you reject both of them as representing your argument?"
In effect, the attendance case requires a dialogue approach rather than
an absolute judgment or one-shot outcome.

The phone book case illustrates an enthymeme that may not, when
filled out, produce an argument that is a good one:

> Bob Sturges can't have a telephone, because his name isn't listed in the
> phone book.

The missing premise in this case seems to be the statement "If your
name is not listed in the phone book, then you don't have a telephone."
This assumption is false, as a matter of common knowledge, because
it is possible to have an unlisted number.[8] If the missing assumption is
inserted, the resulting argument is valid, but has a false premise. On the
other hand, you could interpret the missing premise as a conditional
that is not strict, but is a defeasible rule of thumb. Suppose we interpret
the missing premise as saying that it's a pretty good guess, although it
could be wrong, that if somebody is not listed, he or she does not have a
telephone. This interpretation makes the premise come out as rough but
somewhat plausible, rather than as false. But the argument is no longer
valid. Instead, it is a plausible argument that could carry some weight,
but is defeasible.

It could be noted that the form of the argument in this case is also
that of argument from ignorance. As described in the Silver Blaze case,
such arguments can have a deductive form if the requirement of epistemic closure is met, licensing the closed world assumption, that is, if
the knowledge base is complete, and no relevant facts are assumed to be
missing (Walton, 1996, p. 112). But more typically, they are arguments

---

[8] Once again, the notion of common knowledge comes into play here.

used in cases where knowledge is incomplete. In such cases, even when they are reasonable, they tend to be plausible kinds of argument that can be used tentatively as a basis for action or further investigation.

7. LIMITATIONS OF SCHEMES

The following two cases do not offer a panacea for argument analysis. The first is adapted from an example used by Peirce.[9] The argument in the fossilized fish case is the following:

> Fossilized remains of fish were found on Mount Lemmon; therefore, Mount Lemmon was under water at one time.

There seem to be several implicit assumptions in this argument. One is that Mount Lemmon is not presently under water. But there are three others.

1. If fossilized remains of fish were found on Mount Lemmon, then there were fish on Mount Lemmon at one time.
2. Fish can survive only in water.
3. If there were fish on Mount Lemmon at one time, then Mount Lemmon was under water at that time.

How the argument works can be analyzed as follows. The given premise, along with missing premise one, implies by *modus ponens* that there were fish at Mount Lemmon at one time. This statement, along with missing premise three, implies the conclusion of the given argument. Premise two functions as support for three. Two could perhaps be rephrased as "All places where fish survive are places that are under water." So construed, two deductively implies three. But does the argument have to be reconstructed in a way that makes the warrant it is based on an absolute generalization or conditional? Maybe not. Perhaps the missing premise could be expressed using the following generalization: "Anywhere fossilized remains of fish are found is a place that was under water at one time." This generalization could be seen as one that is subject to exceptions, depending on what kind of explanations of the fossilized remains are possible or plausible.

---

[9] Peirce (1965, p. 375), in his paper "Deduction, Induction and Hypothesis," originally published in 1878, cited this example in the following words. "Fossils are found; say, remains like those of fishes, but far in the interior of the country. To explain the phenomenon, we suppose the sea once washed over this land."

Peirce used this very case (or one like it) to illustrate abductive inference. Along Peircean lines, the argument can be reconstructed as an instance of inference to the best explanation. The given datum is the finding of fossilized remains of fish on Mount Lemmon. But then, as Peirce might ask, how could such a finding be explained? A possible explanation is that fish were living in their natural state in the region of Mount Lemmon at one time. That hypothesis would explain how the fossils got there. But how could it be that fish survived in that location, considering the additional fact there is no body of water near Mount Lemmon at the present time? For fish can survive (under natural conditions) only in water. A plausible explanation would be that Mount Lemmon was under water at one time, when the fish were there. Of course, there could be other explanations. The fish could have been transported there, for example. Or they could have been blown there by a hurricane, or by some other major disturbance. But it could be that the best explanation is that Mount Lemmon was under water at one time.

What the fossilized fish case shows is that there seem to be two ways of reconstructing this kind of argument. One way is the usual method of enthymemes or incomplete arguments. This way is to fill in the missing premises in the usual way, and then to show how the conclusion follows by a chain of reasoning from the given premises along with the missing premises. The other way is the method of abduction. This way is to reconstruct the argument as an instance of inference to the best explanation. According to this method, you start from the given data and then construct a hypothesis that seems best to explain the data. Then, from that hypothesis, you may construct a further hypothesis needed to explain the initial hypothesis. Using this method, you get a chain of inferences to the best explanation from the given data. These two ways seem to be equivalent. In practice, they pretty much seem to amount to the same general method. A chain of inferences is used to fill in the gaps in the line of reasoning between the given premises and the conclusion to be proved.

One interesting aspect of this case is that it looks as if the first reconstruction, using the method of enthymemes, is based on deductive argument. The reason is that *modus ponens* is used to derive the conclusions needed in the chain of reasoning. Yet the abductive reconstruction uses inference to the best explanation, which is not a deductive form of argument. The resolution of the puzzle of a deductive form being used in nondeductive reasoning comes through the realization that premises one and three in the fossilized fish case are really not conditionals of the kind that support *modus ponens* inferences of the deductively valid sort. They are defeasible conditionals, of the sort often called "defeasible rules" in AI (Verheij, 1996). They are not universally true, but hold only with a kind of generality that is subject to default. They represent hypotheses that are plausible, but that can be defeated by incoming new information describing a particular finding in the given case. It would not be too surprising to find a case like the fossilized fish case treated in logic textbooks as being an enthymeme that can be filled in using rules of deductive logic. But a more careful analysis of the case should reveal that the argument is abductive.

The last case is one of a series of cases studied by Walton (2001). These are cases of enthymemes based on a kind of missing assumption that is often characterized by the phrase 'common knowledge'. In connection with enthymemes, a number of definitions of this phrase have been given. According to Govier (1992, p. 120), an implicit premise in an argument is based on common knowledge if it states something known by virtually everyone, depending on audience, context, time, and place. As examples, Govier (1992, p. 120) cites the statements "Human beings have hearts" and "Many millions of civilians have been killed in twentieth-century wars." Freeman has what appears to be a less strict and more variable definition of 'common knowledge' that is more accommodating to defeasible statements. Freeman (1995, p. 269) stipulates that to claim that a statement is common knowledge is to claim that many, most, or all people accept that claim. But Freeman (p. 269) adds the warning, "popularity is never sufficient to warrant acceptance," based on the danger posed by the *argumentum ad populum* or appeal to popular opinion, known to be fallacious in some instances. Freeman describes common knowledge as a form of presumption, rather than knowledge, based on the shared "lived experience" of a speaker and hearer (p. 272). In their account of the kind of common knowledge characteristic of enthymemes, Jackson and Jacobs (1980, p. 263) emphasize Gricean postulates based on rules of conversation that allow participants to participate collaboratively in a dialogue by making assumptions about what the other party can reasonably be expected to know. The literature in AI on scripts and plan recognition (Carberry, 1990) is full of examples of this kind of assumption making.

A glance through the logic textbooks reveals many examples of arguments with missing premises based on assumptions that come under the heading of common knowledge. One common type of example comes from everyday human experience of the way things can generally be expected to go in common situations that both speaker and hearer can be assumed to be stereotypically familiar with. The next case, also cited by Walton (2001), is a textbook exercise taken from Copi (1986, p. 246). Let's call it the textbook case.

> Although these textbooks purport to be a universal guide to learning of great worth and importance – there is a single clue that points to another direction. In the six years I taught in city and country schools, no one ever stole a textbook.[10]

The three nonexplicit assumptions in this case cited in the analysis presented by Walton (2001) are the following statements.

1. Anything that is a universal guide to learning of great worth and importance would be regarded as highly valuable.
2. Anything that is regarded as highly valuable, and would not be too difficult to steal, would likely be stolen.
3. These textbooks would not be too difficult to steal.

The gist of the argument in this case can now be reconstructed by the following expansion of what the writer is presumably telling us. Because no one has ever stolen a textbook, in the writer's experience, the assumption that these textbooks are regarded as highly valuable is refuted. This assumption is shown to be false. From that conclusion, another is then suggested. This conclusion is that these textbooks are not the universal guide to learning of great worth and importance they are taken to be.

Statements one, two, and three are based on what is called common knowledge. They are assumptions about the way things generally work, about familiar human institutions and values, and about the way we can normally expect most people to react. For example, statement three is based on common knowledge about how textbooks are used in schools. In a typical situation, the textbooks have to be distributed to the students, and this need for distribution makes them easy to steal. The reader is assumed to be familiar with how textbooks are normally used in the schools, and with the fact that theft is a common occurrence in that setting that is hard to prevent. In the literature on planning in AI (Carberry, 1990), these assumptions would be classified as domain-dependent knowledge, and they are notoriously difficult to capture in

[10] W. Ron Jones, "How to Write and Publish Your Own Dick and Jane," *Changing Education* 5(4) (Winter-Spring 1974): 17–19.

a principled way (Lenat, 1995). But they are not based on specialized expert knowledge. They represent common knowledge about the way things can normally be expected to work in a typical situation known to both sides in a conversation.

The textbook case shows even more clearly how incomplete arguments can rest on what is called common knowledge shared by the proponent of the argument and the intended recipient. Thus the study of this case has indicated the limits of argumentation schemes as a tool for the identification of implicit premises in incomplete arguments. This limitation has already been remarked upon by Gerritsen (2001). She observes (p. 73) that the identification of an argumentation scheme in an argument gives only a general clue to the unexpressed premise, "while the problems of identifying unexpressed premises are often about details and peculiarities." For example, the abortion case rests on an argument from analogy, and the identification of the argumentation scheme of argument from analogy is a general clue that helps to identify the unexpressed premise. But the textbook case shows very clearly how identification of argumentation schemes is not sufficient to fill in all the details and peculiarities needed to identify the unexpressed premises.

8. DISCUSSION OF CASES

The cases fall roughly into a pattern suggesting two types. In the one type, a strict (absolute) universal generalization appears to be the missing assumption needed to complete the argument. Deductive logic can be used to furnish the right kind of structure needed to make the argument valid. In the other type of case, a nonstrict (defeasible) generalization appears to be the missing assumption needed to complete the argument. Plugging in a strict generalization in these cases would produce a false, or easily refutable, premise (or conclusion, as the case may be). Such a reconstruction would therefore not fit with optimal methods for dealing with enthymemes. It would violate the negative principle that you shouldn't reconstruct an argument in such a way as to impute to it a false or implausible unstated assumption, if there is a more plausible (or true) statement that would also fit into the slot for the missing assumption.

The cases can plausibly be taken to indicate that deductive logic alone is not sufficient as the structural tool of inference needed to aid in the filling in of incomplete arguments. It is shown that deductive logic is the right tool in some cases. But the weaker standards of appraisal are more appropriate in others, as shown by Ennis (2001). In many of these other cases – and these kinds of cases seem to be very common in natural language argumentation – presumptive argumentation schemes are the right tool for the job. This in itself is an important finding, given the traditional tendency in logic to advocate deductive logic as the right tool for enthymemes, even applying it to cases where argumentation schemes are clearly more appropriate and would do a much better job of locating the missing assumptions. In general, different standards or structural models of rational argument need to be applied to different cases. In some cases – for example, where the generalization that is the warrant of the inference is strict – deductive logic is the right standard. In other cases (though none are included in the case studies here), inductive logic of the modern statistical kind could be the right standard. But in many cases, presumptive argumentation schemes provide the right structual models. In these cases, the standard of argument is neither deductive nor inductive, but falls into a third category. It could be called the category of plausible arguments, based on argumentation schemes and defeasible generalizations. Some of the cases involve an argument that would nowadays often be classified as abductive. This is not to belittle the problem of classification. There are substantial practical challenges in carrying this out.[11] Following Mann (1987), we suggest a pragmatic approach based on plausibility judgments. When analyzing a given text of discourse, there may be more than one way of reasonably interpreting the text, and the job of the analyst may be to consider alternative interpretations as hypotheses.

Groarke has argued (1999, 2001) that argumentation schemes can be captured in a deductivist framework, with deductive logic propagating a level of certainty or presumption from the premises to the conclusion. He gives the following example:

Jones is a politician, so he is not to be trusted.

He suggests that the missing premise is the generalization "No politicians can be trusted." He suggests that there are other possible premises that would result in deductive validity (e.g., "If Jones is a politician, then he is not to be trusted," which could work in the rather peculiar context in which Jones has been claiming not to be a politician), but that the "No politicians . . . " premise is the most appropriate (1999, p. 6).

> In the absence of some explicit indication that this idiosyncratic assumption is the basis of the conclusion, it is reasonable to assume that it is the latter

---

[11] Some argue that there is no reasonable way to classify reasonable arguments at all (Hitchcock, 1981; Ennis, 2001).

generalization about politicians which drives the inference. It can therefore be designated the pragmatic optimum.

We are concerned here about how this designation is to be made in practice – and, perhaps, automatically. The answer lies in the guidance afforded by the structure of argumentation schemes – if we have a scheme of a particular type in play, then we know what missing premises are to be expected.

There is also a potential problem with the deductivisation of schemes. There seem to be two approaches to analysing Groarke's example. First, Groarke's own:

(1) Jones is a politician.
(2) No politicians can be trusted.
(3) Jones is not to be trusted.

This is a clear deductive argument. Groarke would argue that (2) is only plausible (whereas (1) is certain), and that therefore this plausibility is transferred to the conclusion, (3). The result is that the conclusion is plausible. An alternative reading is

(1) Jones is a politician.
(2) Usually, politicians cannot be trusted.
(3) Jones is not to be trusted.

This is based on some sort of scheme (perhaps the circumstantial *ad hominem*, or a specialization of it) that says

(1) Person X is an A.
(2) As usually have some feature F.
(3) X has feature F.

and, as an inbuilt part of the scheme (or rather, as a feature of the scheme), that the conclusion is only plausible; it is only defeasible. So, with the argumentation scheme approach, we have the same conclusion, again marked as plausible. The only difference between the Groarke approach and the scheme approach is that in the former the implicit premise is a universal generalization that is only plausible, rather than certain, while in the latter the implicit premise is a nondeductive, nonuniversal generalization that admits exceptions. We argue that it is a generalization that admits exceptions, rather than a universal generalization that might be wrong, that is driving such argumentation. For a single exception to a universal generalization would demonstrate it to be wrong, and yet intuition leads us to view an exception to a generalization as just that – an exception to a generalization that still holds. We conclude from this example, therefore, that although deductive logic has a role to play, it can function best when complemented by an approach based on nondeductive argumentation schemes.

The case studies, as analyzed and discussed here, bring out the importance of argumentation schemes as a supplement to deductive logic. But they do not, even so, provide anything like a complete solution to the problem of incomplete arguments. They do move the discussion of the problem along, however. What is made clear is that a formal, inferential component is a necessary part of the construction of an enthymeme machine, and that argumentation schemes should be an important part of that component. To conclude, it is well to comment on some of the general issues concerning enthymemes that still remain to be dealt with. It is useful to see that, in order to construct the enthymeme machine, a contextual component needs to be used alongside the formal inferential component.

9. THE ATTRIBUTION PROBLEM

The most general issue in dealing with incomplete arguments is how a statement can be attributed to an arguer as part of her argument if she has never gone on record as making that exact statement explicitly. It could be called the problem of attribution. The problem of attribution is one of interpreting a claim supposedly made, based on a quotation, or given text of discourse, that records what the arguer actually said or wrote. Some would say that you can never attribute a claim to someone unless they actually made that exact claim. For after all, it may be said, you can never really look into the other person's mind and see what they meant, or intended to say. All attributions other than exact quotations of claims made are, many would say, "subjective." There is something to this line of argument. It is often made by students who are reluctant to take on the task of analysis of argumentation in a text of discourse because they fear that the whole project is dangerously "subjective." Many philosophers, especially of the postmodernist stripe, have voiced the same objection. It has to be admitted that there is something to this objection, and it should be taken seriously.

The topic of a recent discussion with Bart Verheij was the argument, "John is a thief; therefore, John is punishable."[12] Verheij took the

---

[12] This discussion took place in several e-mail exchanges on the subject of defeasible reasoning in November 2000.

position that this argument is a different argument from the following valid argument: "If John is a thief, he is punishable; John is a thief; therefore, John is punishable."[13] The point of view on incomplete arguments (enthymemes) typically expressed in the logic textbooks is that the second argument is an analysis or reconstruction of the first óne. In other words, the assumption is that the two arguments are equivalent, in some sense, or that at least the second one represents the argument underlying the first. But Verheij took the point of view that the two arguments are not equivalent. As conceded earlier, there is something to be said for this point of view, and there could be various reasons for maintaining it. One of these reasons frequently surfaces when you try to teach students how to analyze arguments by filling in missing premises. Students can be highly skeptical when asked to fill in missing premises or conclusions in texts of discourse they are analyzing. Some students ask how you can attribute a statement to someone if they didn't explicitly say it. After all, isn't making such an interpretation potentially unfair, if the proponent of the argument might not agree with it? How can you tell what such a person was really thinking anyhow? Such doubts are legitimate. In order to respond to them appropriately, the defender of the doctrine of enthymemes is rightly put in the position of having to show why it should be intellectually acceptable for a critic to insert missing premises into an argument analysis.

This dispute about enthymemes is not new. Burnyeat (1994) has brought out a number of interesting historical points relating to enthymemes arising from the texts of Aristotle and other Greek philosophers. One interesting point (1994, p. 46) concerns two views that were expressed by ancient philosophers. Antipater of Tarsus, head of the Stoic school about 159-130 B.C., defended one-premised arguments like "If you breathe, you are alive" (Kneale and Kneale, 1962, p. 163). But this posed a problem in relation to Stoic logic, because the five basic types of arguments recognized by the Stoics all have two premises. It also posed a problem for the Aristotelian commentator Alexander of Aphrodisias, because an Aristotelian syllogism must have two premises. To justify syllogistic logic, Alexander argued that the kind of argument cited by Antipater must be incomplete. To fill it out, you must add a missing premise like "All who breathe are alive."[14] There is another

---

[13] See the analysis of defeasible legal reasoning given by Verheij (1999).

[14] A modern candidate for a counterexample to this generalization would be the case of a brain-dead patient whose breathing is artificially sustained by a respirator.

example that Burnyeat (1994, p. 46) ascribes to Antipater: "This man deserves punishment, for he is a traitor." The issue is whether this argument has as a missing or unstated assumption the premise "All traitors deserve punishment." It would seem that there are two viewpoints on this issue. According to the standard view of the enthymeme found in the logic textbooks, and according to Alexander of Aphrodisias as well, the two arguments are equivalent. Or at least the second one represents a more explicit version of the first. According to Verheij and Antipater as well as many skeptical students of argumentation, the two arguments are not equivalent. One should be seen as quite different from the other.

Arguments can be given on both sides of this issue. But the case studies given earlier, along with the Araucaria system of marking up an argument, suggest a way of approaching the attribution problem. That way involves seeing the new argument reconstructed from an incomplete argument as being closely related to, but different from, the original argument. As Burnyeat (1994) showed, there is considerable persuasive evidence that Aristotle's notion of enthymeme has been systematically misinterpreted by commentators, and by mainstream logic, for over two thousand years. The version of Aristotle that Burnyeat brings out sees Aristotle as viewing an enthymeme as essentially a type of argument based on a defeasible argumentation scheme.

Araucaria is equipped with a set of argumentation schemes. When a user constructs an argument diagram, she can identify the scheme that fits a given set of premises and conclusion that she has identified as an argument in a given text. Araucaria can then fit the scheme to the specified parts of the argument, and identify the missing premises required by that scheme. This part of the process is straightforward, but other issues need to be resolved. One issue is the identification of additional nonexplicit premises. This task seems best accomplished through the use of critical questions. Each argumentation scheme has a matching set of critical questions. The best approach is to use the critical questions corresponding to a scheme (and available in Araucaria) to extend the process of identifying additional nonexplicit premises. But critical questions can have critical subquestions. For example, appeal to expert opinion has six basic critical questions (Walton, 1997, p. 223). But it also has critical subquestions under each of these main critical questions. This poses the practical problem of how long the process of identifying nonexplicit assumptions should go on. It may not be useful to list all critical questions for all schemes. And thus it may be best to make the assumption that normally the main critical questions are invoked only when identifying nonexplicit premises.

A further problem is that schemes are related to each other. In many cases, some schemes are subsumed under others. For example, the scheme for appeal to expert opinion is treated by Walton (1996, 1997) as a subtype of argument from position to know. Another issue is whether deductive and inductive forms of argument can be included along with defeasible argumentation schemes, as part of the sets of structures used by the system to mark up arguments. The best approach to these issues is to classify the schemes into a taxonomic hierarchy with a threefold root representing reasoning types – deductive, inductive, and plausible. Thus, as things turn out, this approach to enthymemes does take some steps toward solving the problem of attribution in incomplete arguments. According to this approach, the system can use argumentation schemes and critical questions to identify nonexplicit premises in an incomplete argument identified by a user in a given text of discourse. But so far, most of the attention has been directed to the inferential component, relating to the use of arguments forms. The textbook case, in particular, suggests that there are also important assumptions about common knowledge. We now turn to a discussion of an additional component of enthymemes relating to the context of dialogue in a given case.

10. THE DIALECTICAL COMPONENT OF THE ENTHYMEME MACHINE

The case study discussions presented here have shown that, in addition to the inferential component, there is another factor necessary for the reconstruction of incomplete arguments. In addition to generating an argument that is structurally correct by some standard of inference, the machine should give preference to missing premise (or conclusion) candidates that are true, that represent common knowledge, or that are at least plausible, in context. The principle of charity tells the machine to pick missing parts that make the argument strongest. But as indicated in the case studies, this approach isn't always right. The distinction between used and needed premises also seemed to be a helpful tool. But in fact what the machine needs to do, as indicated by the cases, is to combine both aspects. To solve the attribution problem, some new approach is needed.

This new approach takes a line of reasoning that supports a version of Alexander's and Verheij's point of view. According to this point of view, the reconstructed argument and the original incomplete argument are not exactly identical. Yet it can be argued that they are closely related in a pragmatic way. It can be argued that when a critic analyzes an argument in a given text of discourse, she can analyze it only by bringing out the implicit assumptions in the argument. To do this, she should see herself as engaging in a critical discussion with the proponent of the argument. A critical discussion type of dialogue can be maximally successful only if the strongest possible arguments on both sides are brought forward. How should missing premises or conclusions be inserted by the enthymeme machine in the context of such a critical discussion? In a typical critical discussion, each party has a thesis to defend, and the thesis of one party is opposed to the thesis of the other party (Van Eemeren and Grootendorst, 1992). To make the critical discussion successful, each party should put forward the strongest and most convincing arguments possible in support of his or her thesis. This aspect seems to support the principle of charity. What makes an argument strong? The reply suggested earlier is that the critic needs to put in the assumptions required to make an argument come out valid, or at least structurally correct according to whatever the accepted standards are. After all, if the argument really depends on this assumption, and it is not stated as part of the argument, the other side in the critical discussion could ask critical questions about this gap. Asking such critical questions would reveal a weakness in the argument, showing a missing premise in it. Therefore, from a viewpoint of having a good critical discussion, putting in such a missing premise can be justified. In a sense, Gilbert's (1991) dialogic algorithm for refining the principle of charity to get at exactly what implicit premise the protagonist is working from provides a generic set of critical questions, or rather, a single generic critical question aimed at refining universal generalizations. This would form one part of the critical questioning that probes the inferential link in the argument.

But it was shown in the case studies that the inferential component is not enough. The enthymeme machine should select missing assumptions that are true, or at least seem to be true, or that represent common knowledge or the arguer's position as indicated by the discourse. Often, such missing assumptions are statements that would generally appear to be acceptable as common knowledge to the audience, or statements that seem to be based on the arguer's position (commitments), as far as the text and context of discourse indicates. This observation suggests that dialectical criteria are important for constructing the enthymeme machine – for example, in cases where the context of the given argument is that of a critical discussion. In a critical discussion, the arguer's goal is to convince the audience (other party) that her (the arguer's) thesis is true (Van Eemeren and Grootendorst, 1992). For this purpose, the arguer ideally needs not only strong arguments (structurally), but also arguments based on premises that the audience will accept, or can be led to accept. She also needs arguments that will support her own position, and especially her thesis to be proved. Here, then, is the pragmatic component. Incomplete arguments should be filled in with missing assumptions (a) that are plausible to the intended audience or recipient of the argument, and (b) that appear to fit in with the position advocated by the arguer, as far as the evidence of the text indicates (Gilbert, 1991). Here, then, is an approach to solving the attribution problem. In addition to the argumentation scheme component, a dialectical component is needed representing other relevant factors of the text and context of dialogue in a given case.

But some other clarifications and qualifications are needed. The inserted premise must be marked clearly as an assumption put in by the critic, and distinguished from other statements explicitly stated as premises or conclusions. Such a marker should indicate that the inserted statement is merely a hypothesis inserted by the analyst.[15] As long as that requirement is met, and provided there is evidence that the inserted statement does seem to represent what we take to be the arguer's position (commitments), putting in missing premises can be pragmatically justified. This way of building an enthymeme machine fits with the view of Antipater and Verheij. The two arguments about John being a thief, according to that view, are not equivalent (the same argument). One argument is a kind of artificial construct made up of the other argument plus contextual information about the type of dialogue the arguments are supposedly embedded in. In some cases, you could justify replacing one with the other. But the two arguments are not equivalent in the sense that they are substitutable for each other in all contexts of argumentation. The justification for replacing one with the other depends on the purpose of the exchange. And yet there is another sense in which the two arguments are, if not equivalent in a context-free way, at least equivalent in a more limited sense. In this sense, one can be substituted for the other in the right context, and under the right conditions, for a purpose.

---

[15] Araucaria explicitly marks implicit premises. The AML tag marking a proposition has an attribute *missing* that is set to *No*, in the case of original components of the argument, or to *Yes*, for analyst-added parts.

Speaking of the logician's dream of constructing an enthymeme machine, one might ask what its anticipated uses are. Of course, the traditional need for such a machine is evident in existing methods used in applied logic and critical thinking, where the concept of the enthymeme is already an important and well-established part of the curriculum. But there are other significant potential uses as well. One is in the development of critical thinking tutorial software for educational applications. A software system that helps a user to identify implicit premises by using argumentation schemes, and possibly other dialectical clues, could be useful in guiding a user to probe logical gaps in an argument and to ask appropriate critical questions. The user numbers the component statements in an argument and identifies an argumentation scheme, or type of argument, linking the given premises to the conclusion. The machine then applies the scheme and inserts a missing assumption that completes the argument. The machine then engages in a critical dialogue with the user. For example, it might give hints by asking the user about other possible choices. A second anticipated use is in the field of communication, especially in rhetoric, where the analysis of incompletely expressed arguments is centrally important. A third application is to computing, and especially to multi-agent systems, where software agents often need to communicate using argumentation (Reed, 1998). Agents often need to communicate by asking questions of other agents, and by acting on the basis of practical arguments and directives expressed by another agent. In this application, as well as in the first two, filling in the missing parts of an incomplete argument is an important preliminary to efficient communication and carrying out of practical tasks based on information and directives expressed either in natural language or in the artificially constructed discourse of computer languages. An enthymeme machine would be a valuable tool for carrying out such tasks in an automated and efficient way. A fourth possible use is in legal argumentation, where many typical kinds of legal arguments can be analyzed as based on assumptions in the form of unstated premises or conclusions. An enthymeme machine would be a valuable tool that could be used in many computer systems of the kind now being developed in artificial intelligence and law (Prakken, 2002). Araucaria is a tool for supporting a human analyst that goes some way toward meeting the challenges of enthymemes. It is a useful and valuable first step. But much work remains to be done in tackling the difficult task of truly automating the process of enthymeme reconstruction.

7

Attack, Opposition, Rebuttal, and Refutation

In this chapter we develop a pragmatic theory of refutation in which a refutation is defined as a sequence of dialogue moves in which an argument is used by one party to attack and defeat an opposed argument put forward previously by the other party. A fundamental distinction is drawn between refutation and rebuttal. A rebuttal is an argument that is opposed to another argument. It stands against the argument it is opposed to. But it does not necessarily refute that argument. Refutation is something more powerful. A refutation knocks down the original argument. It not only is opposed to the original argument, but also has enough strength itself as an argument that it overpowers the original argument and knocks it down (defeats it). This distinction is not a very firm or precise one in everyday conversational usage.[1] An attempted refutation is, after all, a refutation in conversational English, even if it is not strong enough to knock down the argument it is opposed to. But still, as we hope to show, there is something to this distinction. Refutation is stronger than rebuttal. A refutation is something like a strong rebuttal, or a rebuttal that has active force in successfully attacking the argument it is aimed at.

A parallel distinction that has been very influential in recent work in artificial intelligence can be drawn between attacking and defeating an argument (Dung, 1995; Prakken, 1997). One argument attacks another if the second is addressed to the first one, or meets it, and if it is opposed to it in such a way that it goes against it. One argument defeats another

---

[1] The way Toulmin (1958) uses the term 'rebuttal', it refers to a way of attacking the inferential link between the premises and conclusion of an argument by arguing that a rule is subject to an exception. The way the term 'refutation' is used in Araucaria, it is a stronger notion akin to classical negation in logic. For dialogue analysis of Toulmin's model that spells out details of how a Toulmin rebuttal works, see Bench-Capon (1998).

if it not only meets it, but is also strong enough that it removes the support that the first argument had or seemed to have originally (Prakken,
2001). The notions of attack and defeat seem so fundamental that it is
hard to try to define them in other terms that don't simply repeat or
depend on the notions to be defined. Is the pair 'attack and defeat' the
same as the pair 'rebuttal and refutation'? And what about the distinction between undercutters and defeaters, also widely known and used
in AI (Bex et al., 1998)? Is this distinction the same as the other two or
not? The problem has led into a terminological swamp that makes one
hesitate to use any of these terms without getting bogged down.[2] The
first step in clearing this forest of verbiage will be to define the notion of
an argument and then to try to define, or at least to clarify, notions like
attack, defeat, refutation, and rebuttal in terms of this original notion.
The framework is one in which you have one argument to begin with,
and then a second argument directed against the first one. But what does
'against' mean in this context? This question is one of how to define the
notion of opposition. In this chapter we will see that there are different
ways to define opposition, some deeper than others, and that there is
much that can be found by going back to ancient and medieval sources,
including a distinctive argumentation scheme for argument for opposition.

1. ATTACKING, QUESTIONING, REBUTTING, AND REFUTING

Opposition and refutation are fundamental logical notions basic to critical argumentation. A critical discussion arises from a conflict of opinion
between two parties. There can be stronger or weaker kinds of conflict
of opinion. In the stronger kind of conflict, one party has a thesis to be
proved, and the other has a thesis that is the opposite of the first thesis.
In the weaker kind of conflict, one party has a thesis to be proved, and
the other doubts that thesis, but has no positive thesis of his own. In the
stronger kind of conflict, each party must refute the thesis of the other
in order to win. In the weaker, a party can refute the other and win by
showing that her thesis is doubtful. But maybe the term 'refutation' is too
strong here; perhaps a distinction should be drawn between a refutation
and a rebuttal. In addition to, and perhaps underlying, this dialectical

---

[2] The notions of argument defeat and rebuttal in the Toulmin model of argument are
cloely related to a notion of warrant that has proved notoriously difficult to interpret
(Hitchcock, 2005).

view of 'conflict of opinion', there is also a logical view, long known and appealed to in logic. In formal logic, we have various concepts that represent the notion of opposition. One is negation. In classical deductive logic, the negation of a proposition is its opposite in a strong sense of opposition. Such pairs of propositions are said to be contradictories. "This pen is black" and "This pen is not black" are contradictories. The one proposition is true if and only if the other is false. But there is also a weaker sense of opposition, that of contraries. Two propositions are contraries if and only if it is not possible for both to be true, even though it is possible for both to be false. "This pen is black" and "This pen is green" (all over) are contraries. Questions are raised by these traditional notions about the relationship between opposition and refutation, questions that have rarely been explored in any depth.

These questions have become especially acute with the rise of the notion of defeasibility in recent work in computing. A distinction is often drawn between two types of opposition or refutation called defeaters and undercutters (Pollock, 1995). At a first pass, this distinction can be roughly characterized as follows. A defeater refutes an argument it is aimed at, whereas an undercutter merely removes the support given to the conclusion by the premises. Recent work in the emerging field of computational dialectics has come to be based on the primitive notions of attack and defeat. But what do 'attack' and 'defeat' really mean in the sphere of arguments? Let's begin with the notion of attack.

According to a hypothesis that is plausible, there are only three ways of attacking (and refuting) an argument. The first way is to argue that the premises are not true, or at any rate have not been shown to be true. The second way is to argue that the conclusion does not follow from the premises. The third way is to argue that the conclusion is false, or at any rate, that there are reasons to think so. This triad of attacks, when said to be the only ways to attack an argument, could be called the three-ways hypothesis. The third way can be carried out by presenting a counterargument that is stronger than the original argument, and that has a conclusion opposite to that of the original argument. In the case of a deductively valid argument, however, all three options are not open in all cases. If the premises are true and the conclusion follows from them, the conclusion has to be true, too. Thus in the case of such an argument, finding a stronger counterargument with the opposite conclusion is not possible. The argument cannot be defeated. If its conclusion is true, the opposite conclusion (its negation) cannot be true. A deductively valid argument with premises is called a sound argument in logic. A sound argument is not only conclusive, but also incontrovertible, meaning that no argument can refute it. However, it may be that we can rarely find incontrovertible arguments, since an argument for a proposition is not useful or even appropriate unless there is some reason to doubt that the proposition is true. For one might well argue that the purpose of putting forward an argument is to respond to, and to try to remove, doubts that have been expressed.

Many common arguments are defeasible (fallible), meaning that even if the premises give reasons to support the conclusion as true, new evidence may later come in that shows that the conclusion is false. Such arguments hold only tentatively, and may have to be retracted later if new evidence shows that the original argument is no longer tenable. Thus defeasible arguments are not incontrovertible. In the case of this kind of argument, all three options are open. Even though the premises are true (or are tenable as commitments now), and the conclusion follows from the premises by an argumentation scheme (a form of argument that is not necessarily deductively valid), the conclusion could possibly be refuted by questioning. It could even be cast in doubt or refuted by a stronger opposed argument.

Is the three-ways hypothesis true? There are two grounds for doubting it. The first is the observation that an argument can be refuted by showing that it commits the fallacy of begging the question. But this form of refutation is a rather special one that applies only to arguments that are circular. It appears to have to do with a failure of the premises to prove the conclusion, even if the argument is valid and the premises are true. These grounds have to do with burden of proof. There is still another form of refutation that has to do with burden of proof that especially applies in the case of a defeasible argument, and that is even simpler. It may happen in a trial, for example, that the prosecution comes forward with an argument that is supposed to show that the defendant is guilty of the crime alleged. The defense might not challenge the argument on any of the four grounds of attack already cited. But still, they might attack the argument by claiming that it is too weak to fulfill burden of proof. The defense might admit that the argument has some probative weight, but claim that it does not have enough to prove what it is supposed to prove. This response to an argument is a kind of attack on it. It does not claim that the argument is altogether worthless as evidence. It claims only that it does present enough evidence, by itself, to successfully prove its conclusion. In such a case, the defense might not produce any counterargument with the opposite conclusion, or attack the inferential link between the premises and the conclusion. But while the defense might not attack any of the premises, they still may be attacking the argument.

Is the defense really attacking the argument, though, or are they conceding that it is a good argument, but one that is not good enough to meet some standard of goodness that is being claimed for it, or that it might be seen as meeting? This question is controversial. It depends on what you mean by 'attack'. Also, alleging begging the question, the fourth form of attack noted earlier, is a rather special form of refutation that applies only when an argument is circular, a relatively rare form of argument failure. Moreover, it might be possible to analyze begging the question as a way of attacking a premise in an argument, arguing that the failure is one of finding the right sort of premise. Of course, the failure lies not in the truth of the premise, but in how the premise is supported by evidence independent of the conclusion. Some other fallacies, like the straw man fallacy, might also count as attacks, but again, only in special cases. One might still argue that there are only three main ways of attacking an argument, barring such special cases. Thus it remains controversial whether the three-ways hypothesis is true or not, depending on what is meant by 'attack', and on whether the hypothesis can admit of some exceptions in special cases, and on how these exceptional types of cases, corresponding to informal fallacies, might be defined and analyzed.

What is the distinction between attacking an argument and refuting it? Presumably, an argument that attacks another is one that is opposed to it, but that does not necessarily defeat it. An argument that defeats another is one that shows that the other argument has to be given up. But when does an argument have to be given up – when its conclusion is shown to be false, or when doubt can be raised about whether the evidence given for it is sufficient? These questions are harder to answer.[3] They seem to depend on matters of burden of proof. In some cases, merely raising serious doubt about an argument is enough to warrant giving it up as insufficient for proving its conclusion. In other cases, an argument has to be given up only if a stronger counterargument can be given. In law, for example, raising doubt is enough to refute a criminal allegation made

---

[3] In the critical discussion model of rational argumentation (van Eemeren and Grootendorst, 1984, 1987, 1992), a participant (he) must concede defeat if the other party (she) can prove her thesis by means of rational argument, thus showing that she wins and that his viewpoint has not been successfully defended. This notion of having to give up and concede defeat when confronted with a superior argument at the closing stage of a discussion is fundamental to the notion of rational argumentation in a critical discussion.

in a trial, while in a civil case, a claim can be refuted only if the opposed side brings forward a stronger argument against the claim. It seems, then, that the distinction between attacking an argument and refuting it is not an easy one to draw clearly in all cases by any simple criterion. It seems to depend on the context of use of the argument – that is, what the argument is supposed to prove – and on how strong it has to be to succeed in carrying out its goal of proving what it is supposed to prove.

To try and find a basis for this distinction, maybe we can retreat to one that is more fundamental. This is the distinction between asserting a proposition and questioning it. The notion of denial also needs to be brought in here. To deny a proposition is to claim that the proposition is not true. Thus denial of a proposition is assertion of its negation. When you assert a proposition, you are making a positive claim that it is true. You are making a commitment to the proposition asserted, and thus you are also committed to backing it up by giving a reason why it should be taken as true if anyone questions it. Questioning, as opposed to asserting as well as to denying, is a different speech act. When you question a proposition, you are not necessarily making a claim that the proposition is false or true. Questioning does not necessarily imply commitment. When you make an assertion, you are staking out a positive claim. When you assert something, you are saying you are for it, but you are free to question a proposition without being for or against it. You can be agnostic, but still have doubts and raise questions.

This basic distinction between asserting and questioning has implications concerning attack and refutation. Let's take a common sort of example in a conversational exchange between two parties, White and Black. White asserts that Uranus is the seventh planet in orbit around the sun, because she has read that in the *National Enquirer*. Black then questions White's argument by asking the question, "Is the *National Enquirer* a reliable source?" Black is not refuting White's argument, only asking a question about it that raises doubt about it. Suppose, by contrast, that Black puts forward an opposed argument like, "It says in my astronomy textbook that Venus is the seventh planet in orbit around the sun." The conclusion of this argument, "Venus is the seventh planet in orbit around the sun," is opposed to the conclusion of White's argument, "Uranus is the seventh planet in orbit around the sun." To say that they are opposed means that both cannot be true. If one proposition is true, the other is false.[4] Since an astronomy textbook is generally a more reliable source

4 The reason is the implicit assumption that there can be only one planet that is the seventh in orbit around the sun.

than a newspaper, Black's argument is the stronger of the two. So we could say in this instance that it is a refutation of White's argument. Drawing such a distinction in this simple example rests on the distinction between merely questioning a claim and refuting it by offering some opposing argument stronger than the one for it.

This discussion suggests that there is an important epistemological distinction between refuting an argument and merely raising questions about its tenability. In the literature on argumentation, it is common to draw a distinction between two types of conflicts of opinion (Walton, 1998). In the stronger kind of conflict, which could be called a dispute, one party has a thesis (or ultimate *probandum,* as it is called in law), a proposition she holds to be true, while the other party has a thesis that is opposed to the thesis of the first party. There is a contrasting type of conflict that could be called a dissent, in which one party has a thesis to be proved, while the other doubts that this proposition is true, but does not contend that it is false. The problem here is to determine what is meant by 'opposed'. Does one proposition have to be the negation of the other, or could they be said to be opposed if they exhibit some weaker relationship to each other? For example, if White claims that tipping is good, while Black claims that tipping is bad, does their conflict represent a dispute? The answer would be yes if it can be assumed when you argue for the thesis "Tipping is bad," it means or implies that your thesis is opposed to the thesis that tipping is good. One proposition is not the negation of the other, but presumably they are opposed in that, given suitable assumptions about the meanings of 'good' and 'bad', the two propositions can be reduced to propositions that are negations of each other. But how does such a reduction take place? It seems to depend on semantic categories, and on implicit assumptions about one category being opposed to, or being the opposite of, another. In computing, an ontology, or mapping of key concepts in a domain of knowledge, would be the basis on which such semantic assumptions rest. But the importance of semantic assumptions about the meanings of key words in determining whether one proposition is the opposite of another has long been known by philosophers. As shown in the next section, Aristotle based his theory of opposition on making such assumptions explicit.

What is the relationship between opposition and refutation? Can one notion be defined in terms of the other? Is one notion logical and the other dialectical, and if so, can one be reduced to, or defined in terms of, the other? Are there stronger and weaker notions of refutation that can be distinguished? What is the difference between a refutation and a rebuttal? These questions are important, because opposition and refutation (or rebuttal) seem to be the fundamental notions on which logic and argumentation are based. These questions cry out for clarification, and for some attempt at answering them.

To examine some examples of refutation, let's consider the following dialogues. Each is an example of a refutation, or at least an attempted refutation.

Dialogue #1

White: Flax oil reduces cholesterol.

Black: How do you know?

White: Dr. Phil said so, and he is an expert.

Black: Yes, but he's not an expert in nutrition. He's a psychologist.

In this dialogue, Black does not deny that Dr. Phil said that flax oil reduces cholesterol. And he does not deny that Dr. Phil is an expert. Thus he does not deny either of White's premises. But he does deny that White's conclusion follows from these premises. Because Dr. Phil's expertise is not in the appropriate field for the claim made, White's argument from expert opinion fails to hold. By pointing out that Dr. Phil is not an expert in nutrition, Black has refuted White's argument from expert opinion.

Now consider another dialogue.

Dialogue #2

White: Flax oil reduces cholesterol.

Black: How do you know?

White: Dr. Phil said so, and he is an expert.

Black: Experts can be wrong.

White: Yes, but they tend to be right, because they have knowledge of a field.

In this dialogue, Black attempts to refute White's argument from expert opinion by saying that experts can be wrong. This statement is true, for argument from expert opinion is defeasible.[5] But White's reply to the refutation attempt backs up her original argument in the right way, giving

---

[5] We could express these claims using the notion of a generalization, as follows. First, as a generalization, all else being equal, we can presume that experts tend to be right. At least, we can take their opinions to have a special status as evidence bearing on a disputed issue, unless we have reason to think the contrary, in a case in point.

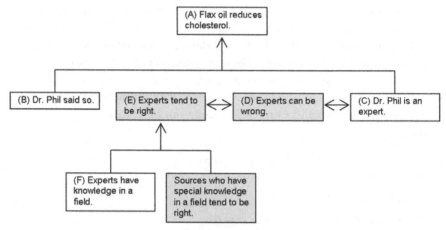

FIGURE 7.1. Araucaria diagram for Dr. Phil Dialogue # 2.

a reason to support it. Thus we could say that White has refuted Black's attempted refutation.

The argumentation in Dialogue #2 can be broken down into a set of propositions as shown on the following key list.

Key List for Dr. Phil Dialogue #2

(A) Flax oil reduces cholesterol.
(B) Dr. Phil said so.
(C) Dr. Phil is an expert.
(D) Experts can be wrong.
(E) Experts tend to be right.
(F) Experts have knowledge in a field.

Using this key list, the Araucaria diagram in Figure 7.1 can be constructed to represent the argumentation in Dialogue #2.

In this diagram, D is shown as a refutation of C. But then E is shown as a refutation of D. Hence, as noted earlier, the case is one in which there is a refutation of a refutation. F is shown as one premise in a linked argument supporting E. Along with F, a missing premise expressing a defeasible generalization has been inserted.

In a third dialogue, Black refutes, or at least challenges, White's argument in a different way.

Dialogue #3

White: Flax oil reduces cholesterol.

Black: How do you know?

White: Dr. Phil said so, and he is an expert.

Black: Where did Dr. Phil say that?

In Dialogue #3, Black questions the truth of White's premise that Dr. Phil said that flax oil reduces cholesterol. Black is not attacking the argument as a whole, by trying to argue that the conclusion does not follow from the premises. He is questioning the truth of the premise that made a claim about what Dr. Phil said. But if the premise does not hold, and cannot be proved to be true, White's whole argument falls down. Thus Black's reply in this sequence of dialogue can be classified as a refutation. Perhaps it could be classified as a weak rather than a strong refutation, as it merely asks a question. If White can answer the question by citing a passage in one of Dr. Phil's books where he made the statement in question, Black's objection could be overcome. However, even a weak refutation by questioning can defeat an argument. For unless White can cite the source of Dr. Phil's alleged statement, his argument about cholesterol is no longer convincing.

Finally, let's consider a sequence of argumentation that is an extension of Dialogue #1.

Dialogue #4

White: Flax oil reduces cholesterol.

Black: How do you know?

White: Dr. Phil said so, and he is an expert.

Black: Yes, but he's not an expert in nutrition. He's a psychologist.

White: OK, but he is still a scientist.

Black: Dr. Wendy said that flax oil does not reduce cholesterol, and she is an expert in nutrition. She is a medical doctor and has a Ph.D. in food science.

In Dialogue #1, Black refuted White's argument, or at least weakened its force, by pointing out that Dr. Phil is not an expert in the appropriate field for the claim made. In dialogue #4, White replies to this objection with some effect by arguing that Dr. Phil does have some scientific knowledge. But then Black comes back with an even stronger refutation. He cites an expert in the appropriate field for the claim, and this expert made a claim that is opposed to the one made by Dr. Phil. Black refutes White's argument by putting forward a rebutting counterargument of the very same type as the one originally used. This argument represents the kind of attack in which the second party presents a counterargument that attacks the conclusion argued for by the first party. The second party is not attacking a premise, or attacking the link between the premises and the conclusion of the target argument. She is refuting the argument by putting forward a stronger argument proving the negation of the conclusion of the target argument.

To get at least the beginnings of a theory of refutation, we will use these four dialogues as representing some common kinds of refutations familiar in everyday conversational argumentation. We will not be concerned with other kinds of refutations or procedural objections, like the objection that an argument is irrelevant. These kinds of attacks have been dealt with extensively elsewhere (Walton, 2004), and it would take us too far afield to try to deal with them here. As indicated by these four paradigm dialogues, our primary focus will be on cases in which one argument is used to attack, rebut, or refute another argument. We will not be centrally concerned with cases of refutation of a claim or hypothesis, or cases of attacking a question. On the other hand, all these notions turn out to be involved. For example, one way to attack an argument is to attack a statement that is one of its premises. In order to begin to understand the notion of refutation represented by these four dialogues, we will start by surveying the main theories that have been influential in the past, and then examine some recent theories.

2. OLDER THEORIES OF REFUTATION

It is well known that Aristotle's distinction between contraries and contradictories was an important foundational notion for his formal logic. Nearly every logic textbook that treats of syllogisms includes a representation of the square of opposition, setting this distinction in terms of relations between pairs of categorical propositions. What is less well known is that Aristotle also had set out a much broader notion of opposition in his *Rhetoric*, based on his notion of the enthymeme. An enthymeme is commonly taken in logic textbooks to be an argument with an implicit premise (or conclusion), but Burnyeat (1994) is a leading exponent of the view that Aristotle meant something else by 'enthymeme'. Aristotle characterized enthymemes as arguments based on premises that are true "only for the most part," contrasting them with the universal generalizations commonly stressed in deductive systems, like syllogistic. Such defeasible, or enthymematic generalizations are subject to defeat as exceptions to a rule may arise in the future. The warrants for these kinds of imperfect inferences have traditionally been called "topics" or *topoi* (places). An example (*Rhetoric* 2.19.24, 1393a6–7) analyzed by Burnyeat

(1994, p. 26) is the generalization "If the sky is clouded over, it is likely to rain." An argument with this generalization as premise, along with another premise, "Today the sky is clouded over," could be used to argue for the conclusion "Today it is likely to rain." This argument could be countered by an opposing one based on a different generalization: "For the most part, if the barometer is high, the day is not likely to turn out rainy" (Burnyeat, 1994, p. 28). The conclusions of two such opposed arguments "contradict each other," and thus each is opposed to the other and could defeat or refute it. Citing examples from Aristotle's writings, Burnyeat argues that what Aristotle referred to by the term 'enthymeme' was not an argument with a missing premise, but a defeasible argument of a common type based on a generalization that is true only "for the most part" and is subject to exceptions.

Aristotle, in the *Topica*, developed a theory of contraries based on the use of the linguistic elements. The sense of a word, on this theory, is determined in part by its referential or predicative use. For instance, 'sharp' can be predicated of a note or of a material substance. In the first case, its sense is close to 'high'; in the second one, it is close to 'pointed'. It is impossible to separate the contraries of a term from its use: 'sharp' in the first sense has as a contrary 'flat', in the latter one 'dull'.

It is interesting to observe that the contrary is determined, on this theory, by the sense of a word in its context, not by its literal meaning. There may be, for this reason, as many contraries of a term as there are possible different senses of it. Some senses, however, may have no contraries. For example, 'loving' when used of the mental state has 'hating' as a contrary, but when used to refer to the physical act, it has no contraries at all (*Topica*, I, XV, 106 b, 1–5). Contradictory opposites depend on use as well. For instance, 'not to see' when used in the sense of 'not to possess sight' has as a contradictory proposition 'to possess sight', but if its sense is 'not to exercise the faculty of sight,' the opposite will be 'exercise the faculty of sight' (*Topica*, I, XV, 106 b, 15–20).

The sense, and consequently the contrary, of a word depends on the genus that it comes under when it is used. The adjective 'sharp' can be classified under different genera. For instance, in the examples just mentioned, it can come under the genus of sound quality or material shape.

On this theory, semantic categories of terms used in natural language discourse affect argumentative oppositions. The first topic Aristotle described in his *Rhetoric* is the enthymeme from contraries, that is, argument from contrary qualities inherent in contrary subjects. In argument from contraries, the consequences of the opposed category

are evaluated against a given one. Consider the argument "Temperance
is good, because intemperance is injurious" (*Rhetoric*, II, XXIII, 1). The
scheme in this argument is based on a relation between the two genera grouping the two related contraries. They are mutually dependent,
because each opposition is determined by the other paradigm. The division between temperance and intemperance is made on the basis of
the consequences, and the differing consequences are determined by an
evaluation. The opposition is meaningful only in relation to the other category. In other words, if I oppose temperance to intemperance, I regard
the opposition under the quality of 'self-control of human beings', in its
turn constituted by two exclusive elements. Argument from contraries
is based on a quasi-logical link between the two correlated paradigms:
from the predication of one subject with one of the contrary qualities
proceeds the predication of its contrary with the other quality. The opposition that is the basis of argument from contraries is thus a semantic
classification.

Aristotle's theory of enthymemes is based on the concept of probability, or *eikos*, that is, things that are likely or, in modern terms, plausible. Oppositions to an argument are possible because "many things that
appear likely are opposed to each other" (*Rhetoric*, II, XXV, 2). The focus
here is on the endoxic premises: the propositions constituting the fundamental basis of the enthymemes are based only on general acceptability.
They are subject to defeat when replied to by objections. There are four
ways to object to such an argument and refute it: from the same subject,
from similarity, from opposition, and from points already decided. In
the first case, the contrary enthymeme proceeds from the same subject –
for instance, love. To the enthymematic premise "Love is a good feeling"
there may be two kinds of objections: a general one, like "Every want
is bad," and a particular one, like "The proverbial expression *Caunian
loves* [incestuous relationship between sister and brother] would not have
arisen had there not been some wicked loves" (*Rhetoric*, II, XXV, 4). In
the first case, the refutation stems from an opposite endoxon, while in
the second case it is an exception that is directed against the premise.

When the terms of the argument and its opposition are in a similar
relation – as, for example, 'benefactor' stands to 'benefited' as 'injurer' –
stands to 'injured', the refutation is from similarity. For instance, Aristotle
opposes to the generalization "Those who have been treated ill always
hate" the rebuttal "Those who have been treated well do not always love"
(*Rhetoric*, II, XXV, 6).

The scheme of argument from contraries is a useful way to find and justify oppositions between arguments based on endoxic generalizations. For example, to "The good man benefits all his friends" it is possible to object that "The bad man does not hurt all his" (*Rhetoric*, II, XXV, 5). Thus a refutation may be drawn from premises that are agreed upon and are not in dispute. The force of this counterargument proceeds from the fact that an accepted proposition is endoxic, meaning that it is considered uncontroversial by the majority, the authorities, or opinion leaders. On Aristotle's view of opposition, opposed arguments are always defeasible when directed against arguments based on endoxic premises. Enthymemes are drawn from probability or, in modern terms, plausibility: the conclusion follows from the premises not by a necessary inference, but only by a probable one.[6] Refutation, in the broad meaning of the term indicated earlier, attacks the plausible basis of arguments, starting from plausible sources in its turn. Objections, for this reason, tend to be plausible arguments themselves, and for that reason are defeasible and subject to criticism. The refuter typically opposes one plausible argument to another, and cannot defeat the plausibility of the first argument only by attacking its endoxic premises (*Rhetoric*, II, XXV, 8, 9). It is interesting to observe that Aristotle opposes the form of the reasoning to the premises, where with the label "form" he indicates the endoxic missing premises stemming from the topical consequence. For instance, if we consider the enthymeme "John hates Bob, because Bob treated him badly," the refutation against the form is focused on the missing premise "Those who have been treated badly always hate." The rebuttal is grounded on the instantiation of argument from consequences and focuses on the implicit endoxic generalization.

We now turn to two quite different ancient approaches to refutation. Both of them are based on legal argumentation and are therefore different in emphasis and approach from Aristotle's. Cicero's approach was influenced by Stoic propositional logic and can thus be contrasted with Aristotle's theory, which took syllogistic logic as its point of departure. But before examining Cicero's theory, it is worth briefly examining Quintilian's account of refutation. The Dr. Phil example we began with in section 1 is a defeasible and plausibilistic argument that is not deductively valid, and that depends on the trustworthiness or credibility of a

---

[6] 'Probability' is not used here in the modern statistical sense, but in an older sense referring to what seems to be true or is generally accepted as true.

source. Such arguments are, of course, quite common in law, and are especially interesting in relation to the study of rebuttal and refutation.

Quintilian's account of refutation is closely related to his conception of credibility. Enthymemes and *epicheiremes* are different from syllogisms with respect to the nature of the premises. While syllogistic premises are considered to be true or false, enthymematic premises derive from things that are perceived by the senses. These include things about which there is general agreement; things established by law, or passed into current usage; and things admitted by either party, or already proved, or not disputed by the adversary (*Institutio Oratoria*, V, 10, 12–14). Enthymemes are, for this reason, defeasible. The credibility of premises can be always questioned or refuted by more credible arguments. Credibility is a concept involving persons or agents as sources of testimonial evidence, a factor that needs to be considered in addition to the reasoning in the argument.

The argument from opposition studied by Quintilian, based on trust, reputation, and the sincerity of the arguers, is now the basis of argumentation in multi-agent computing. Information on the worldwide web is collected by software devices called agents. An agent must not only carry out actions and collect information, but also evaluate the reliability of information before acting on it. For example, an agent may need to judge whether another agent should be considered an expert source of information, on the basis of referrals. Referrals are based on information, but also on trust. Technology for automated reputation management (Yu and Singh, 2000, p. 1) needs to evaluate argumentation based on assumptions about the perceived reliability and reputation of an agent. According to Yu and Singh (2000, p. 4), an agent a assigns a reputation rating to an agent b based on a's direct observations of b, the ratings of b given by b's neighbors, and a's ratings of these neighbors. This rating is a measure of the trust a has in b as a source of reliable information. The trust rating can then be updated as an exchange of messages on the internet proceeds. If a encounters a "bad partner," it can decrease its rating and inform its neighbors (p. 6). For example, suppose a appeals to expert source b in order to support his argument used against c, but c shows that b has been untrustworthy in the past. The trust rating of b could then be downgraded, and the argument from expert opinion could be attacked or even refuted.

Whether such an argument is judged strong or weak depends on the credibility of an agent. Refutation strategies, in Quintilian's view, are similarly focused on the person. In Quintilian's account of refutations in trials, attacks on the person have an important role in situations where they are used to make the accused appear not credible and his argument implausible.

> ... counsel for the defence may deny or justify the facts, raise the question of competence, make excuses, plead for mercy, soften, extenuate or divert the charge, express contempt or derision. (*Institutio Oratoria*, V, 13, 2)

However, the strategies of denying the facts and of raising the question of the competency of the testifier are not the only forms of rebuttal available in the courts. Rebuttal can be performed in two ways: either absolutely, or relative to the nature of the allegation. For example, to the proposition "The property is mine, for I am the only son of the deceased" may be opposed the rebuttal "You are not his son" (*Institutio Oratoria*, V, 10, 12–14). On the other hand, an attack on witness testimony can be launched by discrediting the witness. Still another form of rebuttal of a comparable kind is the argument that the court is not competent to try the case. In many such cases, the notions of rebuttal and refutation are agent-based, and involve reputation management.

Quintilian connected the phenomenon of refutation to a kind of opposition based on the analysis of inconsistencies. For example: Clodia asserts, on the one hand, that she lent Caelius money, which is an indication of great intimacy, but, on the other hand, that he obtained poison to murder her, a sign of violent hatred (*Institutio Oratoria*, V, 13, 30). Another kind of reasoning is a strong form of argument from consequences, called argument from incompatibles. For example, it is possible to argue: "Can money be a good thing when it is possible to put it to a bad use?" (*Institutio Oratoria*, V, 14, 25). In this example, we can notice, the absolute quality of a subject "to be good" is opposed to the quality of its possible use. Therefore, the opposition is based on a missing premise, such as "The nature of a thing cannot be distinguished from its uses," allowing the two different predicates to be considered under the same category.

The Aristotelian account of argument from contraries is founded on its theory of predication and on a plausible kind of reasoning associated with topics and enthymemes, as indicated earlier. Cicero, in his book *Topica*, described the same kind of argumentation from a different point of view, determined by a Stoic, instead of Aristotelian, logical system. Argument from contraries is related to two main figures (*Topica*, XIII, 54).

When you deny that certain things are associated and assume the truth of one or more, so that the remaining statement must be excluded:

Either this or that is true; but this is true; therefore, that is not true.
Either this or that is true; but this is not true; therefore, that is true.

The second form of argument appears to be an instance of the form called disjunctive syllogism in modern deductive logic: either p or q; not-p; therefore q. Cicero offered the following examples for this type of argument (*Topica*, XIII, 55):

Fear this, and not dread the other!

You condemn the woman whom you accuse of nothing, and do you assert that the one deserves punishment whom you believe to deserve reward?

The Ciceronean argument from contraries has one pivotal difference from the Aristotelian one: the opposition is not between the elements belonging to two reciprocally defined categories, but between two contrary properties attributed to the same subject. The logical form of the argument is different from that in Aristotle's account. His version is not based simply on the impossibility of the same subject being predicated of contrary qualities, but, in addition, on the plausible rule that the same quality cannot be attributed to contrary subjects. Moreover, Cicero describes the contradictory elements as a linguistic reality, not as a linguistic practice. These two elements of his theory aim to develop the form of a necessary, not just a plausible, inference from the premises to the conclusion. Cicero's focus is more on the form of the opposition than on its underlying rationale. He studied the forms from which a necessary conclusion is derived. In the case of contraries, the necessity of the deduction stems from the property of contradiction: "whatever is contradictory has such a nature that can never be connected with it" (*Topica*, XII, 53). Cicero distinguishes among different kinds of contraries. They may be opposites, when the things belong to the same genus but differ absolutely. For instance, slowness and speed are adverse and are located under the same genus, motion. Privatives are terms that have the prefix 'in' that removes from them a quality, as inhumanity is the privation of humanity. In Cicero's account, however, privative terms such as 'blind', without the prefix 'in', are considered simply adverse contraries. For instance, 'blind' is considered a contrary of 'sight'. Oppositions can be originated by comparison with something – for example, double and half, father and son. And there are negative inferences based on contrary expressions, such as "If this is so, that is not" or "If it is day, it is not night."

It is interesting to observe that Cicero's treatment of argument from contraries is based on the logical form of disjunctive syllogism, not described in Aristotle. The latter analyzed argument from contraries in relation to the predication in a proposition, and not in relation to the truth values of simple or disjunctive propositions. Moreover, the disjunctive syllogism stems from a particular meaning of the disjunction operator ∨. For Cicero, it indicates that the component propositions cannot both be false: in other words, it has the same truth-value structure as the connector *vel*, representing the inclusive 'or'. The disjunction "It is either black or white" can be true in Cicero's logic, even if both component propositions are false – for example, if the subject is green. The argument form of the disjunctive syllogism ($p \lor q$; $\neg p$, therefore q), consequently, is not valid, because it is possible for the proposition q to be false and the disjunction to be true (Stump, 1988, p. 235). The distinction between contradictory propositions and contrary or opposite ones can be brought out by the study of Boethius on the nature of contraries. His description of the logical operator ∨ as an *aut... aut...* connector is at the basis of a different theory.

The first important distinction Boethius draws is that between exclusive contrariety and contrariety admitting an intermediate. In the first case, contraries cannot exist at the same time, that is, the existence of one of them excludes the existence of the other, and vice versa. On the contrary, if an intermediate is admitted, this mutual exclusion is not valid anymore. If we analyze the derived propositions, the truth-value conditions are different: while in the first case the disjunct propositions cannot both be false, in the second case they can. Similar to the exclusive contraries are the privative and relative opposites. Terms like 'blindness' and 'sight' are mutually exclusive, as contraries, and therefore have the same values in case of disjunction. On the other hand, relatives like 'father' and 'son' cannot be separated, and from this it follows that the affirmation of one relative excludes the negation of the other. They must be distinguished from the contraries, because they are not defined in relation to a paradigm, that is, to a genus. The privative opposites are defined in relation to the quality they are deprived of, while the relative ones are reciprocally determined. The last class of incompatibilities Boethius examines is the contradictories, that is, the relation between the affirmation and negation of the same proposition. The relation is not between terms, but between sentences (Stump, 1988, p. 119).

TABLE 7.1. *Peter of Spain's four types of opposition*

Topic	Maxim	Example
Relative opposition	Given one correlative, the other is given as well; when one correlative is rejected, the other is rejected as well.	There is a father, therefore there is a son. There is no father; therefore, there is no son.
Contraries	If one immediate contrary is excluded, we must posit the other if its subject persists.	This animal is not well; therefore, it is ill.
Privative opposition	Given one of opposed privations, the other is removed from the same thing.	He is sighted; therefore, he is not blind.
Contradiction	If one contradictory opposite is true, then the other is false, and the converse.	That Socrates is sitting is true; therefore, that he is not sitting is false.

Peter of Spain distinguished four kinds of oppositions: relative opposition, contrariety, privative opposition, and contradiction (Peter of Spain, 1990, pp. 62–64).

One can see how Peter's way of classifying different kinds of opposition fits with the Aristotelian kind of framework that typified the medieval approach.

In light of the recent concern with defeasibility, rebuttal, and undercutting in AI, as well as with reputation management in multi-agent computing, it is remarkable that so little is widely known about older theories of contraries and opposition. This neglect is especially remarkable in light of the fact that Aristotle, Cicero, Boethius, and Peter of Spain regarded these notions as fundamental to the study of rational argumentation of the most common kinds.

3. NEWER THEORIES OF REFUTATION

Modern approaches to opposition carry over the central Aristotelian basis of the notion, but also follow Cicero in that they see opposition as partially defined by the type of argument involved. Petrus Ramus's system of classification of the topics from opposition, for example, is different from the medieval one. He defines the opposition in terms of what he calls "gaynesettes," organizing them in subcategories according to the nature of the opposition and its logical form. Gaynesettes are arguments that always disagree: it is impossible, in other words, to attribute them to the same part of the subject, in the same respect, at the same time. For instance, Socrates cannot be white and black on one part, or father and son of one man; however, he can be white on one part and black on another, or son of one man and father of another.

Ramus categorized the oppositions under two different criteria: the paradigm and the logical form (Ramus, 1969, pp. 21–24). On his theory, the opposition between black and white is different from the one between father and son. Moreover, pairs like beast-man do not have the same logical form as beast–not beast and, similarly, as sober-drunk. These arguments are called *disparates*. Contraries can be positive or negative. Positive contraries are divided into relative and repugning. A relative contrary is defined by its mutual relation with the other – for example, the word 'father' can exist only relative to 'son'. Repugning arguments, on the other hand, are contraries that are opposed between contrasting pairs, like men and beasts, virtue and vice (p. 23). Negative contraries derive from the affirmation and negation at the same time of the same term, like beast–not beast, or from the negation in one term of a semantic property present in the other, like rest and movement, blindness and sight.

The division of opposites is not considered in the logic of Port Royal, which instead adopts the medieval classification of them. The general categorization of the *topoi* is, however, different from the medieval tradition. Arguments from opposition, like topics from cause to effect and whole to part, are considered metaphysical topics that reflect the order of the world. For this reason, their account of the oppositions, even if following a traditional classification, is deeply different from the Aristotelian one. Oppositions are not a property of the words, of the linguistic elements, but a characteristic of the reality itself. For this reason, topics were reduced to the function of classifying consequences, instead of being their fundamentals.

Perelman's theory of refutation is, like the ancient views, based on the concepts of incompatibility and contradiction. Contradictions or incompatibilities may be shown in the opponent's thesis itself, or between his argument and a rebuttal. While contradictions are propositions differing only in their truth values, incompatibility holds in relation to circumstances, or point of view. Perelman uses the expression "to create incompatibility" to indicate the process of pointing out "circumstances which make unavoidable a choice between the two propositions involved"

(Perelman and Olbrechts-Tyteca, 1969, p. 201). The reason for incompatibility lies in the unity of the system the elements are shown to belong to. For example, two statements about the same person at different times may be presented as incompatible if all his statements are regarded as forming a single system. Incompatibilities depend on the perspective from which the speaker looks at reality, on his choice of the categories used to classify two elements. Thus this theory is very close to the Aristotelian logic of classes. The main strategies used to create incompatibilities are based on the notion of relevance: the objection must be relevant to the matter in order for the speaker to be able to classify the two propositions under a unique system.

In *The New Rhetoric* (Perelman and Olbrechts-Tyteca, 1969, p. 462), the authors examined strategies of refutation in relation to the particular modes of argumentation. Argumentation may be based on connections between elements: independent elements are to be shown interdependent through connecting links, that is, argumentation schemes. A refutation is based on dissociating the associated elements. If the argument proceeds from example, a contrary example may be found. If the argument proceeds from a rule, an exception to the rule may be opposed. If the conclusion is supported by an analogy, a refutation may consist in extending the analogy until a contradiction is shown, as in this example (Perelman and Olbrechts-Tyteca, 1969, p. 387):

> Berriat Saint Prix, arguing against a jurist who, scorning any reference to Roman law and early jurisprudence, claimed in a work on the Civil Code, to describe "the veins, muscles, features and soul of the law" expressed regret that the writer did not "follow his metaphor to the end" for "he would have soon perceived that every living being receives its organization from an earlier being from which it was begotten."

If the argument is from dissociation, the dissociated pair may be rejected on the basis of another pair. For example, a dissociation may be objected to as illusory, by relying on the couple verbal/real (Perelman and Olbrechts-Tyteca, 1969, p. 427). The theory of refutation in *The New Rhetoric* is an Aristotelian one, but consciously departs from any previous rhetorical tradition based on the rather complex notion of dissociation.

In *The New Rhetoric*, many argumentation schemes are identified. When we turn to recent work on refutation in epistemology, the approach is quite different. It is typically based on the notion of an agent as a repository of beliefs, changing its beliefs when an argument is refuted. Such a change of beliefs is based on logical reasoning, including a few defeasible forms of argument that have been acknowledged. Pollock (1974) distinguishes between two kinds of refutations: defeaters and undercutters. His theory is based on two underlying concepts: prima facie reasons and justification. Prima facie reasons are logical reasons that are defeasible, meaning that they may cease to be good reasons when associated with some additional belief (Pollock, 1974, p. 40). Justification, on the other hand, is a nonmonotonic consequence, opposite to overruling and defeasibility, definable in terms of victory or loss in a dispute. With justified arguments, a dispute can be won; overruled ones lead to loss; and defeasible ones leave the dispute undecided (Prakken, 2003, p. 2). The notion of justification is related to the concept of belief: therefore, the reasons supporting a conclusion are simply reasons to believe a proposition. The justification link is constituted by both contingent (I consider my informant reliable, etc.) and inductive (e.g., past reliability of the informant) reasons.

Defeaters are precisely the propositions that, added to prima facie reasons, defeat the justification. On the other hand, prima facie reasons are logical reasons for which there exist defeaters (Pollock, 1974, p. 42). In an example employed by Pollock (p. 42), given the prima facie reason "x looks red to me; therefore, x is red," the proposition "Jones told me that x is not red, and Jones is generally reliable" is one of its possible defeaters. We can think of defeaters (or Defeaters I, in Pollock's terminology) as attacking the conclusion of the argument directly. In the example just given, moreover, we can observe that a whole line of argument is developed to support a conclusion contrary to the one proceeding from the prima facie reason. Another kind of attack can be moved against a process of justification. Instead of focusing on the conclusion, it is possible to undermine its connection with its premises. The inferential link is the target of the Type II defeater, or undercutter.

For example (Pollock, 1974, p. 43), given the prima facie reason "x looks red to person S; therefore, x is red," a possible undercutter might be "There are red lights shining on x.' Knowing that x looks red to me is not a conclusive basis for determining whether x is red, because red lights can make a white object look red. The undercutter is a reason to believe that the premises do not justify the conclusion, but it does not follow that the conclusion is wrong. In other words, the undercutter is a reason *not to believe* the conclusion, it is not a reason to believe that the conclusion is *not believable*. It may be true, or believable, despite the undercutter. It is the justification link between P and Q (premise and conclusion) that is defeated, not Q (the conclusion). After the attack, it is not reasonable anymore to believe in the truth of Q on the basis of P. In addition, a distinction is drawn between personal and impersonal defeaters. For example, while the undercutter in the example just given is an impersonal one, the proposition "Person S is hypnotized" is a personal defeater of the justification "x looks red to person S; therefore, x is red." It is a distinction between the kind of circumstances attacked, and they may be general or related to a subject S: that is, between "x *looks red* to person S" and "x looks red *to person S*."

Defeaters are parasitic on undercutters. It is possible to defeat the conclusion "x is red" only if the perceptual reason justifying it is assumed to be undermined. For this reason, the process of defeating the conclusion presupposes reasons to believe that its justification can be ignored. They derive from the inductive generalization proceeding from the undercutter. For example, in the argument "x looks red to me; therefore, x is red," the defeater "Jones told me that x is not red, and Jones is generally reliable" can defeat the conclusion only because of the fact that Jones's reliability is superior to my own perception. This follows from presupposed undercutters. I can believe that Jones is more reliable than my perception because I have reasons to believe that my perception cannot justify the conclusion. From perception and induction I can understand undermining reasons (i.e., I am wearing red glasses) and from that infer that the opposite conclusion is justified.

The problem is posed of figuring out how the undercutters work. The undercutter is supposed to attack the link between the premises and the conclusion of a given argument. But how does such an attack work? In the red light example, the undercutter "there are red lights shining on x" is information that, if known to be true in a given case, is an exception to the generalization on which the inference is based. This generalization can be formulated as a conditional: "If x looks red to person S, then generally (but subject to exceptions) x is red." The conditional can be seen as a premise in a defeasible *modus ponens* argument.

> If x looks red to person S, then generally (but subject to exceptions) x is red.
> x looks red to person S in this case.
> Therefore, x is red in this case.

Suppose that, in this case, new information comes in that there is a red light shining on x. Since this proposition serves as an exception to the defeasible generalization in the first premise, the argument defaults.

But is the undercutter attacking the link between the premises and the conclusion, or is it merely defeating the first premise, as applied to the given case? Presumably it is defeating the applicability of the premise to the case, by citing a fact in the case that is an exception to the general rule stated in the first premise.

Thus it is seemingly hard to grasp Pollock's notion that the undercutter attacks the inferential link. According to the three-ways hypothesis stated earlier, there are three ways to attack an argument. You can attack one or more of the premises; you can attack the conclusion, by posing a counterargument; or you can attack the inferential link between the premises and the conclusion. But how is this third way carried out? Is it really any different from attacking a premise? To try to answer these questions, we return to considering the case of argument from expert opinion, a defeasible argumentation scheme.

4. ARGUMENTATION SCHEMES AND CRITICAL QUESTIONS

As we have already shown, refutation can take three forms, or even four. In many cases, one argument is rebutted by another argument paired to it. The second argument attacks the first one, but the attack itself can be in the form of a different kind of argument, and it can attack the first argument in various ways. As Aristotle showed, such attacks are based on topics, or common types of arguments, nowadays called argumentation schemes. The refutation we began with in Dialogue #1 was based on an appeal to expert opinion, sometimes called an argument from expert opinion. We can represent the structure of such an argument in the following way, showing it is based on a defeasible generalization. In the format of Toulmin (1958), this generalization represents the warrant on which the inference is based.

The premise takes the form of a defeasible generalization, "If expert E asserts A, then A is plausibly true." A second premise takes the form of a pronouncement of an expert, "Expert E asserts that A is known to be true." The conclusion says, "Plausibly A." An Example would be the argument, "Dr. Bob told me that the leg is broken; therefore, the leg is broken." The implicit assumption is that E (in this instance, Dr. Bob) is an expert.

How would an undercutter be used to attack this argument? As shown in the last section, it could present data known in a special case that make the generalization premise default. But this is merely to attack a premise. How would the undercutter attack the argumentation scheme for argument from expert opinion, that is, the inferential link between premises and conclusion?

To address this question we need to review the argumentation scheme for argument from expert opinion, which has the following inferential structure (Walton, 1997, p. 210).

Argumentation Scheme for Argument from Expert Opinion

E is an expert in domain *D*.
E asserts that *A* is known to be true.
A is within *D*.
Therefore, *A* may plausibly be taken to be true.

How could an undercutter be used to attack an argument that fits this argumentation scheme by somehow attacking the inferential link between its premises and conclusion? To answer this question, we return to the four example dialogues of attacks on argument from expert opinion.

First, let's return to Dialogue #1, and diagram it using Araucaria. We begin by listing the propositions (statements) that appear as premises or conclusions in the argumentation.

(A) Flax oil reduces cholesterol.
(B) Dr. Phil said so.
(C) Dr. Phil is an expert.
(D) Dr. Phil is not an expert an nutrition.
(E) Dr. Phil is a psychologist.

In Figure 7.2, we see that both B and C are premises required to support A, based on the argumentation scheme for argument from expert opinion. Statement D appears in a leftmost box, and is joined to A by a double arrow, indicating that D is a refutation of A. Since E is a premise supporting D, the whole structure shown in the diagram represents the argument from E to D as a refutation or rebuttal of the argument that appears on the right (from B and C to A). The so-called owners of each of the arguments are also represented on the diagram. White is the party arguing for A, while Black argues for D in order to refute A. Here we have a very clear case of a refutation that can be visualized as argumentation displayed on an *Araucaria* diagram. We began with the set of statements shown in the key list. Then we used the diagram to show how one argument is used to rebut, or to try to refute, the other.

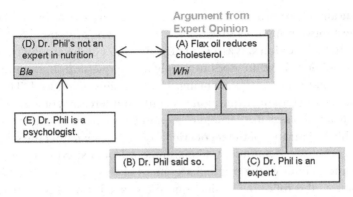

FIGURE 7.2. Key list and diagram for Dialogue #1.

But there is something not quite right with this way of diagramming the argument. D is supposedly an undercutter, meaning that D attacks the inferential link. But the way it is pictured on the diagram, D is refuting A, meaning that D is supposed to be an opposite or negation of A. But D is not then an undercutter in Pollock's sense, but a defeater. Also, it is unclear how D is an attack on A at all. What D really attacks is the assumption in the argument that E is an expert in the field into which the argument falls, namely, nutrition. But does this mean that D attacks an implicit premise, or that D attacks by raising an appropriate critical question?

When we turn to the refutations in the next two dialogues, the straightforward kind of analysis of undercutting one would attempt using *Araucaria* is also problematic. In Dialogue #2, Black, when he replied "Experts can be wrong," appeared to be attacking the argumentation scheme itself, using an undercutter. It doesn't appear that he was attacking any specific premise of the argument from expert opinion that had just been put forward by White. It appears that he was attacking the inferential link between these premises and the conclusion that White drew from them. But now look back to the *Araucaria* diagram for Dr. Phil Dialogue #2. The generalization "Experts can be wrong" is shown as a refutation of the premise "Dr. Phil is an expert." The generalization "Experts tend to be right" is shown as a refutation of the generalization "Experts can be wrong." These refutations are thus not shown as undercutters in Pollock's sense. As indicated, the case is one of a refutation of a premise of the argument from expert opinion.

In Dialogue #3, Black asks where Dr. Phil said that flax oil reduces cholesterol. In this case, Black is only asking a question. He is not putting forward a counterargument that attacks the original argument. Nor is he making any assertion at all. There is no statement that can be inserted into a key list for drawing up a diagram for the attacking argument. Still, the asking of the question acts as a kind of undercutter of the argument from expert opinion. If White does not reply to the question, her argument defaults, and is not very convincing. Its plausibility is reduced. Thus in this case, merely asking a question is a kind of undercutter, or acts as one.

To deal with cases like Dialogues #2 and #3, we need to better understand how asking one of the set of critical questions matching the scheme for a given type of argument, like argument from expert opinion, acts as an undercutter of the original argument of this type. To review from Chapter 1, the following critical questions match the argument from expert opinion (Walton, 1997, p. 25).

1. *Expertise Question*: How credible (knowledgeable) is E as an expert source?
2. *Field Question*: Is E an expert in the field that A is in?
3. *Opinion Question*: What did E assert that implies A?
4. *Trustworthiness Question*: Is E personally reliable as a source – for example, is E biased?
5. *Consistency Question*: Is A consistent with what other experts assert?
6. *Backup Evidence Question*: Is E's assertion based on evidence?

Critical question one presupposes that E is a knowledgeable source, and thus can be seen as a presupposition of the premise "Expert E asserts that A is true." The predicate 'being an expert' means having knowledge in a particular field. On the other hand, the reliability and consistency factors are not presuppositions of any premise. Even if an expert is biased, or if what she says is not consistent with what the other experts say, it does not follow that she is not an expert or that she has not claimed A. Thus there appear to be differences in how the critical questions need to be treated. Some can be seen as attacking a premise of the argument from expert opinion, while others do not appear to have such a function.

There is a difference between developing an opposed line of argument and destroying the proponent's reasoning without reaching any contrary conclusion. This difference was the basis of Pollock's distinction between undercutters and defeaters. In Dr. Phil Dialogue #1, the reply defeats the original argument, whereas in Dialogues #2 and #3, the reply appears merely to undercut the original argument. In Dialogue #3, Black tries to rebut White's original argument by undercutting the inference on which White's argument is based. But then White restores the argument by rebutting Black's rebuttal. Here White undercuts, or perhaps even defeats, Black's attack. So how useful is the distinction between defeaters and undercutters in dealing with such cases? It seems to be helpful up to a point, but its results are different from those of an approach based on argumentation schemes and critical questions. In the latter approach, the focus is not just on the plausibility of the conclusion, but also on the type of argumentative move. This approach concerns dialogues, and not adjustments of the belief values of a single rational agent.

Finally, we come to Dialogue #4. This case is an interesting one, because it involves a counterargument used to refute an original argument, and the counterargument is stronger than the original argument. Thus it is a case of a successful refutation.

To represent refutation in Figure 7.3, an implicit premise, "Flax oil does not reduce cholesterol," has been inserted. It is the opposite of the conclusion of the original argument from expert opinion, and the double arrow between it and A represents refutation. Thus Black's argument rebuts White's because it has the opposite conclusion. So far, so good. But how can we represent the notion that when Black says that Dr. Phil is not an expert in nutrition, he is refuting White's original argument from expert opinion? This can't be represented on the diagram because, like the case of Dialogue #1, the counterargument is an undercutter that attacks the inference from the premises to the conclusion. This kind of attack can't be represented on *Araucaria*. So what has been done here is to present D as a refutation of C, and then to represent E as support for D.

An interesting aspect of Dialogue #4 is that it represents a refutation (a successful one, that is) as opposed to a rebuttal or counterattack, because the counterargument is stronger than the original one it attacks. The argument based on Dr. Phil's expertise, although not worthless, turned out to be weaker than the one based on Dr. Wendy's expertise. To represent this aspect of the dialogue, argument evaluations were added, as indicated in the diagram. This aspect will turn out to be important, because the distinction between a rebuttal and a refutation turns on an evaluation of the strength of the counterargument in relation to the strength of the original argument.

5. DIFFERENT KINDS OF OPPOSITION

Problems for the pragmatic theory start to arise as soon as we encounter refutations of the kind commonly found in everyday examples of argumentation. Using an example from Cicero (*De Inventione*, I, 63), suppose that a man is accused by prosecutors of having committed a murder known to have taken place in Rome on a certain day. Let's say he puts

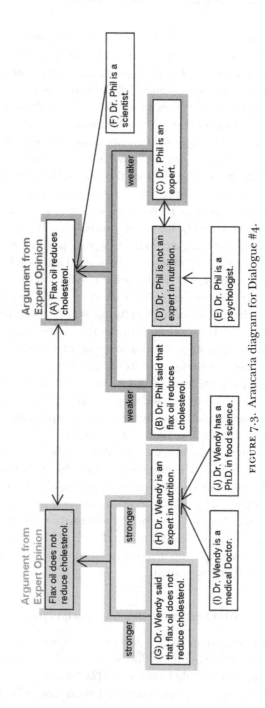

FIGURE 7.3. Araucaria diagram for Dialogue #4.

forward an alibi: "I was in Athens on that day." If his claim to being in Athens on that day can be substantiated, he has refuted the argument in the charge made against him by the prosecutors. It is a refutation because the man's being in Athens on that day is opposed to his being in Rome on that day, given the travel conditions in ancient times. But the meaning of opposition in such a case cannot be reduced to the kind of propositional negations that we know in logic, either classical negation or negation as failure. At least it cannot be so reduced without making some additional assumptions in the form of missing premises based on *endoxa.* It was perhaps for this reason that the older theories of refutation outlined earlier were based on a much broader and richer notion of opposition than those we are currently familiar with.

Oppositions were defined with respect to a paradigm established by the use of a term, and based on a specification of the sense of the term. For instance, the adjective 'white' may be used to denote a color absorbing a determinate wavelength of light, or a yellow-appearing wine, or grey-appearing meat, or pinkish skin. The meaning depends on the paradigmatic alternatives; it can be conceived as a semantic area defined by the other elements and by the reality it refers to. White is opposite to red in regard to meat, or to wine, but not to skin. Rome was incompatible with Athens under the aspect 'physical presence on the same day', but not under the point of view of 'beautiful city', under the conditions of the ancient world. These kinds of oppositions are created by the uses of terms, and they are not linguistic necessities. Moreover, they presuppose supporting reasons of their incompatibility. The argument "I was in Athens the day the murder was committed; therefore, I wasn't in Rome" takes its support from the missing premise "It is impossible to be both in Athens and in Rome on the same day." The nature of the opposition between being in Athens and in Rome on the same day derives from the impossibility, in the first century A.D., of reaching Athens from Rome in one day of travelling. The defeasibility of the opposition depends on the contextual assumption that what was impossible two millennia ago is normal nowadays.

The first topic Aristotle described in his *Rhetoric* was the enthymeme from contraries, that is, from contrary qualities inherent in contrary subjects. The example presented earlier was the argument "Temperance is good, because intemperance is injurious" (*Rhetoric,* II, XXIII, 1). This argument works somewhat like *reductio ad absurdum,* but is not a deductive type of argument. It is based on the argumentation scheme for argument from consequences. It argues for the proposition "Temperance is a good policy" by refuting the opposed proposition "Intemperance is a good policy." The refutation works because the negative consequences of intemperance are cited as a reason why it is not a good policy. This example raises the question of whether argument from opposition is itself an argumentation scheme. In this scheme, an argument is supported by arguing that the opposed argument has reasons that go against it. Such a counterargument or attack can be a refutation when the counterargument has stronger reasons behind it than the original argument. The kind of *reductio ad absurdum* we are familiar with in modern logic can be classified as a species of argument from opposition applicable to deductive arguments. What this ancient example shows is that argument from opposition also applies to defeasible arguments based on argumentation schemes that are not deductive in nature.

These considerations lead to some others on whether certain instances of arguments from opposition have already been recognized in classical deductive logic under the form of the disjunctive syllogism. This form of argument was known to the Stoics, as in the example, "The fox went over the hill or into the hole; he did not go over the hill; therefore, he went into the hole." Cicero's theory of refutation was based on Stoic propositional logic. His argumentation scheme for argument from contraries follows the logical form of disjunctive syllogism: $A \lor B, \neg A$; therefore, B. But in modern propositional calculus, this form of argument can be valid even if the two propositions A and B are not really opposed.

Either Socrates is running or Plato is green.
Socrates is not running.
Plato is green.

This argument is a valid disjunctive syllogism. The terms 'either... or...' are simply a translation in natural language of the logical operator $\lor$ (in its value of *aut... aut...*, the exclusive 'or'). There are differences between disjunctive syllogism in modern proposition logic and the argumentation scheme based on opposition. In natural language examples of argumentation from opposition, the argument depends on an incompatibility not just between truth values, but between semantic properties. In a real argument from opposition, the disjunctive terms must be opposed to each other and must be semantically related.

In the medieval tradition, it is interesting to notice that the consequences from oppositions were divided into two categories of loci: the topics from opposites (relatively and privatively), like arguments from contraries and contradictories, were listed under extrinsic topics, while the locus from division was thought to be separate. The difference lies in the nature of the predication. In the first class, the referred fragment of reality is categorized by the property, and the meaning of the referring term is not relevant in determining the opposition. In the second case, the definition of the subject term is taken into consideration. The categorization, in other words, is partially derived from the meaning of the term.

In choosing the predicate 'black', we exclude that 'white', 'yellow', 'red'...apply to the subject in question. Suppose White argues that Cleitomachus was white because he was a Greek philosopher who was a student of Carneades, and it is often said that the Greek philosophers were white males. Suppose Black then argues that Cleitomachus was black because he came from Africa. Neither argument is conclusive, but Black's argument can rightly be classified as a refutation, or at least as an attempted refutation, of White's argument. Yet the two conclusions, "Cleitomachus was white" and "Cleitomachus was black," are not opposites in the sense that one is the classical negation or the contradictory of the other. They are opposed in that both can't be true. They are contraries of each other. But what is contrariety in this sense? It means that it is not logically possible for both statements to be true. But how can this be proved? In order to prove it, a missing linguistic premise has to be added – a statement that if something is white (all over), then the same thing cannot be black (all over, at the same time). Moreover, some explanation may have to be added to the effect that 'white' and 'black', in this sense, refer to skin color, according to conventionally accepted ways of making such distinctions in common speech. Lexical oppositions of this sort, therefore, must be distinguished from contradictories like "Cleitomachus is white" and "Cleitomachus is not white."

6. UNDERCUTTING AND SCHEMES

In order to help us better understand the relationship between opposition and refutation, it is useful to consider the Dr. Phil dialogues once again. In Dialogue #1, the original argument from expert opinion was refuted by denying that the source is a specialist in the field in question. In Dialogue #2, there was a refutation of a refutation. In Dialogue #3, it was the opinion of the expert that was questioned. In these three cases, the refutation was directed against a premise of the argument. In Dialogue #1, the target of the attack was the assumption that *A* is within *D*. In Dialogue # 2, the attack was against a generalization that could be seen as an implicit premise on which the argument rested. In Dialogue #3, the

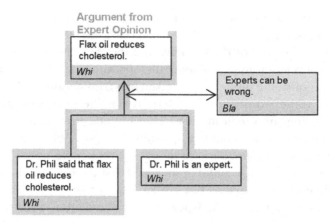

FIGURE 7.4. Araucaria-style representation of an undercutter.

attack was against the explicit premise that *E* asserts that *A* is known to be true. In these cases, the attack can be seen as directed against a premise. All three cases of undercutters can be seen as attacks on a premise of the original argument. The arguer can attack the undercutter by arguing that there is additional evidence that his premise is true, or at least plausible. If he cannot defend the premise in this way, his argument collapses. By way of contrast, in Dialogue #4, the respondent refutes the original argument by mounting another argument, saying " Dr. Wendy said that flax oil does not reduce cholesterol, and she is an expert in nutrition. She is a medical doctor and has a Ph.D. in food science." In this case, Black has presented stronger evidence for a contradictory conclusion. White's argument, however, is not defeated as no longer having argumentative force. He could go on to reinforce it with additional evidence or with another argument, increasing its acceptability and overcoming the rebuttal. He could reply to the refutation by finding an even stronger argument for his conclusion.

What then is an undercutter, or attack directed against the inferential link between premises and conclusion? The problem is to represent on a diagram the sort of attack that is directed against the inferential link, as might be found, for example, in Dr. Phil Dialogue #2. To visualize it on an Araucaria diagram, something like Figure 7.4 needs to be drawn.

In this diagram, the assertion "Experts can be wrong" is not shown as a refutation of another proposition (node) in the diagram. It is shown as an undercutter that attacks the inferential link between the two premises and the conclusion of the argument. The problem is that this type of diagram cannot be drawn in Araucaria. It represents a kind of argument structure called entanglement (Verhiej, 2003b) in which an arrow is drawn from a node representing an attack on an argument to the arrow linking the inference between the premises and conclusion of the argument. This notation represents Pollock's notion of an undercutter on an argument diagram.

But is the entanglement really needed on argument diagrams, or can undercutters be represented as attacks on premises, sometimes on defeasible generalizations or on implicit premises? For example, what about the Araucaria-style representation of an undercutter in the diagram above? The way to represent it may be to add an additional premise to the original argument from expert opinion. It is both an implicit premise and a generalization.

> *Conditional Premise*: If source E is an expert in a subject domain S containing proposition A, and E asserts that proposition A is true (false), then A may plausibly be taken to be true (false).

If this premise is seen as a material conditional of the kind familiar in deductive logic, it would make the argument from expert opinion purport to be deductively valid. It would have the deductive *modus ponens* form. However, White's rebuttal suggests that the missing premise should be represented as a conditional premise that expresses a defeasible generalization, making the argument as a whole defeasible. Thus seen, it has the form of defeasible *modus ponens*. Using the device of enthymemes, there is perhaps a way of representing the kind of refutation exemplified by Dialogue #2 as an attack on a premise. The undercutter could be diagrammed as a statement expressing special conditions that refute the conditional premise, when that generalization is taken as an implicit premise of the argument from expert opinion. By this means the notion of an undercutter could be represented in the following argument diagram.

In the pair of diagrams in Figure 7.5, in the diagram on the left, Black's attack on White's argument is represented as an attack on the implicit conditional on which White's argument rests. With the claim that "Experts can be wrong," Black attacks White's plausible implicit proposition.

In the diagram on the right, White replies to Black's attack by using the premise "Experts tend to be right" to support the premise that Black attacked. But can this way of diagramming the undercutter be used to represent Pollack's view of defeasibility?

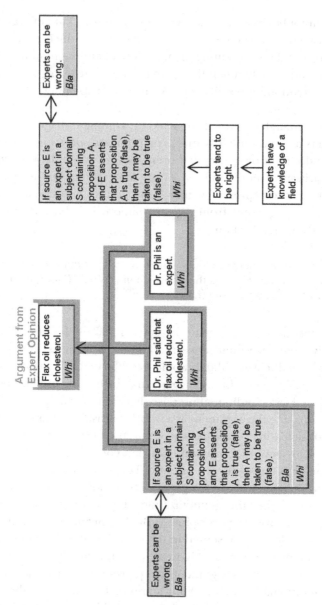

FIGURE 7.5. Pair of Araucaria diagrams representing undercutting.

254

Let's go back to Pollock's example of an undercutter. We suggest that it can be said to rest on an argumentation scheme from appearance. Part of this scheme could be put in the form of a conditional: if something appears to have a certain property, then it really has that property. For example, if an object looks red, then it is red. But such a generalization is defeasible, for, as Pollock points out, we can consider a special case in which the object is illuminated by a red light. In such circumstances it will not appear red to an observer. The argument has a defeasible *modus ponens* form as used in a given case.

Modus Ponens Form of the Red Light Example

If something appears to have a certain property, then it really has that property.
This object appears to have the property of being red.
Therefore, this object is red.

Now suppose that this argument defaults in a given case, because we have the additonal premise that a red light is shining on the object. Citing these circumstances is an undercutter. It shows that the generalization stated in the conditional premise does not apply to this particular kind of case. It defaults, as we say. Now consider a comparable kind of case in which the conditional defaults.

Prakken (2003) has shown that the argument in this example has a characteristic argumentation scheme. It is called argument from appearance (Walton, 2006b).

Argumentation Scheme for Argument from Appearance

If something looks like an x, then it is an x.
This object looks like an x.
Therefore this object is an x.

Prakken (2003) cites another example of argument from appearance. It is a legal argument, and provides the theme of his paper.

The Affidavit Argument

This object looks like an affidavit.
Therefore, this object is an affidavit.

The affidavit argument is defeasible, and is similar to Pollock's red light argument in that both are arguments from appearance. But the affidavit

argument suggests convincingly how common and important defeasible arguments are in legal evidence – and, by the way, how important
argumentation schemes are as well.

As an argumentation scheme, argument from appearance needs to be
evaluated in a dialogue format by asking critical questions. Each critical
question is a potential defeater of the argument. Three critical questions
matching the argument from appearance are the following.

> CQ1: Is there some reason why this object might look like an x but not really
> be one?
> CQ2: Can the hypothesis that the object is an x be tested by collecting more
> data?
> CQ3: Are there counterbalancing reasons for accepting the hypothesis that
> the object is something else, as opposed to being an x?

The first critical question can be illustrated by Pollock's example. If an
object looks red to me, I might quite reasonably conclude that it is red,
at least as a hypothesis. But then I might find out that the object was
illuminated by a red light at the time, making it appear red. This finding
defeats the argument, because a red light will make any object look red
to a normal human observer. Such a finding is a reason of the kind
mentioned in CQ1, because the object might look red, but not really
be red.

Pollock's theory is that the additional premise indicated in the *modus
ponens* form of the red light example is an undercutter because it attacks
the inferential link from the premises to the conclusion. But does it
really? Or does it just attack and refute the first premise by providing
an exception to the general rule? If the latter interpretation of how the
argument from appearance is attacked is right, an Araucaria diagram
can be used to represent undercutting argumentation. The technical
problem is that we can't represent a statement refuting an inferential
link (arrow) on an Araucaria diagram, as indicated in the Araucaria-
style representation of an undercutter diagram shown earlier. But can we
look on the statement of the special circumstances as refuting a missing
premise in the form of a conditional that does appear as a node in the
diagram? This solution seems to justify Pollock's view that the undercutter
refutes the inferential link between the premises and the conclusion of
the argument from expert opinion. The reason is that it refutes the
conditional premise on which the scheme depends. But is this view of
the matter the proper way to view defeasibility? We suspect it is not,
because what really happens is not that the conditional is refuted. It still
holds. What has happened is that it defaults, because the case is shown

to be one to which it does not apply. Thus the conditional should not be said to have been refuted. What has been shown is that it doesn't apply to this particular case.

In short, we have some reservations about Pollock's view that the undercutter attacks the inferential link (the argumentation scheme) on which the argument is based in a given case. We propose that it is better to think of undercutters in a different way. They perform the role of critical questions that do not, at least irretrievably, defeat an argument they attack. But they do temporarily defeat the argument unless the critical question can be answered satisfactorily by the proponent of the argument. In this light, let us review the list of critical questions matching the argument from expert opinion (Walton, 1997, p. 211–225) to see how they might be used to attack an argument and undercut or defeat it.

Expertise Question: How credible is *E* as an expert source?
Field Question: Is *E* an expert in the field that *A* is in?
Opinion Question: What did *E* assert that implies *A*?
Trustworthiness Question: Is *E* personally reliable as a source?
Consistency Question: Is *A* consistent with what other experts assert?
Backup Evidence Question: Is *E*'s assertion based on evidence?

The first two questions impinge on the first premise, "Source *E* is an expert in subject domain *S* containing proposition *A*." The expertise question is focused on the categorization of the source as an expert, that is, "an individual with specialized skills that can be tapped for information to which she has privileged access" (Walton, 1997, pp. 112–113). Therefore, the attacks will question the evidence for this classification - that is, her job, her qualifications, the testimony of peer experts on her competence, as well as her experience, skills, and publications. The field question is directed against the correspondence between the domain of expertise and the proposition: the domain itself may be examined from the point of view of type (knowledge or technical skill), of its relation to other fields, or of its development. The opinion and trustworthiness questions impinge on the second premise of the scheme: "*E* asserts that proposition *A* is true (false)." The former questions the correspondence between the source words and the reported proposition and attacks the inferential and interpretative links between them. The trustworthiness question raises problems concerning bias, honesty, or conscientiousness of the expert, matters that can undermine the tenability of her statement in favor of the truth of the conclusion. The last two questions represent other possible critical points of attack. The passage from expert opinion to the plausibility of the conclusion could be weakened by the observation that "experts may be wrong." The consistency and backup evidence questions examine two other possible points of attack. Her assertion may conflict with the generally accepted opinion shared by the majority of experts. Or her opinion may be not based on strong relevant evidence that she can cite, and therefore, it could be rejected as a strong reason to conclude that the proposition at issue is highly plausible.

What we observe generally here is that critical questions matching a scheme are aimed at attacking the argument by undercutting its applicability to a given case that can be shown to have special features to which the scheme does not properly apply. In other words, the critical questions undercut the argument, not by defeating the scheme itself – that is, by attacking the link between the premises and the conclusion – but by showing that the scheme does not apply to the particulars of the given case. Thus we recommend that the critical questions matching a scheme be seen as devices for attack that work by arguing that the requirements of the scheme are not met. In our view, the defeater/undercutter distinction is best seen as a device for referring to different kinds of attacks.

We begin by reviewing the three-way hypothesis with our addition to it. There are three ways of attacking an argument. The first way is to argue that the premises are not true, or have not been shown to be true. The second way is to argue that the conclusion does not follow from the premises. The third way is to argue that the conclusion is false, or at any rate, that there are reasons to think so. In addition, we have noted that there is a fourth way, in which the attacker argues that the argument given does not meet the burden of proof. To this four-ways hypothesis we have to add the distinction between rebuttal and refutation. A rebuttal should be defined as an attack on a prior argument in which the second argument is opposed to the first. A refutation is a successful rebuttal, that is, an attack that knocks down (successfully defeats) the prior argument.

It is the second way, representing the undercutter, that is the most controversial. How can you argue that the conclusion does not follow from the premises? For if a given argument in a particular case can be classified under a certain scheme, then it does (presumably) meet the requirement for that scheme. For example, if a given argument is identified as having the *modus ponens* form, then the conclusion does follow from the premises. So how is undercutting possible at all? To examine this question, let's look again at argument from expert opinion. If an argument really is of this type, and has this scheme, then there is an inferential link between the premises and the conclusion. How could that be attacked? One way is that the attacker could argue that the given argument is not really an argument from expert opinion, even though it may look like one or the audience may be taking it to be one. For example, suppose the source cited in an ad is a Hollywood celebrity who claims that a diet is a healthy way to lose weight. This argument could be attacked by arguing that the celebrity is not an expert on matters of healthy diet. But how should we classify such an attack in light of the four-ways hypothesis? Is it claiming that the argument is not really an argument from expert opinion? That is, is it claiming that the scheme for argument from expert opinion does not fit the argument in the ad? For example, is it claiming that the argument used in the ad is really only an appeal to popular opinion, or some other type of argument, as opposed to an appeal to expert opinion? This claim could be an undercutter, perhaps, but not of the kind that Pollock described in the example of the red light. Or is there another way to interpret the attack? Could it be an attack on the premise that the source is an expert? If so, it falls under the first way and not the second way. It is an attack on a premise. But this way of attack can be diagrammed without resorting to entanglement.

What about the claim that the argument is not really an argument from expert opinion? Could this form of attack be represented diagrammatically without using entanglement? Suppose it is a linked argument with two premises, for example, that looks like it might be an appeal to expert opinion but really is not. It seems that this kind of failure can be represented in an Araucaria diagram simply by not using the argumentation scheme for argument from expert opinion to mark the linked argument going from the premises to the conclusion.

7. A CASE STUDY OF COMBINED REBUTTALS

The structure of a refutation can be quite complex in some instances of realistic argumentation, because the refutation can be made up of many component arguments. Each of the component arguments may be relatively weak or inconclusive by itself, but when taken together they can form a mass of evidence, made up of complex argumentation, that is relatively strong. Even if such a mass of evidence is not a conclusive proof of the ultimate thesis being advocated, it may be relatively strong, and it may be strong enough to fulfill a burden of proof. On controversial issues, one may not be able to expect a conclusive proof of an ultimate thesis or a conclusive refutation of it. Instead, many defeasible arguments can be used to build up a mass of evidence that is fairly persuasive. For example, take a public policy issue like euthanasia or the legalization of marijuana. You are never going to settle such an issue by finding conclusive arguments on one side or the other. What you do find is that there are many inconclusive arguments used by both sides. For example, if a policy of allowing euthanasia or of legalizing marijuana has been established in a country, you will find lots of argument from consequences. For example, someone opposed to the policy of legalization of marijuana may argue that when this policy was carried out, it led to negative consequences. Such arguments, in turn, will need to be supported by arguments from expert opinion; documented scientific studies of such supposedly harmful effects, when carried out by scientific experts, can be used as persuasive arguments. In such cases, then, one might expect a mass of complex argumentation made up of arguments from expert opinion. No single argument refutes the claim it is used to attack, by proving that it is false. But we do not expect that. We are nonetheless impressed if a number of such inconclusive arguments can be formed into a complex network of argumentation in which many separate rebuttals fuse together to form a strong argument.

The following passage is from the article "Marijuana"a Loss of Innocence' (Gerlof Leistra and Simon Rozendaal, *Reader's Digest*, April 2004, pp. 97–102). The argument of the article is that a Dutch drug policy designed to tolerate the use of cannabis, started in 1976, has been a failure. The following passage (pp. 99–100) is part of the article.

> The Dutch drug policy is based on a distinction between soft drugs and hard drugs. It is believed that hard drugs are dangerous, while soft drugs are relatively harmless. But this is a shaky hypothesis. Over the past two decades, research published in scientific journals such as *Nature*, the *British Medical Journal* and *The Journal of the American Medical Association* indicates that long-term cannabis use is more dangerous than previously assumed. There appears to be a clear relationship with schizophrenia. "It cannot be denied that over the course of time, this link has become more pronounced," says Harald Wychgel, a drug expert at Utrecht's Trimbos Institute, which monitors mental health and addiction in the Netherlands.
>
> Experts such as Don Linszen, professor of psychiatry at the Academic Medical Centre in Amsterdam, and Jim van Os, professor of psychiatry in Maastricht, believe that cannabis use can increase the risk of psychotic disorders and can result in a poor prognosis for those with a tendency to psychosis.

The article goes on to offer other arguments against the hypothesis that soft drugs are harmless. For example, it argues that cannabis is more carcinogenic than tobacco, and that long-term use of cannabis lowers the IQ (p. 100).

The main thesis of the article is that the Dutch policy is "wrongheaded" (p. 100). But the part of the article just quoted is meant to be a rebuttal, part of a refutation of the hypothesis that soft drugs are relatively harmless, compared to hard drugs. What is interesting is that the article gives a number of arguments against the hypothesis that soft drugs are relatively safe. These arguments are meant to be rebuttals of the hypothesis, and when taken together they are meant to be a refutation of it. Both arguments just quoted, and many other arguments in the article as well, are arguments from expert opinion.

The argumentation in this quotation can be analyzed using Araucaria. We begin by identifying the explicit statements in the article using a key list.

Key List for the Dutch Policy Case

(A) It is believed that hard drugs are dangerous, while soft drugs are relatively harmless.

(B) Over the past two decades, research published in scientific journals such as *Nature,* the *British Medical Journal,* and *The Journal of the American Medical Association* indicates that long-term cannabis use is more dangerous than previously assumed.

(C) There appears to be a clear relationship with schizophrenia.

(D) Harald Wychgel says that it cannot be denied that over the course of time, this link has become more pronounced.

(E) Harald Wychgel is a drug expert at Utrecht's Trimbos Institute, which monitors mental health and addiction in the Netherlands.

(F) Don Linszen, a professor of psychiatry at the Academic Medical Centre in Amsterdam, is an expert on the dangers of drug use.

(G) Don Linszen believes that cannabis use can increase the risk of psychotic disorders and can result in a poor prognosis for those with a tendency to psychosis.

(H) Jim van Os, a professor of psychiatry in Maastricht, is an expert on the dangers of drug use.

(I) Jim van Os believes that cannabis use can increase the risk of psychotic disorders and can result in a poor prognosis for those with a tendency to psychosis.

To this list of explicit statements we add the unstated or missing premise (J) that long-term cannabis use is more dangerous than previously assumed. As Figure 7.6 shows, the three arguments from expert opinion support this statement, and it is meant to be a rebuttal of the hypothesis

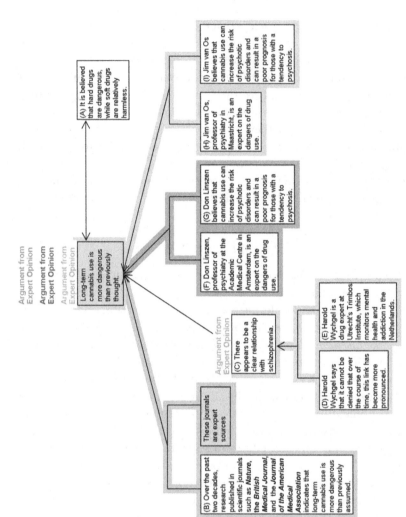

FIGURE 7.6. Araucaria argument diagram for the Dutch policy case.

represented by statement A. Statement J, supported by several arguments that appeal to expert opinions, is meant as a refutation of A. These different arguments from expert opinion are presented in rebuttal of hypothesis A. One such argument is based on premises B and the unstated premise that the journals cited are expert sources.

What this case shows is how rebuttals can come in packages. A strong rebuttal can be made up of several rebuttals that, by themselves, are weaker, but that when taken together can make up an attempted refutation that combines them in an argumentation structure. We see that a single refutation by itself – for example, based on argument from expert opinion – may not be very strong, but it can still have some strength as a rebuttal. When two or more such relatively weak rebuttals are combined in an argumentation structure, there can be an additive effect. The mass of evidence offered forms a refutation that can be stronger than any of its component rebuttals.

8. ARGUMENT FROM OPPOSITES

The subject of refutation has taken us into even deeper problems relating to the notions of negation, opposites, and contraries. These notions are fundamental to logic and argumentation theory, and were judged to be so in the ancient world, as shown in the brief history of the subject outlined earlier. We don't really have solutions to these problems at present, because we do not have any new theory of negation or contrariety to offer at this point. Still, it is worth noting that argument from contraries was recognized as an argumentation topic by ancient writers.

Aristotle (*Rhetoric*, Book II, Chapter 23, 1397a) postulated a topic he called argument from opposites (translation quoted from p. 297 of work cited in the Bibliography).

> One topic of demonstrative enthymemes is derived from opposites; for it is necessary to consider whether one opposite is predicable of the other, as a means of destroying an argument, if it is not, as a means of constructing one.

He offered four examples of this type of argument (1397a).

Self-control is good, for lack of self-control is harmful.
If the war is responsible for the present evils, one must repair them with the aid of peace.
For if it is unfair to be angry with those who have done wrong unintentionally, it is not fitting to feel beholden to one who is forced to do us good.

If men are in the habit of gaining credit for false statements, you must
also admit the contrary, that men often disbelieve what is true.

Aristotelian topics are argumentation schemes that generally represent
arguments that are reasonable, but that can also be used fallaciously in
some cases. So it is not entirely to be taken for granted that Aristotle is
endorsing argument from opposites as a reasonable form of argumentation. Still, the examples just offered look fairly reasonable.[7]

Cicero, in *Ad Herennium*, Book IV, Chapter 25, also recognized argument from opposites. At least, he postulated an argumentation scheme
he called reasoning from contraries that looks to be the same as the form
of argument that Aristotle called argument from opposites.

> Reasoning by contraries is the figure which, of two opposite statements, uses
> one so as neatly and directly to prove the other.

He offered the following examples (translation quoted from p. 293 of
work cited in the Bibliography).

> Now how should you expect one who has ever been hostile to his own
> interest to be friendly to another's?
> Now why should you think that one who is, as you have learned, a faithless
> friend, can be an honourable enemy? Or how should you expect a person
> whose arrogance has been insufferable in private life, to be agreeable and
> not forget himself when in power, and one who in ordinary conversation
> and among friends has never spoken the truth, to refrain from lies before
> public assemblies?
> Do we fear to fight them on the level plain when we have hurled them
> down from the hills? When they outnumbered us, they were no match for
> us; now that we outnumber them, do we fear that they will conquer us?

Cicero thought this a highly plausible and useful form of argument, for
several reasons. First, it is "agreeable to the ear on account of its brief and
complete rounding-off." Second, "by means of the contrary statement it
also forcibly proves what the speaker needs to prove." Third, "from a
statement which is not open to question it draws a thought which is in
question, in such a way that the inference cannot be refuted, or can be
refuted only with the greatest difficulty."

It is not too clear, perhaps, whether Cicero regarded reasoning by
contraries as a reasonable form of argument in general, or whether he

---

[7] Still, each argument has its special features. We have already commented that the first
one appears to involve negative argumentation from consequences, for example.

was putting it forward only as a useful means of persuading an audience. Still, judging from his examples, it looks like it could be a reasonable form of argument in some instances, even if it is one that should be regarded as defeasible.

So far, so good. It looks as if argument from opposites was recognized as an argumentation scheme in the ancient world, and it appears from the examples cited that it should be cited as an argumentation scheme representing a form of argument that is defeasible. It looks like a fairly standard argumentation scheme of a presumptive sort, in other words. But there is another ancient source that needs to be considered in this connection. Plato (*Phaedo*, 69c-72d) used what appears to be this very form of argument as a proof of the claim that living men are produced from the dead (quoted from Plato, *Phaedo*, translated by C. Stanford [New York, Hurst and Company, 1835], pp. 27, 28).

"We are sufficiently assured of this, that all things are so produced, contraries from contraries?"

"Sufficiently so."

"But further; is there something of this nature in them; for instance, two stages of generation between each pair, as all contraries imply two extremes, from the one to the other, and from the other back again to it? For between the greater and the less intervenes the process of increase and diminution; and do we, therefore, call the one the act of increasing, and the other the act of diminishing?"

"Yes," said Cebes.

"So, therefore, with the act of separating and of mixing, of growing cold and growing warm, and all things similarly, even though we should not have names to designate them by at times, still must they not in fact be at all times so disposed as to be produced from each other and that their generation should be reciprocal?"

"By all means," replied Cebes.

"What then," said he, "has life any contrary as sleeping has its contrary, waking?"

"Certainly."

"What is it?"

"Death."

"Are not these then produced from each other, since they are contraries, and the stages of generation between them are two, since they are two themselves?"

"How should it be otherwise?"

"I shall tell you then," said Socrates, "one combination of contraries amongst those which I mentioned just now, both itself and its stages of generation, but do you tell me the other. I say then, that sleeping is one thing, and waking another, and that waking is produced from sleeping and sleeping from waking, and that the stages of their generation are, the one falling asleep, and the other awakening. Is this sufficiently clear or not?"

"Quite so indeed."

"Do you now tell me likewise in regard to life and death. Do you not say that death is the contrary of life?"

"I say so."

"And that they are produced from each other?"

"Yes."

"What then is that which is produced from life?"

"Death," said Cebes.

"And that which is produced from death?"

"I must allow," said Cebes, "to be life."

"Then, Cebes, from the dead are living things, and living men produced?"

The conclusion of this argument, as stated in the last sentence, is not one that is generally accepted. And the form of argument it is based on does not seem to hold generally. It seems easy to find a counterexample, "Socrates is white; white is the contrary of black; therefore, Socrates will be black."

Argument from opposites is a defeasible argumentation scheme that has the following form (see Chapter 3).

Positive Form of Argument from Opposites

The opposite of subject S has the property P.
Therefore, S has the property not-P (the opposite of property P).

In some instances, the argument will have the following negative form.

Negative Form of Argument from Opposites

The opposite of subject S has the property not-P (the opposite of property P).
Therefore, S has the property P.

It is somewhat arbitrary which form is called positive and which negative, as both contain negated expressions.

But should argument from opposites be more widely recognized as an important defeasible argumentation scheme? And if so, under what conditions is it used in a way that is faulty or fallacious? In order to answer this question, the examples put forward by Plato, Aristotle, and Cicero have to be examined, and other cases studied in depth. We do not try to solve the problem here.

9. A PRAGMATIC THEORY OF REFUTATION

What has been considered so far leads to asking whether it is possible to have a theory of refutation based on the concept of negation. In logic, two types of negation have been recognized. Classical negation refers to the opposite truth-value of a proposition. In other words, a proposition is true (false) if and only if its negation is false (true). Negation as failure means that if there is a failure to prove that a proposition is true, then, by default, that proposition is not true (Clark, 1978). In other words, its negation is true. These two types of negation appear to correspond to the two types of refutation of a proposition. A proposition is strongly refuted if proven to be false. A proposition is weakly refuted if not shown to be true. This model of refutation fits the kind of refutation called *reductio ad absurdum*, or indirect proof, very well. This kind of argumentation shows that a proposition is true by showing that the assumption that it is false leads to a contradiction. Since a contradiction has to be false, by *modus tollens* the assumption in question has to be true. Here the notion of refutation of a proposition based on classical negation seems to work. However, we have been concerned not with refutation of propositions, but with refutation of arguments. A refutation is taken as a species of argument used to attack another argument.

Some might say that these notions are connected, because to refute an argument is basically to attack the conclusion of that argument by mounting an argument for its opposite (classical negation). Dialogue #4 fits this model. To diagram the Dr. Wendy argument, showing it to be a refutation of the Dr. Phil argument, we inserted an implicit conclusion that is the negation of the conclusion of the Dr. Phil argument. The Dr. Wendy argument, because its conclusion is the negation of the prior Dr. Phil argument, fits the model of a refutation. But this model doesn't fit the refutations represented in Dialogues #1, #2, and #3 very well. These are weak rather than strong refutations. The arguments are undercutters rather than defeaters, in Pollock's language. This kind of refutation, or perhaps rebuttal, doesn't seem to fit the model of propositional refutation based on classical negation. Small wonder, because we are dealing with defeasible arguments here. A refutation is typically the use of one defeasible argument to attack another. This can take place, as in Dialogue #4, by the use of a defeasible counterargument that is opposed to the conclusion of the original defeasible argument. Or, as in the cases of the other three dialogues, it can proceed by the asking of critical questions that attack and weaken the inferential linkage between the premises and the conclusion, or by attacking one of the premises by expressing doubts about it.

To consider the argumentation displayed in all four Dr. Phil dialogues as refutations, or at least as attempted refutations, we have to move to a broader, more pragmatic approach. In such a pragmatic approach, a refutation is defined as something that has three parts. First, it is generally an argument, although it can also be appropriate in some instances to speak of the refutation of a proposition. Second, it is an argument that is opposed to a prior argument. Third, it is a successful refutation if the opposed argument is stronger than the original one. What does this mean? To examine this condition, we have to turn to the case of Dialogue #4. In this case, the Dr. Wendy argument turned out to be stronger than the Dr. Phil argument to which it was opposed. In pragmatic terms, this third condition means that a refutation is an argument that has a certain purpose or function as a speech act. A refutation is meant to fulfil a certain kind of function called the probative function, meaning that it is being used to prove or disprove something. In the case of a refutation, it is used to disprove a prior argument. Such a three-part definition of the notion of refutation is inherently pragmatic, meaning that it is based on how an argument is being used for some conversational purpose. In logic, we are used to semantic definitions of 'argument', whereby an argument is defined as a set of statements (propositions), one of which is designated as the conclusion and the rest as the premises. Contrasted with this semantic approach, there is a pragmatic approach in which an argument is seen as more than just a designated set of statements.

The pragmatic approach follows the definition of Walton (1996, p. 18): an *argument* is a sequence of reasoning made up of a chaining together of inferences that is used to contribute to the settling of an unsettled issue in a dialogue. Some arguments are merely hypothetical, meaning that the premises are merely assumptions, and no attempt is made to prove the conclusion using them as evidence. Other arguments are meant to prove the conclusion by using the premises as evidence. In such a case, we say that the argument is used to fulfil a probative function. The *probative function* of an argument is defined (Walton, 1996, p. 20) as the use of an argument to alter the respondent's commitments so that, although he did not accept the conclusion beforehand, he is now led to accept the conclusion on the basis of the support given to it by the premises, statements he was committed to before the argument was presented. An account of the probative function, something evidently known and discussed in Greek philosophy, was given by Sextus Empiricus in his *Outlines of Pyrrhonism* (Book II, 140–141). According to this account, probative (*apodeiktikoi*) arguments are ones "which deduce something non-evident by means of pre-evident premises." Sextus offered the following famous example: "If sweat pours through the surface, there are insensible pores; but in fact sweat does pour through the surface; therefore there are insensible pores." In such an argument, the conclusion was not evident before the argument was used, but once the argument is presented, the conclusion is made evident by it. This example illustrates very well how the probative function works. You can express this idea by saying that not every argument is meant to be a proof. An argument that is meant to be a proof is one that is put forward to fulfil a probative function in a dialogue setting.

The probative function of an argument refers to the use of an argument by one participant in a dialogue to remove the doubt or disagreement of a respondent, by using premises the respondent is already committed to. Thus the probative function of an argument is a pragmatic characteristic. It relates not just to the truth or falsity of the premises and conclusion, or to the validity of the argument, but to how that argument was used for some purpose by its proponent in a dialogue. The term 'argumentation' is often used, as opposed to 'argument', to indicate this pragmatic approach.

In such a pragmatic approach, argumentation is defined as a chaining together of statements in a sequence of reasoning used for some purpose in a context of dialogue. Reasoning can be used in explanations as well as in arguments. An explanation is something used by one party in a dialogue to help the other come to understand some issue he did not understand before, as indicated by his question. An argument has a different purpose. An argument is the use of reasoning to fulfil a goal of a dialogue, a goal of settling some issue that has two sides. Thus the notion of burden of proof is important. An argument can be attacked on the grounds that it fails to fulfil a burden of proof set for it in a dialogue. This view of argumentation is pragmatic, because every argument is seen as a case of two parties reasoning together for some purpose in order to settle an issue. Each party has a viewpoint, meaning he or she is in favor of or doubts some thesis at issue, and each tries to prove his or her thesis to the other party using the other party's commitments as premises. Commitment refers to the acceptance of a proposition by a participant in a dialogue. Each party has a commitment set, and statements are inserted into it or deleted from it as the dialogue proceeds (Hamblin, 1970). Commitments do not have to be logically consistent with each other. But if one party's commitments are apparently inconsistent, the other can, and in some instance should, challenge the inconsistency by attacking the argument.

This notion of argument is a dialectical one. An argument is not just a set of statements, but a set of statements of a special kind. One of the statements is a claim made by one party in a dialogue; the other party doubts that claim, or has made an opposed claim. On this pragmatic view, an argument is designed to fulfil what is called a probative function. Thus one party, when using an argument, tries to prove something, some particular statement, to the other. The other has expressed doubt or opposition to that statement. You could say that this pragmatic way of viewing an argument sees an argument in a social way. There are always two parties involved, the pro and the contra, and the argument is something used by one side to try to change the viewpoint or commitments of the other. An argument, so conceived, is a kind of speech act.

A refutation can now be defined as a special kind of argument, one addressed to a prior argument with the aim of opposing it and removing the probative function carried out by that prior argument. In other words, a refutation always involves a pair of arguments, one of which appears after another in a dialogue sequence. One argument is aimed at the other with the purpose of refuting it by showing that the probative function that it seemed to fulfil is not really fulfilled after all. The three-ways hypothesis sets out the three primary ways an argument can be attacked or refuted. However, as already noted, there are other ways that apply to special cases. An argument can be refuted by showing that it begs the question, or even more simply by showing that it fails to meet the burden of proof that is appropriate for it. It was Aristotle's view that the notion of refutation is fundamental to the notion of fallacy. Indeed, on his theory fallacies are defined as different forms of sophistical refutations. On Aristotle's view, as shown earlier, the notion of a refutation is quite a broad and pragmatic one that depends on an underlying notion of opposition. Moreover, this notion of opposition cannot be defined exclusively in terms of classical negation of the kind used in modern propositional logic. It depends on semantic categories.

A refutation, in the pragmatic sense outlined here, is an argument that attacks and rebuts an opposed argument, but in a strong or decisive way that knocks the opposed argument down. There is also another commonly accepted meaning of 'refutation' in which a statement, as opposed to a whole argument, is attacked. This is also called a refutation in ordinary usage, and it can represent a kind of refutation. The usage here is comparable to that of the term 'fallacy'. In everyday usage, it refers to commonly accepted false beliefs, while in logic it refers to fallacious arguments or techniques of argument. Despite common usage, however, the paradigm kind of refutation we are centrally concerned with in argumentation theory is refutation of an argument by another argument. Thus we often take for granted that this sense of 'refutation' is what is meant, even though the other sense can be meant as well, and is sometimes important. There is also a third sense of 'refutation' that involves the asking of critical questions. In many instances, asking a critical question does not refute an argument, but merely casts doubt on it, or suspends it temporarily. However, if the critical question is not answered by the proponent of the argument, the effect is to suspend or bracket the argument temporarily. And this suspension can be a kind of refutation in some instances, as we have seen. Here too it may be appropriate to speak of refutation, even though the refutation is not a case of one argument against another.

10. UNSOLVED PROBLEMS OF REFUTATION

A refutation is more than just an attack on or rebuttal of another argument. It is a successful attack by an opposed argument that is stronger than the original one. A refutation is more than just an attempt at rebuttal. It is a rebuttal that has some effect in knocking down, or at least temporarily cancelling out, the argument it is aimed at. This is our basic account of what a refutation is, for purposes of argumentation theory. But many problems have been raised by our attempt to defend this definition. One is the problem of what is meant by 'opposed'. Is simple classical negation, of the kind we are familiar with, enough? Or should some deeper kind of negation (like complement) that applies to predicates be utilized? Or should the distinction between contraries and contradictories be appealed to? We have not been able to provide definitive answers to these questions. They scarcely seem to have been asked in recent times. Still, we have made some progress toward analyzing refutation, and have shown that studies of rational argumentation in classical and medieval times have seen these questions as central.

A refutation offers another argument, with premises and a conclusion, set up in opposition to the one that is the target of the refutation. As shown, this can be done in the following main ways. One way is to attack a premise of the original argument. This in turn can be done in two ways. First, it can be argued that the premise is false. Second, doubt can be cast on whether the premise has been proved by questioning it. Another way is to cast doubt on whether the inferential link between the premises and conclusion of the original argument holds, by arguing that the argument in question does not fit the scheme it was alleged to. Another way is to attack the conclusion, casting doubt on it by questioning it, or by posing a counterargument designed to prove the opposite conclusion. Still another way is to argue that the burden of proof is not fulfilled by examining the probative function. Still another way is to ask a critical question that shifts the burden of proof to the proponent by making the argument default temporarily unless an appropriate answer to the question is offered.

One problem concerns the difference between two kinds of critical questions. Some attacks posed by critical questions, as seen in the dialogues already presented, can reverse the burden of proof. Whether such a reversal occurs depends, however, on the strength of the refutation itself. For example, consider the accusation of expert bias. In this kind of case, the opponent may question the expert's objectivity, but an allegation of bias normally has to be backed by sufficient proof if it is to be effective in raising doubt about the worth of the argument from expert opinion. It must satisfy, in other words, the burden of proof itself, to be considered a countermove that successfully shifts the initiative to the other side. Unless it does so, the burden of proof remains on the critic. On the other hand, the field and expertise questions usually shift the burden of proof back to the proponent, once asked, even if not backed up by additional evidence right away. The difference between these two dialectical phenomena may be explained by connecting critical questions with premise refutation. The attack may be direct or indirect. That is, the truth of the premise may be directly questioned, or critical questions may be asked that cast doubt on the argument only indirectly. The expert's trustworthiness does not directly depend on lack of bias. An expert may be biased, but her opinion might still be true, or at any rate, well worth taking into account. Thus in order to refute an argument based on her expert opinion, the critical question needs to be backed up with evidence. By contrast, when the mere statement that the expert is not a real expert or is not a specialist in the field in question is made, refutation occurs, because the argument is held in abeyance until the proponent can offer some evidence to back up these claims.

The same kind of problem applies to inductive arguments of the kind commonly used in science. Suppose you do a statistical study, and a critic argues that your database was too small, or that you picked a biased sample. At one time, before statistical methods of data collection became standardized in their present form, it might have been up to the critic to prove the sample size was too small, or that it was biased. Now it is up to the proponent to prove these things, or at least to prove, if challenged, that the sample size is adequate and free of bias – for example, by using a double-blind method of testing, or whatever method might be appropriate to the investigation.

Another problem concerns rebuttals of endoxic premises. Aristotle, in his account of different strategies of refutation, described a type of rebuttal directed against endoxic premises. In order to understand the mechanism of such an objection, it is necessary to distinguish between two different kinds of *endoxa*. *Endoxa* are propositions considered true by everybody, or by the majority, or by the wise. Under this label, however, are grouped both "universal truths," like "The earth is round" and "Birds fly," and proverbial expressions or social values and moral judgements considered truths, like "One can learn from mistakes," "Richness is one of the most important targets of life," and "Love is a good feeling." The first class of *endoxa* has a nature different from that of the second. Generalizations grouped under the former category represent situations we take to be standard or normal, and they default in special cases. They organize objective reality based on the way we use words in everyday discourse and as applied to kinds of situations we routinely deal with in everyday life. On the other hand, proverbial and social *endoxa* are forms of organization that tie agents to a given set of facts of the kind encountered in deliberation. Proverbs are illuminative general rules of conduct and judgment, considered to hold in standard circumstances, while social *endoxa* are focused on practical conduct in special circumstances. Both types of endoxic premises can be classified as defeasible generalizations, but not as those associated with any particular argumentation scheme. Anderson and Twining have recognized the importance of these kinds of generalizations in legal argumentation.

In many cases, refutations proceed by attacking an endoxic premise of this sort that is either an explicit or an implicit premise in an argument. Much the same remarks as those already made about conditional warrants and critical questions can be applied to these refutations. The generalization is typically attacked, and possibly defeated, by arguing that it does not fit the particulars of the given case. The language of undercutters could be used to apply to the argumentation in such cases. But according to our approach, the rebuttal can simply be seen as a kind of attack on a premise.

8

The History of Schemes

Topics (*topoi*), in a long tradition stemming from Aristotle's rhetoric and early writings on argumentation and logic, are the places where arguments can be found to make a case, and the warrants that can back a logical inference leading from premises to a conclusion. Argumentation schemes are tools of modern argumentation theory that have been developed to fulfil the latter function, but may also be useful to fulfill the former one as well. In this chapter we will outline the varied developments of the *topoi* in both the logical and rhetorical traditions, starting with Aristotle, the first to describe them. We will examine some leading accounts of them given in the Middle Ages, when they were studied in relation to logical consequences.

Aristotle's *Topics* contains accounts of many commonly used types of arguments he calls topics (*topoi*, or places). There are some 300–400 of these topics, depending on how you count them, according to Kienpointner (1997, p. 227). Many topics can also be found in Aristotle's *Rhetoric*. What these topics supposedly represent has been subject to many different interpretations over the centuries. Many have interpreted the topic as a device to help an arguer search around to find a useful argument she can use, for example, in a debate or in a court of law. Other have taken the topic to have a guaranteeing or warranting function that enables rational inferences to be drawn from a set of premises to a conclusion. The second interpretation makes the topic perform what seems to be a logical function, while the first interpretation views the topic as having a rhetorical function. Since Aristotle's time, topics have been a subject of interest in both rhetoric and logic, but no systematic theory of them has ever taken hold. Commentators have interpreted the topics in a wide variety of ways, but it is fair to say that no single theory of them, or systematic account of them, has proved to be widely useful.

The question is what relationship the modern schemes have to the ancient topics. Ultimately, this question is one for specialists in ancient Greek philosophy and rhetoric, as well as for medieval scholars. Still, for those of us who are working on argumentation schemes as a tool for modern argumentation theory and informal logic, there is some need to look back at the history of the subject in light of our current concerns. Some work on the subject has begun. Warnick (2000) compares twenty-eight topics identified in Aristotle's *Rhetoric* to the thirteen comparable argument schemes identified in *The New Rhetoric*. Warnick shows (p. 123) that some of the forms of argument recognized as schemes in *The New Rhetoric* are the same as those presented as topics in Aristotle's *Rhetoric*. The work so far raises many questions. Are schemes the same as topics, or would it be a mistake to try to fit Aristotle's Greek notion to the *topos* into the modern theory of argumentation schemes? Certainly, whatever one might say here, the *topoi* are the historical forerunners of the schemes. Those of us working on schemes, while we do not wish to be side-tracked into specialized historical matters of Greek philosophy and rhetoric, would like to encourage such specialists to recognize the study of schemes and to undertake studies of how they relate to the historical *topoi*. We expect plenty of controversy, as many fields and factions are involved and have issues at stake.

1. ARISTOTLE ON THE TOPICS

The notion of *topos* is implicit in the methods of argumentation used by the Sophists, but its systematic employment can be traced back to Plato, who, for the first time, gave a systematic explanation of dialectical topics. They were the means to understand the Ideas: *topoi* represent the relations that link or divide them, their structural connections, their hierarchy. Dialectic, based on the *topoi*, is considered the way to ascend to the intelligible world (Robinson, 1962, pp. 61–93). Plato delineates a set of dialectical procedures, never called *topoi*, but closely related to the Aristotelian concept. The first procedure is division, that is, the organization of sensible reality reflecting the order of the Ideas. Other rules of organization are the ones from the more and the less, from contraries, from similitude, from definition, and from relation. Under these general labels are gathered what Aristotle later defined as *topoi*. Another way in which *topoi* are represented in Plato's writings is as the examples of kinds of argumentation used by Socrates in the Platonic dialogues. We typically find Socrates using his method of *elenchus* to question the opinions of experts, to probe into contradictions in what has been said, to use argument from analogy, and so forth (Robinson, 1962, pp. 7–19). Socrates provides the model of the working dialectician using common forms of argumentation to raise questions and to probe into the weak points of an opponent's argumentation.

The Aristotelian work on *topoi* is based on a conception of dialectic that is different from Plato's. For Aristotle, dialectic does not lead to the discovery of the first principles, but only prepares for their intuition. It is intellectual intuition, not dialectic (as in Plato's philosophy) that reaches the first principles. Aristotelian dialectic is only a method of finding arguments through the *topoi*, or "points of view." It is the method for reaching a probable conclusion from the *endoxa*, the opinions commonly accepted by the majority and the wise. From this set of given premises, *topoi* have the function of supporting the desired conclusion by warrants, and of finding premises needed to support the conclusion as well. This double role in the argumentative process was in previous studies considered to be covered respectively by the two different kinds of *topoi* Aristotle pointed to: the proper (*idoi*) and the common (*koinoi*) topoi.

Another ancient account of argumentation schemes is that given in the *Rhetoric to Alexander*, thought to be authored by Anaximenes of Lampsacus. The argumentation schemes, called *pisteis*, or proofs, are based on five principles (Braet, 2004, p. 129):

- argumentation on the basis of classification
- argumentation on the basis of analogous acts
- argumentation on the basis of opposite acts
- argumentation on the basis of authoritative statements on previous comparable acts
- argumentation on the basis of significance criteria (on principles of amplification and minimizing, i.e., consequences)

The argumentation schemes are classified in two main groups: intrinsic and extrinsic *pisteis* (Braet, 2004, p. 130): extrinsic *pisteis* are those not derived from the case itself.

Aristotle's account of *topoi* probably predates that of the *Rhetoric to Alexander*. The *Topics* is a very early work. However, we make no specific claims about matters of dating, leaving that to the experts.

Aristotle's theory of *topoi* can be better understood if located within his general theory of logical and rhetorical argumentation. The main distinction between these two types of argumentation is that the first is the methodology dealing with certainties and establishing truth, while

TABLE 8.1. *Intrinsic and extrinsic pisteis*

Intrinsic *Pisteis*	Extrinsic *Pisteis* (from Authority)
1. From causal probability	1. The view of the speaker himself
2. From examples	2. Witness testimony (voluntary)
3. From sign of contradiction	3. Testimony under torture
4. From contradiction	4. Statement under oath
5. From personal judgement in general (*gnomé*)	
6. From sign	
7. Refutation	

the second is the domain of plausible argumentation, comprising methods used to establish beliefs. Rhetoric and dialectic are two different disciplines characterized by common forms of argumentation: both are based on *endoxa*, opinions held by men in general, or by the majority, or by the wise, and on the *topoi*, principles of probable and plausible reasoning warranting the passage from the premise(s) of an argument to its conclusion. Table 8.2 shows the general scheme of the Aristotelian subdivision of the realm of arguments (McBurney, 1936, p. 60).

In this scheme the term 'syllogism' (*sullogismos*, which could include kinds of reasoning in addition to those cited in the formal theory of syllogisms) is used to refer to both dialectical and demonstrative (*apodeictic*) reasoning. As Burnyeat (1996, p. 96) states, Aristotle used the term *sullogismos* to refer to a valid argument, and not only to what we nowadays commonly think of as a syllogism. It includes the meaning of both necessary and plausible proof. Furthermore, Aristotle defined the enthymeme as a syllogism of a special kind. Burnyeat (1996) has raised doubts about the traditionally accepted view that an enthymeme is a syllogism with a missing (unstated) premise, however. It could refer to what we will call an argumentation scheme, especially one of the presumptive type. Enthymeme and dialectical proof are, however, deeply distinct from demonstration.

Another important distinction that has to be made is the one between dialectic and rhetoric. As Burnyeat (1996, p. 98) writes, dialectic is the art of question and answer, based on premises that are posited because they have been accepted by the other party or have been conceded for the argument's sake. Rhetoric, by contrast, is based on what is the case: the premises are assertive. They proceed from what are purported to be the facts. Moreover, the conclusion in a rhetorical argument does not follow necessarily from the premises. It can follow always or for

TABLE 8.2. *Aristotelian subdivision of arguments*

Argumentative Inquiry and Proof
(establishing beliefs)
(deals with probabilities)

Dialectic

Rhetoric

Scientific Demonstration
(establishing truth)
(deals with certainties)

Induction — Complete enumeration

Deduction — The apodeictic syllogism

Induction — Admission by respondent or group consensus

Deduction — The dialectical syllogism

Nonartistic proofs

Artistic proofs

Ethos   Pathos   Proof by argument

Example   Enthymeme

the most part. This is the main difference between the two kinds of consequences. Enthymemes, rhetorical *sullogismoi*, are characterized by *probabilitas consequentiae* and not, as dialectical and demonstrative ones are, by *necessitas consequentiae* (Burnyeat, 1994, p. 27). The modality is not referred to the consequences, but to the inferential link. The passage between premises and conclusion in an enthymeme is, unlike that of a dialectical proof, not a necessary one.

The *topoi* have a double function characterized by the same name. In one guise, they are points of view from which a conclusion is proved true or false, while in another guise they function as the reason why the search for the argument is directed in a specific way (De Pater, 1965, p. 116). The term *topos* unifies the double function of proof and invention, according to how the givens, the premises already known, are selected in relation to the kind of warrant chosen to support the conclusion (De Pater, 1965, p. 134). Common and proper *topoi* have, therefore, the same double function, but there is a difference between the two in their status. Common *topoi* are general rules, and represent the genus under which proper *topoi* can be found. These general rules are *endoxa*, similar to common knowledge from which the specific reasons warranting the conclusion (premises) are derived. But, in reality, these general rules are a higher kind of *endoxa*, which are sometimes necessary. They are never questioned, as they are axioms, elements laid down in the sense of primary, basic propositions presupposed (*arché*). The function and character of *topoi* of this sort must be distinguished from the instruments.[1] *Topoi* select the reason, or the premise, among a set of given facts, serving as instruments in an encyclopaedic search concerning how each subject can be presented (De Pater, 1965, p. 133). The following example illustrates the specific role of *topoi* as distinct from other kinds of enquiry (De Pater, 1965, p. 133):

> S'il faut louer quelqu'un auprès de gens qui estiment le plus la science, on prendra comme lieu propre que le fait d'être savant est digne d'éloge, et de tout ce qu'on sait de cette personne, on retiendra comme donnée privilégiée le fait qu'elle est savante (car elle a écrit des livres savants, etc.). Mais si l'on adresse son discours à des gens qui n'ont pas d'estime pour la science, mais bien pour la force corporelle, on prendra un lieu propre

---

[1] Aristotle pointed out four instruments. The first is a set of rules to find, order, and multiply the facts that may be subjects of discussion. The second has the function of discovering the different meanings of an expression. The third instrument is used to discover the differences, that is, to help in the discovery of the essence of the things. Through the fourth instrument what is similar is examined. (De Pater, 1965, pp. 150–159)

TABLE 8.3. *Role of topos in proof*

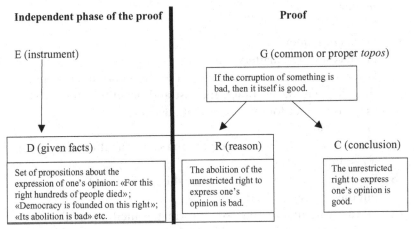

comme "celui qui sait mettre la main à la pâte, fait preuve de caractère," et on choisira parmi les données celles qui en sont des cas et des signes.

The argumentative process can be diagrammed as in Table 8.3 (Kienpointnter, 1986, p. 281).

Aristotle enumerated about 400 proper *topoi*, organized into four major classes: definition, genus, property, and accident. He divided *topoi* into three main categories, according to their function in communication: dialectical *topoi* (which aid in the construction of arguments in dialectical disputations), rhetorical *topoi* (used for the construction of rhetorical arguments), and mnemonic *topoi* (which have the function of recalling memorized things).

Twenty-eight general *topoi* are listed (Aristotle, 1991, pp. 190–203):

1. From opposites
2. From different grammatical forms of the same word
3. From correlatives
4. From the more and less
5. From looking at the time
6. From turning what has been said against oneself upon the one who said it
7. From definition
8. From varied meanings
9. From division
10. From induction

11. From previous judgment about the same or a similar or opposite matter
12. From the parts
13. From consequences
14. From contrasted matters
15. When one's opponents do not praise the same things openly as they do secretly, but praise the just and beautiful while privately they wish rather for what is to their advantage
16. From analogy
17. From arguing that if some result is the same, the things from which it resulted are also
18. From not always choosing the same thing before and after an event, but the reverse
19. From saying that the purpose for which something might exist or might happen is the cause for which it does exist or has happened
20. From what turns the mind in favor and what turns the mind against something
21. From things that are thought to have taken place but yet are implausible
22. From contradictions
23. From cause of false impression
24. From cause and effect
25. From a better plan of a different sort from what is advised or is being done or has been done
26. From looking together at something that is about to be done and something contrary to it that has been done
27. Accusation or defence from mistakes that have been made
28. From the meaning of a name

The work of Aristotle on dialectic is connected with philosophical speculation on the first principles. Even if they do not represent, as for Plato, the way to reach the Ideas, they are instruments to get close to them. *Topoi* are related to the predicables and consequently to the order of reality, and deemed by Aristotle to have the nature of axioms or basic principles.

2. CICERO

This philosophical aspect of *topoi* is lost in the Latin works of Cicero and Quintilian. Cicero reduces the number of *topoi* to a small set called *differentiae*, grouping them as general principles or maxims. Cicero's work is an attempt to organize arguments thought to be useful in arguing

TABLE 8.4. *Cicero's system of classification of loci*

Intrinsic		Extrinsic
In eo ipso de quo agitur	Ex iis rebus quae quodam modo affectae sunt ad id de quo quaeritur	

1. definitio
  a. partitio (definition by whole-part relation)
  b. divisio (definition by genus-species relation)
2. notatio (etymological relation)
	1. coniugata (inflectional relations)
	2. genus (genus-species relation)
	3. forma (species-genus relation)
	4. similitudo (similarity relation)
	5. differentia (difference relation)
	6. contraria (four types of opposite relation)
	7. adiuncta (co-occurrence relation)
	8. antecedentia
	9. consequentia
	10. repugnantia (contraries, opposites)
	11. efficentia (cause-effect relation)
	12. effecta (effect-cause relation)
	13. ex comparatione maiorum, minorum, parium
		Auctoritas (authority)

legal cases (Stump, 1988, pp. 8, 9). Cicero reduces the Aristotelian list of *topoi* to twenty *loci*. In his *Topics*, he divides the common places into two main classes: *loci*, which "in eo ipso de quo agitur haerent,"[2] and *loci*, extrinsically taken. While only topics related to definition fall into the first group, most of the *loci* are taken from things somehow related to the thesis.[3] The Aristotelian nonartistic lines of proof (argument from authority) are classified in the latter category. The classification system can be represented as in Table 8.4.

Later, Quintilian (*Institutio Oratoria*, V) divided the intrinsic proofs into two categories of topics "within" and "somehow related" to the subject of discussion (for the scheme, see Kienpointner, 1986, p. 282).

Along with this system of classification, Cicero connects the theory of topics to the division of oratory according to Hermagora's stasis. When the subject at issue is definite, defined in time and place, the inquiry is

[2] Cicero, *Topica*, 8, 3–4.
[3] Ducuntur etiam argumenta ex iis rebus quae quodam modo affectae sunt ad id de quo quaeritur (ibid., 11, 1–2).

TABLE 8.5. *Division of topics by questions*

Conjecture	Definition	Qualification
Cause, effect, circumstances	Definition, description, notatio, divisio, partitio, consequent, antecedent, inconsistencies,[a] cause and effect, adiuncta	Topics from comparison

[a] Cicero, following Aristotle, distinguishes among four kinds of opposition. The partition is similar to the Aristotelian account. The oppositions are the following: 1. contraries, adverse (predicates belonging to the same genus): sapientia-stultitia; 2. privatives (characterised by the privative affix particle): dignitas-indignitas; 3. correlatives: longum-breve; maius-minus; 4. negatives: si hoc est, illud non est (*Topica*, 47).

about a *causa* (*hypothesis*), while when the matter is about infinite things (semantic properties of words, for instance) the inquiry is about a *propositum* (*thesis*) (*Topica*, 79). Inquiries can be theoretical or practical. Every theoretical question is divided into three steps: "aut sitne aut quid sit aut quale sit," that is, conjecture, definition, and qualification (distinction between right and wrong). Topics can be divided according to these questions (*Topica*, 87).

In addition to the classification of dialectical topics according to rhetorical categories, Cicero advances a classification of rhetorical topics. Rhetorical *loci* deal with circumstances, that is, the conditions of an action such as its place and time, its actors, its consequences, and its reasons (*De Inventione*, I, 33). These topics (e.g., from nature, from origins) constitute the roots of medieval rhetorical tradition, distinct from dialectics but treated in the same studies.

In addition to these topics, Cicero discusses the *loci* used in special kinds of orations, such as the *Indignatio* and *Conquestio*, whose goals are, respectively, to move the hearer against the opponent or a fact and to conquer audience attention and benevolence (*De Inventione*, I, 53–54).

3. BOETHIUS

Boethius commented on and organized Cicero's *Topics* in *In Ciceronis topica* and *De differentiis topicis*. The latter work constitutes the roots of the dialectics of the Middle Ages. Two aspects of Boethius's dialectic are extremely relevant for the subsequent medieval development of *loci*: the distinction between necessary and plausible connections and that between dialectical and rhetorical *loci*. The *De differentiis topicis*

TABLE 8.6. *Special kinds of orations*

Indignatio	Conquestio
a. *Auctoritas*: how important or serious this fact was deemed to be by the ancestors	a. From comparison between the previous fortune and the present misery
b. People affected by the fact (i.e.: he damaged the inferiors; therefore, he is arrogant)	b. From showing in which miseries they have been, are, and will be
c. What would happen if everybody else did this	c. Each inconvenience is separately treated to exaggerate
d. From precedent	d. From all the discreditable or low or mean circumstances affecting a person
e. Impossibility of changing his own mind about that decision	e. From the unjust fate
f. The action has been done on purpose; therefore, it cannot be pardoned.	f. From empathy: consider his problems as if they affected yourself
g. Qualifications (nefarious, fool)	g. From injustice (what happened to this person was not just, considering his actions)
h. Exeptionality of gravity	h. Our speech is made to refer to things that are devoid of both language and sense
i. Comparison with terrible cases	i. From desolate condition
j. From scandalous circumstances	j. Recommendation to bury one's children, or one's parents, or one's own body
k. From the least suitable person to do such an action	k. A separation is lamented when you are separated from anyone with whom you have lived most pleasantly.
l. From the fact that, we being the first people to whom this has happened, it has never occurred in any other instance	l. We are ill-treated by those by whom above all others we least ought to be so
m. Insult followed an injury	m. Appeal to pity
n. From empathy: consider our injuries as if they affected yourself	n. We are complaining not only of our own fortunes, but also of those who ought to be dear to us
o. From the fact that these actions would appear scandalous even to an enemy	o. From empathy: our hearts are full of pity for others, but the others forget about the help we gave them

comprehends topics that in Aristotle were treated in the two separate treatises, namely, *Topica* and *Rhetorica*. Boethius's unified account is useful for pointing out the differences between the two kinds of topics and their connection to the syllogism. While dialectical *loci* stem from the rules of prediction and the logical-semantic properties of the predicates,

TABLE 8.7. *Intrinsic, extrinsic and intermediate loci*

Loci intrinseci

1. *From the substance of the thing*

Differentia	Maxim
From the definition	That to what the definition of the genus does not belong, is not a species of the genus defined
From the description	Said to be the same as from the definition, since description acts as a definition
From the explanation of the name	Explanation of the name is said to be a substitute for a defintion.

2. *From things that accompany the substance of the thing*

From the whole (genus)	What belongs to the genus, belongs to the species
From the integral whole	What suits the whole, fits the parts also
From a part (species)	What inheres in the individual parts, must inhere in the whole
From the parts of an integral whole	None is given
From efficient cause	Things whose efficient causes are natural, are themselves natural.
From the matter	Where the matter is lacking, what is made from the matter is lacking.
From the end	That whose end is good, is itself also good.
From the form	Each thing was capable of as much as its natural form allowed.
From the generation (effects)	That whose production is good, is itself also good; and vice versa.
From the corruption	That whose destruction is bad, is itself good; and vice versa.
From uses	That whose use is good, is itself also good.
From associated accidents	What follows from something that does not inhere in a thing cannot inhere in that thing either.

Loci extrinseci

From estimation about a thing	What seems true to everyone or the many or the wise should not be gainsaid.
From similarities	If that which inheres in a similar manner (to the thing in question) is not a property, neither can the thing in question be a property.

Loci extrinseci

From what is more	If what seems the more to inhere does not inhere, neither will that inhere which seems the less to inhere.
From things that are less	If what seems the less to inhere inheres, then what seems the more to inhere will inhere.
From proportion	What occurs in one thing, must occur in what is proportional to that thing.
From contraries	Contrary predicates belong to contrary subjects.
From opposites with reference to privation and possession	Where privation can be present, there the possession is not a property.
From relative opposites	Properties of opposites which are related to each other are themselves related.
From opposites with reference to affirmation and negation	The properties of opposites must be opposite.
From transumption	The *locus* is said to consist in transferring the question to better-known terms than the original ones.

Loci medii

From inflections	The *locus* is said to consist in ascribing something to/denying something from an adverb that is ascribed to/denied from the principal noun.
From coordinates	The *locus* is said to consist in ascribing something to/denying something from a word of one word-class that is ascribed to/denied from a word of another word-class, but from the same stem or root.
From division	1. By affirmation and negation: he has feet or he does not have feet 2. By partition: he is ill or he is well

rhetorical *topoi* represent the possible connections between things having the qualities (1215C).[4]

Some dialectical topics, such as topics from definition or from genus and species, are necessary, while others (for instance, from *adiuncta*)

---

[4] Rhetorical *loci* are similar in form to dialectical ones, but they proceed from frequent connections between things, from stereotypes and not from semantic properties of concepts (for instance: usually people addicted to alcohol are dissolute; this person is alcoholic; therefore, he is dissolute). See Boethius, *De Differentiis Topicis*, 1215b.

represent only frequent connections. This relation between probable and necessary consequence was crucial in the Middle Ages. Garlandus Compotista classified the topics according to their logical (demonstrative) role. Topics from whole, part, and equal are the foundations of categorical syllogism, while all the topics are conceived under the logical forms of topics from antecedent and consequent (Stump, 1982, p. 277).

For Boethius, the function of a *locus* is not properly the *inventio* of the premises supporting the argument, but the discovery of a middle term linking the premise and the conclusion. The Aristotelian *topoi* are interpreted as *maximae propositiones* falling under *differentiae*, genera of these maxims. *Maximae propositiones* are general principles known *per se*. They are also called axioms. They are general (indefinite in respect to particulars) and generic propositions that several arguments can instantiate. The function or place of arguments (or the genus of particular arguments) is secondary to their role of warrant, a device for confirming arguments. The process of argumentation is described as dependent on the choice of a *differentia* (criteria of appropriateness, or the genus of maxims) and the two terms of the conclusion. From the genus of the *maximae propositiones* and the terms, the maxim is found, and from maxim and the end terms, the middle term is derived (Stump, 1988, p. 6).

The structure of a topic can be represented as follows:

First term: Every virtue is advantageous.
Middle term: Justice is a virtue.
Second term: Therefore, justice is advantageous.
Maxim: What belongs to the genus, belongs to the species.
Differentia: From the whole, that is, the genus.

Topoi are divided into three main categories: intrinsic, extrinsic, and intermediate *loci*. While the first two categories are similar to Cicero's organization, the third class is based on different principles. *Loci medii* represent semantic connections of grammatical relations, such as from words stemming from the same root, or semantic relations of division underlying the definition of the word.

Following Cicero's division, Boethius analyzes the rhetorical topics, dealing not with abstract principles of inference, but with singular cases.[5]

---

[5] Distant autem a superioribus, quod superiores loci, uel facta continebant, uel factis ita adhaerebant, ut separari non possint, ut locus, tempus, et caetera quae gestum negotium non relinquunt; haec uero quae sunt adiuncta negotio non adhaerent ipsi negotio sed accidunt circumstantiis, et tunc demum argumentum praestant, cum ad comparationem ueniunt. Sumuntur uero argumenta non ex contrarietate sed ex contrario, et non ex similitudine sed ex simili, ut appareat non ex relatione sumi argumentum sed ex adiunctis

For instance, reasoning from place, name, and time depends on the fact, stems from the factors of the event and not from the logical-semantic relations between concepts. The rhetorical topics are organized into the five classes identified by Cicero (*De differentiis Topicis*, 1212A–1214A):

Who Is It:	What and Why:	Circumstances:	Attributes:	Consequences:
Attributi	*Continentia cum*	*In gestione*	*Adiuncta*	*Quae negotium*
Personae	*ipso negotio*	*negotii*	*negotio*	*consequuntur*
Nomen (*verres*)	*Summa facti*	When: *tempus,*	Species	What it is called
Natura (*barbar*)	(murder of	*occasio*	Genus	Who are the
Victum (friend	a relative)	Where: *locus*	*Contrarium*	authors of it
of nobles)	*Ante factum*	How: *modus*	*Eventus*	What people
Fortuna (rich)	(he stole a	With whom:	*Comparatio*	think is the
Studium	sword)	*facultas*	*Maius et*	nature of
(architect)	*Dum fit*		*minus*	the thing
Casum (exiled)	(slammed)		*Simile*	
Affectio (lover)	*Post factum*			
Habitus (wise)	(buried him			
Consilium	in a hidden			
Facta	place)			
Orationes				

The following argument from place is useful to clarify how these rhetorical *loci* work: How could a man be murdered in a much-frequented place? (*De Inventione*, I, 43).

4. ABELARD AND THE THIRTEENTH CENTURY

Abelard described topics as imperfect inferences, different from valid categorical syllogisms. The *maxima propositio*, expressing a principle of inference, is related to the function of invention. The *maxima* is the general principle that is useful for finding the propositions accepted by everybody or the by the wise (the *endoxa*) relative to the subject dealt with in the argument. From this perspective, the structure of an argument is similar to that of a syllogism. The main difference lies in the nature of the assumptions, the propositions connecting the general principles to the subject of the reasoning. While dialectical inferences depend on the content of the propositions (or, rather, on the terms and their connections), syllogisms depend only on the form. The difference between form

negotio, et ea esse adiuncta negotio, quae sunt ad ipsum de quo agitur negotium affecta. (*De Differentiis Topicis*, 1214C)

and content can be explained using the following examples. A syllogism such as

omnis homo est animal
sed omne animal est animatum
ergo omnis homo est animatus

depends on a rule of inference, that is (*Dialectica*, 262):

posito antecedenti ponitur consequens.

The connection between the terms of the inference depends only on their positions in the propositions. On the other hand, dialectical inferences cannot be resolved only by considering the positions of the terms. These inferences are imperfect, since assumptions are needed in order for the conclusion to follow from the premises. For instance, the consequence

si est homo, est animal

is necessarily valid, since it is known that "animal" is the genus of man and that "whatever is predicated of the species is predicated of the genus as well." The inference depends on the local connection between the terms, on the *habitudo*.[6]

The mechanism of an argument scheme (see also Rigotti, 2006) can be shown by the ancient model of Abelard, in which the assumptions are connected to the axioms, to the maxims the *locus* proceeds from (*Dialectica*, 315).

In this scheme, the dialectical inference is for the first time[7] analyzed into its components. Abelard's work is extremely relevant because it introduces a notion highly influential in the following studies on dialectics. He draws a distinction between syllogistic inferences and dialectical consequences on the basis of the structure of the logical passage. Moreover, he analyzes the structure of a topical consequence by comparing it to a syllogism. After Abelard, in the twelfth century, the notion of form

---

[6] ...unde sit locus, requiritur, a specie respondemus scientes 'hominem' ad 'animal' secundum hoc quod species eius est antecedere; cuius quidem interrogationis sententiam diligenter inquiramus. (*Dialectica*, 264)

Itaque tam ex loca differentia quam ex maxima propositione firmitas inferentiae custoditur, alio tamen et alio modo; ex differentia quidem hoc modo quod ipsa in antecedenti posita uim inferentiae tenet secundum habitudinem ex qua consequenti comparatur. Oportet enim in ipso antecedenti semper de loco differentia agi [qui] secundum habitudinem ex qua ad illatum terminum inferendum adducitur. (*Dialectica*, 263–264).

[7] Kienpointer, 1987, p. 283.

TABLE 8.8. *Mechanism of an argument scheme*

Consequentia	Si Socrates est homo, est animal
Maxima propositio	de quocumque praedicatur species, et genus
Assumptio	sed homo, qui est species animalis, praedicatur de Socrate quare animal, quod uidelicet genus ipsius est
Assumptio 1	homo est species animalis
Syllogism 1	de quocumque praedicatur species, et genus sed homo est species animalis quare si homo praedicatur de aliquo, et animal
Syllogism 2	si homo praedicatur de aliquo, et animal sed Socrates est homo quare Socrates est animal

of inference was developed into a reduction of all topical inferences to syllogisms. Later, in the thirteenth century, analytical consequences were analyzed as following from topics, "*dici de omni*" and "*dici de nullo*" (every A is B; every B is C; therefore, every A is C). Demonstration is for this reason based on a topical relation (from the whole). (Green-Pedersen, 1984, p. 256).

In the twelfth and thirteenth centuries, William of Sherwood, Peter of Spain, Robert Kilwardby, and Boethius of Dacia developed a theory of enthymemes leading to a deep distinction between *loci* and demonstrative consequences. William of Sherwood, in the twelfth century, proposed to reduce all the arguments stemming from topics to syllogisms (Green-Pedersen, 1984, p. 256). He proposed a classification of *loci* slightly different from Cicero's (Kretzmann, 1966), which with minor changes was to be adopted by the scholars in the following century.

In the thirteenth century, Kilwardby, Peter of Spain, and Boethius of Dacia developed three similar approaches to the doctrine of dialectical inferences. In Kilwardby, demonstrative syllogisms are considered to be dependent on topics from cause or definition, while dialectical inferences proceed from a great variety of middle terms; moreover, middle terms and connections between the middle term and the term in the conclusion are necessary in demonstration, but only probable in dialectical syllogisms (Stump, 1984, p. 284). Peter of Spain's theory can easily be compared to Abelard's proposal (see Stump, 1989). Peter describes enthymemes, or rhetorical syllogisms, as syllogisms with a missing premise. For instance, the enthymeme "A stone is not an animal; therefore, a stone is not a man" is reduced to a syllogism by means of the *differentia* from genus supplying the missing premise, "Every man is an animal," the bridge between the premise and conclusion.

TABLE 8.9. *William of Sherwood's classification of topoi*

Internal Grounds

Arguments from Substance	
1. From definition	Whatever is predicated of (separated from) the definition is predicated also of (separated also from) what is defined.
2. From description	Whatever is predicated of (separated from) the description is predicated also of (separated also from) what is described.
3. From the interpretation of a noun	Whatever is predicated of (separated from) the interpretation is predicated also of (separated also from) what is interpreted.

Arguments from the Concomitance of Substance	
4. From genus	Whatever is separated from the genus is separated from the species.
5. From species	Whatever is predicated of a species is predicated of the genus.

Arguments from Wholes	
6. From an integral whole	What goes together with an integral whole with respect to propositional and perceptible parts goes together with a part; what fails to go together with an integral part fails to go together with the whole.
7. From a quantitative whole	Whatever goes together with a quantitative whole either affirmatively or negatively goes together with a part.
8. From a temporal whole	What goes together with something with respect to a temporal whole goes together with it with respect to the parts.
9. From a locational whole	What goes together with something with respect to a locational whole goes together with it with respect to a part.
10. From a modificational whole	What does not go together with something with respect to a modificational whole goes together with it with respect to a part.

Arguments from Causes	
11. From material cause	If the matter is lacking, then what is dependent on the matter is lacking; if what depends on the matter exists, then that matter exists.

Arguments from Causes

12. From formal cause	Everything is capable only as much as its natural form permits.
13. From efficient cause	As the efficient cause is (is not), so is (is not) its effect.
14. From final cause	A thing whose end is good (evil) is good (evil) as a whole.
15. From generation	What goes together with (is separated from) the generation with respect to the nature of what is generated goes (does not go) together with what is generated.
16. From corruption	What is separated from (goes together with) the corruption with respect to the nature of what is corrupted does not go (goes) together with what is corrupted.
17. From uses	What goes together with (is separated from) a use insofar as it is a use goes together with (is separated from) that of which it is the use.
18. From associated accidents	If one associated accident is separated from anything, the remaining ones are separated from it also.

Extrinsic Grounds

Miscellaneous

19. From authority	What is said by many insofar as they are wise men is not to be contradicted.
20. From likeness	Of things that are alike, a like judgement is to be made.
21. From the superior	What does not go together with something with which it seems to go together the more does not go together with something with which it seems to go together the less.
22. From the inferior	What goes together with something with which it seems to go together the less goes together with something with which it seems to go together the more.
23. From proportion	From proportions a proportional judgement is inferred.

Arguments from Opposites

24. From contrary opposites	If one contrary is in a thing, the other is not; if one exhaustive contrary is not in a thing, then the other is, provided that the subject exists.

(continued)

TABLE 8.9 *(continued)*

Arguments from Opposites

25. From privative opposites	If one privative opposite is not in a thing, the other is, provided that the subject exists and the time is determinate.
26. From mutually related opposites	If one mutually related opposite is predicated of another mutually related opposite, the opposite of the first is also predicated of the opposite of the second.
27. From transumption	What goes together with something more evident as such goes together with something less evident.

Mediate Grounds

28. From coordinates	If one coordinate is predicated of (separated from) another coordinate, then a coordinate of the first is also predicated of (separated from) a coordinate of the second.
29. From grammatically related forms	If one grammatically related form is predicated of (absent from) another grammatically related form, then the form related to the first is also predicated of (absent from) the form related to the second.
30. From division	If one member of a division is separated from something, the other goes together with it; what goes together *simpliciter* with the members of a division goes together *simpliciter* with what is divided.

Modists, however, recognized the difference between demonstrative and dialectical syllogisms. In the thirteenth century, Boethius of Dacia distinguished between two kinds of consequences. From the mode of being of things, common concepts are drawn (for instance, one thing is a *genus*). These concepts, however, are relative (for instance, *genus* is not an absolute concept but is relative to its species: nothing is predicable of genus without the relation to its species). For this reason, topics drawn from them are not causes of the consequence, but signs of them (Stump, 1982). Topical consequences are not, unlike demonstrations, the cause of the relation between the terms. The link between the premises and the conclusion lies in the meaning of the terms.

5. FOURTEENTH-CENTURY LOGIC

In the fourteenth century, logicians considered all the consequences (demonstrative and dialectical) to be topic-dependent. Every dialectical syllogism was deemed to hold by virtue of a topical relation. As an anonymous commentator on Aristotle wrote, "every syllogism holds by means of the topic from a quantitative whole to its part, because the minor is part of the major" (Stump, 1989, p. 161). The enthymeme was thus held to be reducible to the first syllogism (every man is running; therefore, Socrates is running) by virtue of the following topical maxim: "Whatever is posited of a quantitative whole is posited also of its part." Every syllogism is seen, consequently, as topic-dependent, as Burleigh stated (Schupp, 1988, p. 79). Among the *loci*, therefore, were included logical rules (like the one stating that "from what is impossible anything follows") reduced to topical consequences (from the lesser, in the example just given). Topical consequences, in other words, were confused with logical rules and rules of thumb to test the validity of a consequence (as "a good consequence holds by a necessary medium"). Consequences (a term covering every passage from premise[s] to conclusion, such as syllogisms, inferences, and enthymemes) were grounded on topical *maximae propositiones*, general rules of consequences (Stump 1984, p. 295). Topical inferences, in their turn, were treated as necessary. The study of *loci* was developed toward a comprehension of dialectical and logical syllogisms under a unique principle: the relationship between antecedent and consequent by virtue of the middle term.

Burleigh and Ockham organized the consequences into classes. The principle of this classification was based on the type of medium: it could be extrinsic (such as the rule of conversion) or intrinsic (a missing premise such as "Every animal is mortal" in the consequence "Every man is an animal; therefore, every man is mortal"), formal or material (if depending on the truth value or modality of the propositions) (Boh, 1984, p. 310). According to Stump (1988, p. 40), in Burleigh's theory[8] "natural (formal) consequences hold by virtue of a term-relation (inclusion), accidental (material) consequences by virtue of a rule regarding propositions." The first consequence holds by virtue of a *locus intrinsecum*, the second by means of a *locus extrinsecum*. This account is comparable to Ockham's theory, stating that material consequences hold by virtue of *medii extrinseci* and the relations between their terms, while formal ones

---

[8] In Burleigh, in natural consequences the consequent is understood in the antecedent, while accidental consequences hold either by virtue of the significance of the terms (these are also called material consequences) or because the antecedent is impossible or the consequent is necessary (Stump, 1989). The accidental consequences, therefore, are divided in two categories, but only the second of them is described in detail. Ockham's account of accidental consequences corresponds to the formal class of accidental consequences.

hold by virtue of term relations only. The two kinds of consequences were distinguished by the two kinds of *medii*: while the intrinsic ones are based on term relationship (from definition or genus, for example), the extrinsic ones are, for instance, logical rules like "Ex impossibili sequitur quodlibet." Both kinds of rules were grouped under the name of *loci*, the unique principle for necessary consequences. In fact, the modality of the conclusion, in these consequences, is dependent on the modality of the intrinsic medium and of the antecedent. This means that the consequence, the logical relation between premise and conclusion (the *medium extrinsecum*), is always true, is always necessary. For this reason, among the topics, some *loci* such as From Authority disappeared. Because of the extension of their role to cover both dialectical and logical consequences, it was considered that *loci* included not only the ones listed by Boethius, but the Aristotelian logical rules as well. Logical rules (as "Quidquid sequitur ad consequens, sequitur ad antecedens"), distinguished by Abelard from dialectical *loci*, were included among the number of maximal propositions. For this reason, relations between propositions and relations between terms in Burleigh were confused, and the original meaning of *loci* was lost (Schupp, 1988, p. 83). Burleigh's Consequence Rules (Stump, 1989, p. 166–170) can be summarized as in Table 8.10.

6. TOPICS IN THE RENAISSANCE AND IN THE PORT ROYAL LOGIC

The study of dialectics as a field of knowledge distinct from logic was rediscovered in the Renaissance, with Rudulphus Agricola. Topics were regarded during this period as means of *inventio*, and for this reason a clear difference between rhetorical and dialectical *loci* disappeared or became much less distinct than it had been during the Middle Ages. In the Renaissance, logic and dialectics were clearly separated. While logic was considered to be dealing with abstract relations between concepts, matters related to dialogue were treated by the field of dialectic (Agricola, 1976, pp.12–13).

The interest in the study of *loci* as a means of discovery pretty well ended in the seventeenth century, with the advent of the Logic of Port-Royal.[9] Arnauld described them not as a means to find arguments, but only as principles for their classification. He wrote (Arnauld, 1964,

---

[9] See Kienpointner, 1997, p. 228.

TABLE 8.10. *Summary of Burleigh's consequence rules*

---

1. Whatever follows from the consequent follows from the antecedent.
2. Whatever the antecedent follows from, the consequent also follows from.
3. Whatever the opposite of the consequent follows from, the opposite of the antecedent also follows from.
4. From an exclusive proposition to a universal proposition with the terms transposed, and vice versa, is a good consequence. For example: Only a man is an animal; therefore, every animal is a man – and vice versa.
5. From an exclusive to its prejacent is a good consequence. For example: Only a man is running; therefore, a man is running.
6. From a superior to an inferior with an exclusive expression is a fallacy of the consequent. For example : Only an animal is running; therefore, only a man is running.
7. From what is impossible anything follows.
8. What is necessary follows from anything.
9. One or the other of a pair of contradictory propositions is compossible with any possible proposition.
10. Every proposition that includes its opposite, such as "You know that you are a stone," is impossible.
11. Every true past tense proposition is necessary.
12. The opposite of the consequent is not compatible with the antecedent.
13. Whatever is incompatible with the consequent is incompatible with the antecedent.
14. A good consequence holds by a necessary medium.
15. Every good consequence has to be reduced to a syllogism.
16. From one premise nothing follows, either enthymematically or syllogistically.
17. If from the opposite of the conclusion of some syllogism and one of the premises there follows the opposite of the other premise, then the original syllogism was good.
18. Whatever does not follow from the antecedent of a consequence with some addition, either syllogistically or in any other way, does not follow from the consequent with the same addition.
19. Whatever follows from the consequent with some addition follows, either necessarily or formally, from the antecedent with the same addition.

---

p. 237): "True, all arguments can be classified under the headings or general terms called Topics; but arguments are not discovered by means of this classification. Common sense, a careful consideration of subject, the knowledge of many truths – these enable us to find arguments; and then skill allows us to classify these arguments under certain headings: the Topics." He grouped topics into three main categories: those taken from grammar (based on etymology or words derived from a common root), those taken from logic (based on genus, species, difference, property, accident, definition, and division), and those taken from metaphysics (based on cause, effect, relation of the whole to a part, and relation of opposition).

The roots of the Port Royal theory can be found in Petrus Ramus, who separated rhetoric, or *elocutio*, from logic, also called *inventio* and *dispositio*. The two complementary arts of rhetoric and dialectic were separated, and specific tasks were assigned to each of them. Topics were considered a means to discover arguments: their role preceded their *dispositio* in a valid consequence (Arnauld, 1964, p. 236). Consequential link and discovery of premises (*inventio*) were separated in such a categorization, and topics were deprived of their original meaning of general rules of plausible reasoning. They became tools, in other words, instead of axioms.

Port Royal made another important distinction by shifting the focus from the process of arguing to the product of arguing: *loci* were not described as tools, but simply as categorizations of arguments. They, in other words, had no function at all in the process of reasoning. The art of arguing was considered primitive, without any need of artificial tools such as topics. This was a consequence of the Ramistic theory. Therefore, *loci* were described as a means of categorizing arguments, a kind of a posteriori operation: "We can, of course, formulate rules for walking by observing what nature has made us do; but we never walk with the help of these rules. Similarly, in ordinary discourse we make use of all the Topics. ... But it is not by any express reflection on the Topics that we arrive at the thoughts we seek" (Arnauld, 1964, p. 238).

If considered simply as tools for discovering thoughts, topics are of no use at all. They become classifications, useful as a way to learn to regard things in their parts and facets (p. 240). The Port Royal account of the *loci* is different from the role we assign to them in the logic of the arguments. However, it is interesting, as it helps us to understand the reasons why the art of rhetoric was abandoned during the following centuries. Another relevant aspect of this theory is the classification of topics. The focus is on the kinds of arguments, not on the nature of consequences. The division is based on the disciplines the premises belong to, the fields of human knowledge.

Arnauld refused to use the Aristotelian logic based on classes: he considered the categories "of very little use; they help but slightly in the formation of judgement, the true end of logic" (Arnauld, 1964, p. 43). Port Royal logic is a propositional logic, distinct from the Aristotelian logic of classes. Categories, organizations of the linguistic means to conceive reality, are ignored; the relation language-reality, consequently, is not seen as a relation of organization by means of linguistic instruments, but as a

TABLE 8.11. *Port Royal topics*

Topics Taken from Grammar	Topics Taken from Logic	Topics Taken from Metaphysics
a. Arguments based on etymology b. Arguments based on words derived from a common root	a. Whatever is affirmed or denied of the genus is affirmed or denied of the species. b. In denying the genus, we also deny the species. c. In denying all of its species, we deny the genus. d. If we affirm or deny a difference, then we affirm or deny the corresponding species. e. If we can affirm or deny the essential attribute, then we can affirm or deny the species. f. We affirm or deny what is expressed by the defined word when we affirm or deny what is expressed by the defining words.	a. *Cause* 1. Final cause 2. Efficient cause 3. Material cause 4. Formal cause b. *Effect* c. *Relation of the whole to part* d. *Relation of opposition* 1. Relative opposition 2. Contrary opposition 3. Privative opposition 4. Contradictory opposition

form of representation. Enthymemes are reduced to syllogisms with an unexpressed premise: the only difference lies in the expression. Topics do not represent the meaning and reason of enthymematic passages, as distinct from logical ones, but simply classifications of arguments, without any function in the consequential link. The focus on the different kinds of relations between classes shifts to the nature or origin of the arguments. They may derive from logic, grammar, or metaphysics. The metaphysical category is particularly interesting for the study of the Port Royal theory of consequences. The fact that oppositions, cause, effects, and part-whole relation derive from metaphysics, and not from the linguistic organization of reality, explains the different way of conceiving argumentative consequences. They reflect the order that exists in the world independently of language. By contrast, the Aristotelian focus on the logic of classes explains the argumentative passages from a linguistic point of view, as relations between linguistic elements. *Topoi* represent the possible reasonable links between terms – for instance, "black" and "white," or "traces" of the animal and its "presence." Language, being an instrument of organization, reflects our ways of conceiving reality: topics are, as a consequence, possible ways of organizing the world, not aspects of the world itself.

7. MODERN THEORIES OF SCHEMES

7.1. Perelman

The study of topics in the modern age begins with Perelman and Olbrechts-Tyteca's *New Rhetoric*. They list a set of thirteen argumentation schemes instead of the Aristotelian twenty-eight common *topoi*. Their accounting of topics as different in number and form from the traditional ones has opened a new perspective on them. *Topoi* are viewed as "catalogs of the habits of mind endemic to a given culture."[10] In other words, they are not seen as general formal or universal principles, as in the Middle Ages, but as dependent on culture and society. Perelman divides quasi-logical arguments, those based on forms of inference, from others that are dependent on material elements. Perelman's system conciliates the division made by the ancient commentators between dialectical, formal *loci* and rhetorical ones. Perelman classifies the topics into two main categories that correspond to the processes of argumentation: *loci* from association and *loci* from dissociation. The first class is composed of processes that bring separate elements together in such a way that it is possible to conceive them as a unity, evaluating or organizing them by means of each other (Perelman, 1969, p. 190). From the example from Quintilian given by Perelman (1969, p. 223), it is clear what this kind of reasoning is taken to be.

"I have accused; you have condemned," is the famous reply of Domitius Afer.

In this example, two different concepts, accusation and condemnation, are associated into a unity through reasoning from reciprocity: they jointly bring about the same effect.

The other class of *loci* groups correspond to processes of dissociation. These techniques aim to dissociate, separate, or disunite elements conceived as a whole. Consider the example given by Perelman (1969, p. 442).

What religion do I profess? None of all those that you mention. – And why none? – For religion's sake!

[10] Warnick, 2000, p. 111.

In this example from Schiller, Perelman highlights the rhetorical dissociation of the unitary concept of religion into two concepts of *apparent religion* versus *true religion*, or *positive religion* versus *natural religion*. The *New Rhetoric* thinks of rhetoric as the complementary process of connection and rejection of connections between concepts. Arguments from association are divided into three main classes: quasi-logical arguments, relations establishing the structure of reality, and arguments based on the structure of reality. The dissociation constitutes a distinct class. Table 8.12 presents Perelman's categorization (Kienpointner, 2002).

7.2. Toulmin

Toulmin's book *The Uses of Argument* describes argumentation as a process and not as a product of the communicative event. He does not mention topics in his work, but he refers to them with his notion of field.[11] Toulmin defines a field as a logical type in the use of argument: the pattern of reasoning representing the link between premises and conclusion. Fields are the places where warrants are found, in the sense that they belong to these general types.

7.3. Hastings

Hastings' work on the modes of reasoning in argumentation may be considered the basis of the most important theories of contemporary argumentation. In particular, he has improved Toulmin's argumentation diagrams, introducing in his theory the systematic organization of warrants into categories of argumentation schemes. He describes nine modes of reasoning, grouped into three classes: the first one includes verbal and semantic procedure, the second one causal connections, and the third one arguments that support either verbal or causal conclusions. The schematic classification from Hastings (1963, p. 139) is shown in Table 8.13.

Hastings' most innovative and original idea, however, lies in his theory of evaluation of the argumentation schemes. There he lists under each scheme the reasons leading us to acceptance of the argument, and, from

---

[11] "The notion of 'argument fields', as it has been defined by Toulmin, is indeed vague.... According to David Zarefsky (1996, 49), the term 'field' was a metaphor for the location of arguments. In this sense Toulmin's argument fields resemble very much Aristotelian *topoi*, which are nothing but the most important means of selecting the arguments for the enthymemes: they are a repository of arguments." (S. Tardini, 2005, p. 283)

TABLE 8.12. *Perelman's classification of loci*

Quasi-logical Arguments

1. Contradiction and incompatibility
2. Identity and definition
3. Analyticity, analysis, and tautology
4. The rule of justice
5. Arguments of reciprocity
6. Arguments of transitivity
7. Inclusion of the part in the whole
8. Division of the whole into its parts
9. Arguments by comparison
10. Argumentation by sacrifice
11. Probabilities

Arguments Based on the Structure of Reality

Sequential Relations	The Relations of Coexistence	Double Hierarchy Argument	Differences of Degree and Order
1. Causal link	1. The person and his acts		
2. Pragmatic argument	2. Argument from authority		
3. Ends and means	3. The Speech as an act of the speaker		
4. Argument of waste	4. The group and its members		
5. Argument of direction	5. Act and essence		
6. Unlimited development	6. Symbolic relation		

The Relations Establishing the Structure of Reality

Establishment through Particular Case	Reasoning by Analogy
1. Example	1. Analogy
2. Illustration	2. Metaphor
3. Model and anti-model	

these reasons, he derives the critical questions, the necessary conditions that must be fulfilled in order for the scheme to be correctly used.

For instance, he describes the argument from sign (Hastings 1963, p. 55) as:

> The assertion of the existence of B on the presence of A is on the basis of an observed or understood correlation between the occurrence of A and B.

TABLE 8.13. *Hastings' classification of schemes*

Verbal Reasoning	Causal Reasoning	Verbal or Causal Reasoning
1. Argument from example 2. Argument from criteria to a verbal classification 3. Argument from definition	4. Argument from sign 5. Argument from cause 6. Argument from circumstantial evidence	7. Argument from comparison 8. Argument from analogy 9. Argument from testimony

It is based on an inverse order of the reasoning process in the causal or correlative relation involved, $p \supset q, p, q$. In this argument, he points out three aspects for evaluation, stated in form of questions:

1. What is the correlation of the sign with the event signified?
2. Are there other events which would more reliably account for the sign?
3. With what certainty [moving from the sign to the cause and from the cause to the other effect] is the effect produced? Is the cause adequate? Would other factors intervene?

Hastings' classification schemes and theory of critical questions have influenced both the pragma-dialectical school and Walton's pragmatic account of argumentation.

7.4. Schellens

Schellens's argument schemes (Schellens, 1985) are mainly drawn from Hastings's. They are grouped into four classes according to their pragmatic function (Kienpointner, 1992, pp. 201–215). An argument can have two functions: advancing a description (Socrates *is* an animal, because he is a man) or a norm (You *must* pay the fine, because you ran the red light). According to this division, arguments can be based on regularities and, accordingly, lead to descriptive conclusions. They can be founded on rules and be normative. Or they can be pragmatic, that is, normative *and* descriptive, or unbound, that is, normative *or* descriptive. Every scheme is associated with a set of evaluation questions, similar to Hastings's critical questions.

7.4. Kienpointner

In *Alltagslogik*, Kienpointner classifies about sixty context-independent argument schemes into three main groups according to their relation to the rule, or generalization (*endoxon*). Argument schemes may be based on rules taken for granted,[12] establish them by means of induction,[13] or illustrate or confirm them. Argument schemes, in their turn, may have descriptive or normative variants and different logical forms (*modus ponens, modus tollens*, disjunctive syllogism, etc.).

The classification can be represented as in Table 8.14 (Kienpointner, 1992, p. 246).

7.5. Grennan

In Grennan's (1997, pp. 163–165) typology, all the structurally valid inductive[14] inference patterns are classified according to nine warrant types (effect to cause, cause to effect, sign, sample to population, parallel case, analogy, population to sample, authority, ends-means), combined with the types of claims the warrant connects (utterance types expressing the minor premise and the conclusion of an argument, such as obligation.)[15] Both the abstract form of the inference and the pragmatic role of the utterances expressing the sentences are taken into consideration.

7.6. Van Eemeren and Kruiger

In the pragma-dialectical school, three main categories of *topoi*, called argumentation types, are pointed out: the sign relationship, the comparative relationship, and the causal relationship. Every type has a set of critical questions and a set of assessment conditions (Van Eemeren and Kruiger, 1986, pp. 73–74).

8. Conclusions

To conclude, it is helpful to point out some distinctions among the argument schemes analyzed in the previous chapters. This partition can

---

[12] The structure of an argument from rule can be understood from the following example: what is predicated of the definition is predicated of the definiens as well; X is predicated of the definition; X is predicated of the definiens as well. In a concrete argumentation, the rule or generalization might be, for instance, "If a rational animal is risible, a man is risible" (see Keinpointner, 1992, p. 250).

[13] In inductive argumentation (if x1 is Y, x2 is Y, x3 is Y... therefore X is Y), the rule is the outcome of the reasoning.

[14] Inferences, from an informal logic perspective, are considered inductive, since argumentation does not deal with deductive validity. The criterion for discriminating between acceptable and unacceptable patterns is provided by a logical intuition.

[15] Grennan considers not only the pragmatic role, but also the different effects descriptive and normative utterances have on the recipient of the message. The claim types are: obligation, supererogatory actuative, prudential actuative, evaluative, physical-world, mental world, constitutive-rule, and regulative-rule.

TABLE 8.14. *Kienpointner's classification of schemes*

| Argument Schemes Using Rules | | | | Argument Schemes Establishing Rules | Argument Schemes Using or Establishing Rules |
Classification Schemes	Comparison Schemes	Opposition Schemes	Causal Schemes		
a. Definition	a. Similarity	a. Contradictory	a. Cause	1. Inductive argumentation from example	1. Illustrative argumentation from example
b. Genus-species	b. Resemblance	b. Contrary	b. Effect		2. Argumentation from analogy
c. Whole-part	c. Difference	c. Relative	c. Reasons		3. Argumentation from authority
	d. *A maiore*	d. Incompatible	d. Consequence		
	e. *A minore*		e. Means		
			f. End		

TABLE 8.15. *Scheme classification system of the Amsterdam School*

Sign Relationship	Comparative Relationship	Casual Relationship
For X, Y is valid because for X, Z is valid, and Y is symptomatic of Z.	For X, Y is valid because for Z, Y is valid, and X is comparable to Z.	For X, Y is valid because for X, Z is valid, and Z leads to Y.
Critical Questions	Critical Questions	Critical Questions
1. Is Z really valid for X?	1. Is Y valid for Z?	1. Is Z really valid for X?
2. Is Y really symptomatic of Z?	2. Is X really comparable to Z?	2. Does Z really lead to Y?
3. Can Z have other symptoms?	3. Is X also in relevant terms incomparable to Z?	3. Can Z lead somewhere else?
4. Can Y be a symptom of something else?	4. Can Z be better compared to something else?	4. Can Y be the result of something else?

be connected to the Aristotelian conception of argumentation and discourse (see also Rigotti, 2006), distinguishing among the three main
components of conversation – the speaker, the hearer, and the discourse
(Rigotti, 2005).

Given this approach, the schemes examined so far can be divided into
schemes directed to the hearer (emotive arguments, such as ethotic arguments, arguments from threat, and appeal to pity arguments) or to the
discourse (such as the reasoning arguments, arguments from verbal classification, arguments from division, and those from opposition). Other
arguments not directly concerned with the components of the discourse
rely on external factors, such as the arguments from expert or popular
opinion. A second criterion of classification may concern the conclusion, whether it is an opinion or advice. A third observation can be made

TABLE 8.16. *Levels of analysis of schemes*

Role of the Schemes		
Support	Refutation	
Role in the Conversation: Type of Conclusion		
Opinion	Advice or command (practical decision)	
Type of Argumentation according to the Components of the Discourse		
Schemes directed to the discourse	Schemes directed to the interlocutor	Schemes based on external factors

about the role of the schemes in the argumentation. Some schemes are refutation schemes; their role is to be the refutation of the argument constructed following a specific scheme. For instance, argument from arbitrariness of a verbal classification is a refutation scheme of the argument from verbal classification. Not all these criteria are compatible. On the contrary, they represent different levels of analysis of the schemes.

In this chapter, a brief summary of the history of topics and of their classification has been traced. Most argument schemes derive from the dialectical and rhetorical common places, the *loci*. Some of them are based on logical-semantic properties and are necessarily true; others are only plausible. The classic heritage on topics was forgotten or ignored for centuries, and only with modern argumentation studies have these works been inquired into again and applied to the new discipline. Argument schemes stem from both dialectical and rhetorical topics. They include not only semantic inferences but also places from circumstances.

The purpose of this panoramic survey of the tradition has been to show how the schemes presented in the previous chapters are connected to ancient logical and rhetorical studies. Moreover, by comparing them to the modern studies on topics, it is possible to understand their originality and their place in the history of *loci*. The schemes presented in the chapters so far include not only dialectical and rhetorical topics, but also *loci* used to persuade by means of appeal to emotions (Cicero).

9

A User's Compendium of Schemes

In this chapter a compendium of all the schemes discussed is given, along with the source of the scheme in the argumentation literature. We have tried to make the compendium most useful to the reader by presenting the schemes that represent the most commonly used forms of argument, including not only those used in everyday discourse, but also certain schemes that are important in legal and scientific reasoning. Many of these schemes have subtypes, and research on the classification of the subtypes, and even more generally, research on determining which schemes are subspecies of other schemes, is not yet at an advanced stage. Thus while we have made occasional remarks on these matters, hoping to provide some insight, we have not generally listed all the known subspecies of the schemes. There are two especially important exceptions that need to be noted.

We have included a fairly comprehensive account of the known subschemes of the argument from popular opinion, to give the reader an idea of how it can be important to recognize many of these different subschemes in dealing with common arguments of the type associated with informal fallacies. The other important exception is the case of the *argumentum ad hominem*. Many subschemes for this type of argument have been recognized, and much work on trying to organize and classify them has been conducted (Walton, 1998). Also, recent research on the project of formalizing argumentation schemes has chosen the case of the *argumentum ad hominem* as a key example (Verheij, 2003). For these reasons, we have chosen only to include the most basic and central types of *argumentum ad hominem* in the compendium, reserving the problem of how to classify the subschemes, and various related schemes for this type of argumentation, for Chapter 10. Thus the reader who is looking for specialized types of ad hominem arguments will need to consult Chapter 10, while the reader who wants to know only the most general schemes for the most common types of arguments can use the compendium given in this chapter as his or her source of information.

In the compendium each scheme is numbered, and the reader can see that sixty primary schemes have been recognized, along with subschemes under many of them. This number, as suggested earlier, is somewhat arbitrary, as we have chosen only what we take to be the most important schemes for the reader to know about. We could have added a lot more schemes, but decided that doing so would not be very helpful, since what is most important at this point is the project of classifying schemes, or to put it in computer terminology, to provide an ontology of schemes. If we had such a device, we could then organize them in a coherent way even if we had a very large number of schemes. A device of this sort could be very useful for automated collection of data about arguments in natural language discourse, for example, if we could build software that could recognize the schemes.

The reader will notice in the compendium that each scheme is listed with its matching set of appropriate critical questions. Also, references are given to the source where the scheme was found in the literature. This documentation should prove helpful to the reader who wants to use a particular scheme in research and needs to document where it can be found, or who wants to search out more details about it. Occasionally, as well, a comment or two has been inserted to guide the reader by providing clarifications and answers to frequently asked questions.

1. ARGUMENT FROM POSITION TO KNOW

Major Premise: Source *a* is in position to know about things in a certain subject domain *S* containing proposition *A*.
Minor Premise: *a* asserts that *A* is true (false).
Conclusion: *A* is true (false).

Critical Questions

CQ1: Is *a* in position to know whether *A* is true (false)?
CQ2: Is *a* an honest (trustworthy, reliable) source?
CQ3: Did *a* assert that *A* is true (false)?

References

Douglas Walton, *Legal Argumentation and Evidence*. University Park: Pennsylvania State University Press, 2002, p. 46.

2. ARGUMENT FROM EXPERT OPINION

Major Premise: Source E is an expert in subject domain S containing proposition A.
Minor Premise: E asserts that proposition A is true (false).
Conclusion: A is true (false).

Critical Questions

CQ1: *Expertise Question*: How credible is E as an expert source?
CQ2: *Field Question*: Is E an expert in the field that A is in?
CQ3: *Opinion Question*: What did E assert that implies A?
CQ4: *Trustworthiness Question*: Is E personally reliable as a source?
CQ5: *Consistency Question*: Is A consistent with what other experts assert?
CQ6: *Backup Evidence Question*: Is E's assertion based on evidence?

References

Douglas Walton, *Legal Argumentation and Evidence*. University Park: Pennsylvania State University Press, 2002, pp. 49–50.
Douglas Walton, *Appeal to Expert Opinion*. University Park: Pennsylvania State University Press, 1997, pp. 211–225.

3. ARGUMENT FROM WITNESS TESTIMONY

Position to Know Premise: Witness W is in a position to know whether A is true or not.
Truth Telling Premise: Witness W is telling the truth (as W knows it).
Statement Premise: Witness W states that A is true (false).
Conclusion: A may be plausibly taken to be true (false).

Critical Questions

CQ1: Is what the witness said internally consistent?
CQ2: Is what the witness said consistent with the known facts of the case (based on evidence apart from what the witness testified to)?
CQ3: Is what the witness said consistent with what other witnesses have (independently) testified to?
CQ4: Is there some kind of bias that can be attributed to the account given by the witness?
CQ5: How plausible is the statement A asserted by the witness?

References

Douglas Walton, Henry Prakken, and Chris Reed, "Argumentation Schemes and Generalisations in Reasoning about Evidence." *Proceedings of the Ninth*

International Conference on Artificial Intelligence and Law, Edinburgh, 2003. New York: ACM Press, 2003, p. 35.
Douglas Walton, *Witness Testimony Evidence.* Cambridge: Cambridge University Press, 2007, p. 60.

4. ARGUMENT FROM POPULAR OPINION

General Acceptance Premise: A is generally accepted as true.
Presumption Premise: If A is generally accepted as true, that gives a reason in favor of A.
Conclusion: There is a reason in favor of A.

Critical Questions

CQ1: What evidence like a poll or an appeal to common knowledge, supports the claim that A is generally accepted as true?
CQ2: Even if A is generally accepted as true, are there any good reasons for doubting that it is true?

References

Douglas Walton, *Fundamentals of Critical Argumentation.* New York: Cambridge University Press, 2006, p. 91.
Douglas Walton, *Informal Logic.* New York: Cambridge University Press, 1989, p. 89.
Douglas Walton, *Appeal to Popular Opinion.* University Park: Pennsylvania State University Press, 1999, pp. 223–226.

Comment: In argument from popular opinion, various other types of argument tend to support the basic pop scheme type of argument, bolstering the original argument, which often tends to be weak by itself. The general problem of classifying subtypes in relation to more general schemes is considered in Chapter 10.

SUBTYPES OF ARGUMENT FROM POPULAR OPINION

4.1. Pop Scheme

Premise: Everybody in a particular reference group G accepts (rejects) A.
Conclusion: A is true (false)/or: you should accept (reject) A.

4.2. Position-to-Know *Ad Populum* Argument

Premise 1: Everybody in this group G accepts A.
Premise 2: This group is in a special position to know that A is true.
Conclusion: Therefore, A is (plausibly) true.

4.3. Expert Opinion *Ad Populum* Argument

Premise 1: Everybody in this group *G* accepts *A*.
Premise 2: *G* is a group of experts in a domain of knowledge.
Conclusion: Therefore, *A* is true.

4.4. Deliberation *Ad Populum* Argument

Premise 1: Everybody in group *G* accepts *A*.
Premise 2: Group *G* has deliberated intelligently and extensively on whether to accept proposition *A* or not.
Conclusion: Therefore, *A* is (plausibly) true.

4.5. Moral Justification *Ad Populum* Argument

Premise 1: Everybody who is good, or who represents a group *G* with good qualities, accepts policy *P*.
Premise 2: Your goal is (or should be) to be a good person, or a member of a group with good qualities.
Conclusion: Therefore, you should accept *P*.

4.6. Moral Justification (Excuse Subtype) *Ad Populum* Argument

Premise 1: Everybody in a group *G* does *x* (or accepts proposition *A* as a policy).
Premise 2: Doing *x* (or accepting *A*) shows that *x* (or policy *A*) is an acceptable norm of conduct for *G*.
Premise 3: I (the speaker) am a member of *G*.
Conclusion: Therefore, my doing *x* (or accepting *A*) is morally justified as an acceptable action (or policy).

4.7. Snob Appeal *Ad Populum* Argument

Premise 1: Everybody in this group *G* accepts *A* (or has some property, or possesses some object).
Premise 2: This group *G* is elite, that is, everyone who belongs to it has prestige.
Premise 3: Prestige is an important goal for you (the respondent).
Premise 4: If you accept *A* (acquire property *P* or buy object *O*), then you will a member of the group *G*.
Conclusion: Therefore, you ought to accept *A* (etc).

4.8. Appeal to Vanity *Ad Populum* Argument

Premise 1: Everybody in this group of admired (popular) people *G* accepts *A* (possesses *P*, etc.)
Premise 2: If you carry out action *x*, then you will belong to this group *G*.
Conclusion: Therefore, you should carry out action *x*.

4.9. Rhetoric of Belonging *Ad Populum Argument*

Premise 1: Everybody in this group *G* accepts *A*.
Premise 2: Being a member of this group *G* is highly valued for you (the respondent).
Premise 3: If you do not accept *A*, you will be out of this group *G*.
Conclusion: Therefore, you should accept *A*.

4.10. Common Folks *Ad Populum* Argument

Premise: I (the speaker) am an ordinary person, that is, I share a common background with you (the audience).
Conclusion: Therefore, you ought to take what I say as being more credible or acceptable.

4.11. Common Folks (Group Subtype) *Ad Populum* Argument

Premise 1: I (the speaker) am an ordinary person, that is, I share a common background with the members of this audience (group *G*).
Premise 2: You (the respondent) are a member of this audience (group *G*).
Conclusion: Therefore, you should accept what I say.

References

Douglas Walton, *Fundamentals of Critical Argumentation*. New York: Cambridge University Press, 2006, p. 91.
Douglas Walton, *Informal Logic*. New York: Cambridge University Press, 1989, p. 89.
Douglas Walton, *Appeal to Popular Opinion*. University Park: Pennsylvania State University Press, 1999, pp. 223–226.

Comment: It would take up too much space, we judged, to offer examples of all these types of arguments, as examples of all of them can be found in the references just cited. However, we do offer one example of the rhetoric of belonging argument:

I think we can all agree that the overwhelming majority of the leadership of the American movement is composed of decent, honest, dedicated people who have made a great contribution involving great personal sacrifice, helping to build a decent American labor movement.... We happen to believe that leadership in the American movement is a sacred trust. We happen to believe that this is no place for people who want to use the labor movement to make a fast buck. (Bailey, 1983, p. 134, quoted in Walton, 1999, p. 217).

From this example, the reader can get an idea of what some of the other subtypes are like as well.

5. ARGUMENT FROM POPULAR PRACTICE

Major Premise: A is a popular practice among those who are familiar with what is acceptable or not in regard to A.

Minor Premise: If A is a popular practice among those familiar with what is acceptable or not with regard to A, that gives a reason to think that A is acceptable.

Conclusion: Therefore, A is acceptable in this case.

Critical Questions

CQ1: What actions or other indications show that a large majority accepts A?

CQ2: Even if large majority accepts A as true, what grounds might here be there for thinking they are justified in accepting A?

References

Douglas Walton, *Fundamentals of Critical Argumentation*. New York: Cambridge University Press, 2006, p. 93.

Comment: This type is obviously a variant on the pop scheme.

6. ARGUMENT FROM EXAMPLE

6.1. Argument from Example

Premise: In this particular case, the individual a has property F and also property G.

Conclusion: Therefore, generally, if x has property F, then it also has property G.

Critical Questions

CQ1: Is the proposition claimed in the premise in fact true?

CQ2: Does the example cited support the generalization it is supposed to be an instance of?

CQ3: Is the example typical of the kinds of cases the generalization covers?

CQ4: How strong is the generalization?

CQ5: Do special circumstances of the example impair its generalizability?

References

Douglas Walton, *A Pragmatic Theory of Fallacy*. Tuscaloosa and London: University of Alabama Press, 1995, p. 135.

6.2. Argument from Illustration

Premise 1: Usually, if x has property F (belongs to class F), x has property G.
Premise 2: In this case, k has property F and property G.
Conclusion: The rule is valid.

6.3. Argument from Model

Premise 1: Individual P is prestigious for (admired by) individuals Qs.
Premise 2: P carries out acts A.
Premise 3: If P carries out A, A are worth being imitated by Qs.
Conclusion: Qs should do A.

6.4. Argument from Anti-Model

Premise 1: Individual P is not prestigious for (not admired by) individuals Qs.
Premise 2: P carries out acts A.
Premise 3: Qs want to be different from P.
Conclusion: Qs should not do A.

References

Ch. Perelman and L. Olbrechts-Tyteca, *The New Rhetoric: A Treatise on Argumentation*. Notre Dame, Ind.: University of Notre Dame Press, 1969, pp. 357–368.

7. ARGUMENT FROM ANALOGY

Similarity Premise: Generally, case C_1 is similar to case C_2.
Base Premise: A is true (false) in case C_1.
Conclusion: A is true (false) in case C_2.

Critical Questions

CQ1: Are there differences between C_1 and C_2 that would tend to undermine the force of the similarity cited?
CQ2: Is A true (false) in C_1?
CQ3: Is there some other case C_3 that is also similar to C_1, but in which A is false (true)?

References

Douglas Walton, *Fundamentals of Critical Argumentation*. New York: Cambridge University Press, 2006, pp. 96–97.

Comment: This scheme was called the core scheme as presented in Chapter 2, section 4, but the reader might recall that version two was also presented as a variant on the core scheme.

8. PRACTICAL REASONING FROM ANALOGY

8.1. Positive Schema

Base Premise: The right thing to do in S_1 was to carry out action x.
Similarity Premise: S_2 is similar to S_1.
Conclusion: Therefore, the right thing to do in S_2 is carry out x.

8.2. Negative Schema

Base Premise: The wrong thing to do in S_1 was to carry out x.
Similarity Premise: S_2 is similar to S_1.
Conclusion: Therefore, the wrong thing to do in S_2 is carry out x.

References

Douglas Walton, *Informal Logic*. New York: Cambridge University Press, 1989, p. 257.

9. ARGUMENT FROM COMPOSITION

9.1. Generic Composition

Premise: All the parts of X have property Y.
Conclusion: Therefore, X has property Y.

Critical Questions

CQ1: Is property Y compositionally hereditary with regard to aggregate X (when X [the whole] has property Y, then every part that composes X has property Y)?

References

Douglas Walton, *Informal Logic*. New York: Cambridge University Press, 1989, p. 130.

9.2. Inclusion of the Part in the Whole

Premise 1: y is a species (part) of X.
Premise 2: X is A.
Conclusion: y is A (is less A than X, because it is a part of it; it is less A than X because it is a smaller part of it).

References

Ch. Perelman and L. Olbrechts-Tyteca, *The New Rhetoric: A Treatise on Argumentation*. Notre Dame, Ind.: University of Notre Dame Press, 1969, pp. 231–233, 234–241.

10. ARGUMENT FROM DIVISION

10.1. Generic Division

Premise: X has property Y.
Conclusion: Therefore, all the parts of X have property Y.

Critical Questions

CQ1: Is property Y divisionally hereditary with regard to aggregate X (when every part that composes X has property Y, then X [the whole] has property Y)?

References

Douglas Walton, *Argument: The Logic of Fallacies*. Toronto: McGraw-Hill Ryerson, 1982, pp. 206–208.

10.2. Division of the Whole into its Parts

Premise 1: X is the whole of $x_1, x_2, \ldots x_n$ ($x_1, x_2, \ldots x_n$ are the parts of the whole X).
Premise 2: Only if x_1, or x_2, or ... x_n is A, X is A.
Premise 3: x_1 is A (no x is A).
Conclusion: X is A (X is not A).

References

Ch. Perelman and L. Olbrechts-Tyteca, *The New Rhetoric: A Treatise on Argumentation*. Notre Dame, Ind.: University of Notre Dame Press, 1969, pp. 231–233, 234–241.

11. ARGUMENT FROM OPPOSITIONS

11.1. Descriptive Schemes

Major Premise: If X presents the predicate P, X cannot present the contradictory (contrary, converse, incompatible) predicate P' at the same time and in the same respect.
Minor Premise: X presents P.
Conclusion: X cannot present P' at the same time and in the same respect.

Major Premise: If X does not present the predicate P, X must present the contradictory predicate P' at the same time and in the same respect.
Minor Premise: X does not present P.
Conclusion: X must present P' at the same time and in the same respect.

Major Premise: If X does not present the predicate P, X must present the complementary contrary predicate P' at the same time and in the same respect.
Minor Premise: X does not present P.
Conclusion: X must present P' at the same time and in the same respect.

11.2. Normative Schemes

Major Premise: If X receives the evaluation W, X cannot receive the opposite evaluation W' at the same time and in the same respect.
Minor Premise: X receives the evaluation W'
Conclusion: X cannot receive the evaluation W' at the same time and in the same respect.

12. RHETORICAL ARGUMENT FROM OPPOSITIONS

12.1. Normative Schemes

Major Premise: If the treatment Y is required for the subject X, for the contrary subject X' the contrary treatment Y' is required.
Minor Premise: The treatment Y is required for the subject X.
Conclusion: The contrary treatment Y' is required for the subject X'.

12.2. Descriptive Schemes

Major Premise: If X is in relation R with Y, Y can be in the converse relation R' with X and contrariwise.
Minor Premise: X is in relation R with Y.
Conclusion: Y can be in the converse relation R' with X.

Comment: The key notions of opposition and rebuttal are discussed in Chapter 7.

13. ARGUMENT FROM ALTERNATIVES

13.1. Cognitive Schemes

Major Premise: Either X or Y can be the case.
Minor Premise: X is plausibly not the case.
Conclusion: Y is plausibly the case.

13.2. Normative Schemes

Major Premise: Either X or Y can be required.
Minor Premise: X is required.
Conclusion: Y is not required.

References

Manfred Kienpointner, *Alltagslogik, Struktur und Funktion von Argumentations-mustern.* Stuttgart–Bad Cannstatt: Frommann-Holzboog, 1992, p. 306.

14. ARGUMENT FROM VERBAL CLASSIFICATION

Individual Premise: *a* has property *F*.
Classification Premise: For all *x*, if x has property *F*, then *x* can be classified as having property *G*.
Conclusion: *a* has property *G*.

Critical Questions

CQ1: What evidence is there that *a* definitely has property *F*, as opposed to evidence indicating room for doubt about whether it should be so classified?
CQ2: Is the verbal classification in the classification premise based merely on an assumption about word usage that is subject to doubt?

References

Douglas Walton, *Fundamentals of Critical Argumentation.* Cambridge: Cambridge University Press, 2006, p. 129.

15. ARGUMENT FROM DEFINITION TO VERBAL CLASSIFICATION

Definition Premise: *a* fits definition *D*.
Classification Premise: For all *x*, if *a* fits definition *D*, then *x* can be classified as having property *G*.
Conclusion: *a* has property *G*.

Critical Questions

CQ1: What evidence is there that *D* is an adequate definition, in light of other possible alternative definitions that might exclude *a*'s having *G*?
CQ2: Is the verbal classification in the classification premise based merely on a stipulative or biased definition that is subject to doubt?

16. ARGUMENT FROM VAGUENESS OF A VERBAL CLASSIFICATION

Premise 1: If an argument, α, occurs in a context of dialogue that requires a certain level of precision, but some property *F* that occurs in α is defined in a way that is too vague to meet the requirements of that level of precision, then α ought to be rejected as deficient.

Premise 2: α occurs in a context of dialogue that requires a certain level of precision that is appropriate for that context.

Premise 3: Some property F that occurs in α is defined in a way that is too vague to meet the requirement of the level of precision appropriate for the context.

Conclusion: Therefore, α ought be rejected as deficient.

Critical Questions

CQ1: Does the context of dialogue in which α occurs demand some particular level of precision in the key terms used?

CQ2: Is some property F that occurs in α too vague to meet the proper level of standard of precision?

CQ3: Why is this degree of vagueness a problem in relation to the dialogue in which α was advanced?

References

Douglas Walton, *Argumentation Schemes for Presumptive Reasoning*. Mahwah, N.J.: Erlbaum, 1996, pp. 102–103.

17. ARGUMENT FROM ARBITRARINESS OF A VERBAL CLASSIFICATION

Premise 1: If an argument, α, occurs in a context of dialogue that requires a certain level of nonarbitrary definition for a key property F that occurs in α, and if F is defined in an arbitrary way in α, then α ought to be rejected as deficient.

Premise 2: α occurs in a context of dialogue that requires a nonarbitrary definition for a key property F that occurs in α.

Premise 3: Some property F that occurs in α is defined in a way that is arbitrary.

Conclusion: Therefore, α ought be rejected as deficient.

Critical Questions

CQ1: Does the context of dialogue in which α occurs require a nonarbitrary definition of F?

CQ2: Is some property F that occurs in α defined in an arbitrary way?

CQ3: Why is arbitrariness of definition a problem in the context of dialogue in which α was advanced?

References

Douglas Walton, *Argumentation Schemes for Presumptive Reasoning*. Mahwah, N.J.: Erlbaum 1996, pp. 104–105

18. ARGUMENTATION FROM INTERACTION OF ACT AND PERSON

18.1. Variant 1

Premise 1: Person P has done acts A.
Premise 2: To acts A is attributed the value V.
Conclusion: Person P is V.

18.2. Variant 2

Premise 1: Person P is V_1 (a judgment value).
Premise 2: If a person is V_1, his acts A will be (are, were) V_2 (judgment value depending on V_1).
Conclusion: P's acts A will be (are, were) V_2.

References

Ch. Perelman, L. Olbrechts-Tyteca, *The New Rhetoric: A Treatise on Argumentation*. Notre Dame, Ind.: University of Notre Dame Press, 1969, pp. 296–305.

19. ARGUMENTATION FROM VALUES

19.1. Variant 1: Positive Value

Premise 1: Value V is *positive* as judged by agent A (judgment value).
Premise 2: The fact that value V is *positive* affects the interpretation and therefore the evaluation of goal G of agent A (if value V is *good*, it supports commitment to goal G).
Conclusion: V is a reason for retaining commitment to goal G.

19.2. Variant 2: Negative Value

Premise 1: Value V is *negative* as judged by agent A (judgment value).
Premise 2: The fact that value V is *negative* affects the interpretation and therefore the evaluation of goal G of agent A (if value V is *bad*, it goes against commitment to goal G).
Conclusion: V is a reason for retracting commitment to goal G.

References

Trevor Bench-Capon, "Persuasion in Practical Argument Using Value-based Argumentation Frameworks." *Journal of Logic and Computation* 13 (2003), pp. 429–448.

Trevor Bench-Capon, "Agreeing to Differ: Modelling Persuasive Dialogue between Parties without a Consensus about Values." *Informal Logic* 22 (2003a), pp. 231–245.

Comment: Bench-Capon (2003, 2003a) offers the case of Hal and Carla. Diabetic Hal needs insulin to survive, but cannot get any in time to save his life except by taking some from Carla's house without her permission. The argument from positive value for preserving life is weighed against the argument from negative value of taking someone's property without his or her permission.

20. ARGUMENTATION FROM SACRIFICE

Premise 1: For the thing x, sacrifice S is made.
Premise 2: If a great sacrifice S has been made for x, then the value V of x will be greater (and vice versa).
Premise 3: A great (small) sacrifice S has been made for x.
Conclusion: x has a great (small) value V.

References

Ch. Perelman and L. Olbrechts-Tyteca, *The New Rhetoric: A Treatise on Argumentation*. Notre Dame, Ind.: University of Notre Dame Press, 1969, pp. 248–254.

21. ARGUMENTATION FROM THE GROUP AND ITS MEMBERS

21.1. Argumentation from the Group and Its Members: Variant 1

Premise 1: Member M of the group G has the quality Q.
Premise 2: If M has Q, G will have Q as well.
Conclusion: G has Q.

21.2. Argumentation from the Group and Its Members: Variant 2

Premise 1: Group G has the quality Q.
Premise 2: M is a member (idea, habit, custom, product, method) of G.
Premise 3: If G has Q, its members M (ideas, habits, customs, products, methods) will have Q as well.
Conclusion: M is Q.

References

Ch. Perelman and L. Olbrechts-Tyteca, *The New Rhetoric: A Treatise on Argumentation*. Notre Dame, Ind.: University of Notre Dame Press, 1969, pp. 321–326.

22. PRACTICAL REASONING

22.1. Practical Inference

Major Premise: I have a goal G.
Minor Premise: Carrying out this action A is a means to realize G.
Conclusion: Therefore, I ought (practically speaking) to carry out this action A.

Critical Questions

CQ1: What other goals that I have that might conflict with G should be considered?

CQ2: What alternative actions to my bringing about A that would also bring about G should be considered?

CQ3: Among bringing about A and these alternative actions, which is arguably the most efficient?

CQ4: What grounds are there for arguing that it is practically possible for me to bring about A?

CQ5: What consequences of my bringing about A should also be taken into account?

22.2. Necessary Condition Schema

Goal Premise: My goal is to bring about A.
Alternatives Premise: I reasonably consider on the given information that bringing about at least one of $[Bo, B1, \ldots, Bn]$ is necessary to bring about A.
Selection Premise: I have selected one member Bi as an acceptable, or as the most acceptable, necessary condition for A.
Practicality Premise: Nothing unchangeable prevents me from bringing about Bi, as far as I know.
Side Effects Premise: Bringing about A is more acceptable to me than not bringing about Bi.
Conclusion: Therefore, it is required that I bring about Bi.

22.3. Sufficient Condition Schema

Goal Premise: My goal is to bring about A.
Alternatives Premise: I reasonably consider on the given information that bringing about at least one of $[Bo, B1, \ldots, Bn]$ *is* necessary to bring about A.
Selection Premise: I have selected one member Bi as an acceptable, or as the most acceptable, sufficient condition for A.
Practicality Premise: Nothing unchangeable prevents me from bringing about Bi, as far as I know.
Side Effects Premise: Bringing about A is more acceptable to me than not bringing about Bi.
Conclusion: Therefore, it is required that I bring about Bi.

Critical Questions

CQ1: Are there alternative means of realizing *A*, other than *B*? [*Alternative Means Question*].

CQ2: Is *B* an acceptable (or the best) alternative? [*Acceptable/Best Option Question*].

CQ3: Is it possible for agent *a* to do *B*? [*Possibility Question*].

CQ4: Are there negative side effects of *a*'s bringing about *B* that ought to be considered? [*Negative Side Effects Question*].

CQ5: Does *a* have goals other than *A*, which have the potential to conflict with *a*'s realizing *A*? [*Conflicting Goals Question*].

22.4. Value-Based Practical Reasoning

Premise 1: I have a goal *G*.

Premise 2: *G* is supported by my set of values, *V*.

Premise 3: Bringing about *A* is necessary (or sufficient) for me to bring about *G*.

Conclusion: Therefore, I should (practically ought to) bring about *A*.

Critical Questions

CQ1: What other goals do I have that might conflict with *G*?

CQ2: How well is *G* supported by (or at least consistent with) my values *V*?

CQ3: What alternative actions to my bringing about *A* that would also bring about *G* should be considered?

CQ4: Among bringing about *A* and these alternative actions, which is arguably the best of the whole set, in light of considerations of efficiency in bringing about *G*?

CQ5: Among bringing about *A* and these alternative actions, which is arguably the best of the whole set, in light of my values *V*?

CQ6: What grounds are there for arguing that it is practically possible for me to bring about *A*?

CQ7: What consequences of my bringing about *A* that might have even greater negative value than the positive value of *G* should be taken into account?

References

Trevor Bench-Capon, "Persuasion in Practical Argument Using Value-based Argumentation Frameworks." *Journal of Logic and Computation* 13 (2003), pp. 429–448.

Douglas Walton, *Practical Reasoning*. Savage, Md.: Rowman and Littlefield, 1990.

Douglas Walton, "Actions and Inconsistency: The Closure Problem of Practical Reasoning." In *Contemporary Action Theory*, vol. 1, ed. Ghita Holmstrom-Hintikka and Raimo Tuomela. Dordrecht: Kluwer, 1997, p. 164.

Douglas Walton, *Slippery Slope Arguments*. Newport News, Va.: Vale Press, 1992, pp. 89–90.

22.5. Argument from Goal

Major Premise: Doing act *A* contributes to goal *G*.
Minor Premise: Person *P* has goal *G*.
Conclusion: Therefore, person *P* should do act *A*.

References

Bart Verheij, "Dialectical Argumentation with Argumentation Schemes: An Approach to Legal Logic." *Artificial Intelligence and Law* 11 (2003), p. 169.

22.6. Argumentation from Ends and Means

22.6.1. Variant 1

Premise 1: *x* is the means to the end *y*.
Premise 2: *y* is good (bad).
Conclusion: *x* is good/less bad (bad).

22.6.2. Variant 2

Premise 1: Means *x* should be regarded as an end (end *x* should be regarded as a means).
Premise 2: Ends are more important than the means used to achieve them.
Conclusion: *x* should (should not) be regarded as highly important.

22.6.3. Variant 3

Premise 1: *x* is the means to the end *y*.
Premise 2: *x* is easy (not easy) to achieve.
Conclusion: *y* is good (bad).

References

Chaim Perelman and Lucie Olbrechts-Tyteca, *The New Rhetoric: A Treatise on Argumentation*. Notre Dame, Ind.: University of Notre Dame Press, 1969, pp. 273–278.

23. TWO-PERSON PRACTICAL REASONING

Premise 1: *X* intends to realize *A*, and tells *Y* this.
Premise 2: As *Y* sees the situation, *B* is a necessary (sufficient) condition for carrying out *A*, and *Y* tells *X* this.

Conclusion: Therefore, X should carry out B, unless he has better reasons not to.

Critical Questions

CQ1: Does X have other goals (of higher priority) that might conflict with the goal of realizing A?
CQ2: Are there alternative means available to X (other than B) for carrying out A?
CQ3: Would carrying out B have known side effects that might conflict with X's other goals?
CQ4: Is it possible for X to bring about B?
CQ5: Are other actions, as well as B, required for X to bring about A?

References

Douglas Walton, *Appeal to Expert Opinion*. University Park: Pennsylvania State University Press, 1997, p. 163.

24. ARGUMENT FROM WASTE

Premise 1: If a stops trying to realize A now, all a's previous efforts to realize A will be wasted.
Premise 2: If all a's previous attempts to realize A are wasted, that would be a bad thing.
Conclusion: Therefore, a ought to continue trying to realize A.

Critical Questions

CQ1: Is bringing about A possible?
CQ2: Forgetting past losses that cannot be recouped, should a reassessment of the cost and benefits of trying to bring about A from this point in time to be made?

References

Douglas Walton, *A Pragmatic Theory of Fallacy*. Tuscaloosa and London: University of Alabama Press, 1995, p. 157.
Douglas Walton, "The Sunk Costs Fallacy or Argument from Waste." *Argumentation* 16 (2002), p. 488.

25. ARGUMENT FROM SUNK COSTS

$t1$: Time of the proponent's commitment to a certain action (pre-commitment)
$t2$: Time of proponent's confrontation with the decision whether carry out the pre-commitment or not.

Premise 1: There is a choice at *t2* between *A* and *not-A*.
Premise 2: At *t2* I am precommitted to *A* because of what I did or committed myself to at *t1*.
Conclusion: Therefore, I should choose *A*.

References

Douglas Walton, "The Sunk Costs Fallacy or Argument from Waste." *Argumentation* 16 (2002), p. 489.

26. ARGUMENT FROM IGNORANCE

Major Premise: If *A* were true, then *A* would be known to be true.
Minor Premise: It is not the case that *A* is known to be true.
Conclusion: Therefore, *A* is not true.

26.1. Negative Reasoning from Normal Expectations

Major Premise: If the situation were normal, *A* would be true.
Minor Premise: It is not the case that *A* is true.
Conclusion: Therefore, the situation is not normal.

26.2. Negative Practical Reasoning

Premise 1: I do not know whether *A* is true or not.
Premise 2: I have to act on the presumption that *A* is true or not true.
Premise 3: If I act on the presumption that *A* is true, and *A* is not true, consequences *B* will follow.
Premise 4: If I act on the presumption that *A* is not true, and *A* is true, consequences *C* will follow
Premise 5: Consequences *B* (*C*) are more serious than consequences *C* (*B*).
Conclusion: Therefore, I act on the presumption that *A* is not true (true).

Critical Questions

CQ1: How far along has the search for evidence progressed?
CQ2: Which side has the burden of proof in the dialogue as a whole? In other words, what is the ultimate *probandum* and who is supposed to prove it?
CQ3: How strong does the proof need to be in order for this party to be successful in fulfilling the burden?

References

Douglas Walton, *Arguments from Ignorance*. University Park: Pennsylvania State University Press, 1996, pp. 84, 86.

Douglas Walton, *A Pragmatic Theory of Fallacy.* Tuscaloosa and London: University of Alabama Press, 1995, p. 150.

27. EPISTEMIC ARGUMENT FROM IGNORANCE

Premise 1: It has not been established that all the true propositions in D are contained in K.

Premise 2: A is a special type of proposition such that if A were true, A would normally or usually be expected to be in K.

Premise 3: A is in D.

Premise 4: A is not in K.

Premise 5: For all A in D, A is either true or false.

Conclusion: Therefore, it is plausible to presume that A is false (subject to further investigations in D).

References

Douglas Walton, "Nonfallacious Arguments from Ignorance." *American Philosophical Quarterly* 29 (1992), p. 386.

28. ARGUMENT FROM CAUSE TO EFFECT

Major Premise: Generally, if A occurs, then B will (might) occur.

Minor Premise: In this case, A occurs (might occur).

Conclusion: Therefore, in this case, B will (might) occur.

Critical Questions

CQ1: How strong is the causal generalization?

CQ2: Is the evidence cited (if there is any) strong enough to warrant the causal generalization?

CQ3: Are there other causal factors that could interfere with the production of the effect in the given case?

References

Douglas Walton, *A Pragmatic Theory of Fallacy.* Tuscaloosa and London: University of Alabama Press, 1995, pp. 140–141.

29. ARGUMENT FROM CORRELATION TO CAUSE

Premise: There is a positive correlation between A and B.

Conclusion: Therefore, A causes B.

Critical Questions

CQ1: Is there really a correlation between *A* and *B*?
CQ2: Is there any reason to think that the correlation is any more than a coincidence?
CQ3: Could there be some third factor, *C*, that is causing both *A* and *B*?

References

Douglas Walton, *A Pragmatic Theory of Fallacy*. Tuscaloosa and London: University of Alabama Press, 1995, p. 142.
Douglas Walton, *Fundamentals of Critical Argumentation*. New York: Cambridge University Press, 2006, pp. 101–103.

30. ARGUMENT FROM SIGN

Specific Premise: *A* (a finding) is true in this situation.
General Premise: *B* is generally indicated as true when its sign, *A*, is true.
Conclusion: *B* is true in this situation.

Critical Questions

CQ1: What is the strength of the correlation of the sign with the event signified?
CQ2: Are there other events that would more reliably account for the sign?

References

Douglas Walton, *Fundamentals of Critical Argumentation*. New York: Cambridge University Press, 2006, pp. 113–114.

31. ABDUCTIVE ARGUMENTATION SCHEME

31.1. Backward Argumentation Scheme

Premise 1: *D* is a set of data or supposed facts in a case.
Premise 2: Each one of a set of accounts $A_1, A_2, \ldots, A_n$ is successful in explaining *D*.
Premise 3: A_i is the account that explains *D* most successfully.
Conclusion: Therefore, A_i is the most plausible hypothesis in the case.

31.2. Forward Argumentation Scheme

Premise 1: *D* is a set of data or supposed facts in a case.
Premise 2: There is a set of argument diagrams $G_1, G_2, \ldots, G_n$, and in each argument diagram *D* represents premises of an argument that, supplemented by

plausible conditionals and other statements that function as missing parts of enthymemes, leads to a respective conclusion $C_1, C_2, \ldots, C_n$.
Premise 3: The most plausible (strongest) argument is represented by G_i.
Conclusion: Therefore, C_i is the most plausible conclusion in the case.

Critical Questions

CQ1: How satisfactory is A_i itself as an explanation of D, apart from the alternative explanations available so far in the dialogue?
CQ2: How much better an explanation is A_i than the alternative explanation so far in the dialogue?
CQ3: How far has the dialogue progressed? If the dialogue is an inquiry, how thorough has the search been in the investigation of the case?
CQ4: Would it be better to continue the dialogue further, instead of drawing a conclusion at this point?

31.3. Abductive Scheme for Argument from Action to Character

Premise: Agent a did something that can be classified as fitting a particular character quality.
Conclusion: Therefore, a has this character quality.

Critical Questions

CQ1: What is the character quality in question?
CQ2: How is the character quality defined?
CQ3: Does the description of the action in question actually fit the definition of the quality?

31.4. Scheme for Argument from Character to Action (Predictive)

Premise: Agent a has a character quality of a kind that has been defined.
Conclusion: Therefore, if a carries out some action in the future, this action is likely to be classifiable as fitting under that character quality.

Critical Questions

CQ1: What is the character quality in question?
CQ2: How is the character quality defined?
CQ3: Does the description of the action in question actually fit the definition of the quality?

Comment: Even though the critical questions are the same for both, the predictive scheme for argument from character to action needs to be distinguished from the retroductive scheme that reasons from character to a particular action, and these two schemes need to be distinguished from the argument from a past action to an agent's character.

31.5. Retroductive Scheme for Identifying an Agent from a Past Action

Factual Premise: An observed event appears to have been brought about by some agent *a*.

Character Premise: The bringing about of this event fits a certain character quality *Q*.

Agent Trait Premise: *a* has *Q*.

Conclusion: *a* brought about the event in question.

Critical Questions

CQ1: What is the quality *Q* in question?

CQ2: How is *Q* defined?

CQ3: Does the description of the action in question actually fit the definition of *Q*?

CQ4: How large is the reference class of other agents who also might have brought about this event and who have the same character quality?

References

Douglas Walton, *Legal Argumentation and Evidence*. University Park: Pennsylvania State University Press, 2002, p. 44.

Douglas Walton, *Abductive Reasoning*. Tuscaloosa: University of Alabama Press, 2004.

Douglas Walton, *Character Evidence: An Abductive Theory*. Dordrecht: Springer, 2006.

32. ARGUMENT FROM EVIDENCE TO A HYPOTHESIS

32.1. Argument from Verification

Major Premise: If *A* (a hypothesis) as true, then *B* (a proposition reporting an event) will be observed to be true.

Minor Premise: *B* has been observed to be true, in a given instance.

Conclusion: Therefore, *A* is true.

32.2. Argument from Falsification

Major Premise: If *A* (a hypothesis) as true, then *B* (a proposition reporting an event) will be observed to be true.

Minor Premise: *B* has been observed to be false, in a given instance.
Conclusion: Therefore, *A* is false.

Critical Questions

CQ1: Is it the case that if *A* is true, then *B* is true?
CQ2: Has *B* been observed to be true (false)?
CQ3: Could there be some reason why *B* is true, other than its being because of *A* being true?

References

Douglas Walton, *Argumentation Schemes for Presumptive Reasoning*. Mahwah, N.J.: Erlbaum, 1996, pp. 67–70.

33. ARGUMENT FROM CONSEQUENCES

33.1. Argument from Positive Consequences

Premise: If *A* is brought about, good consequences will plausibly occur.
Conclusion: Therefore, *A* should be brought about.

33.2. Argument from Negative Consequences

Premise: If *A* is brought about, then bad consequences will occur.
Conclusion: Therefore, *A* should not be brought about.

33.3. Reasoning from Negative Consequences

Premise 1: If I (an agent) bring about (don't bring about) *A*, then *B* will occur.
Premise 2: *B* is a bad outcome (from the point of view of my goals).
Conclusion: Therefore, I should not (practically speaking) bring about *A*.

33.4. Argument from Negative Consequences (Prudential Inference)

Premise 1: You were considering not doing *A*.
Premise 2: But, if you do not do *A*, some consequence *B*, which will be very bad for you, will occur, or is likely to occur.
Conclusion: Therefore, you ought to reconsider and (other things being equal) you ought (prudentially) to do *A*.

Critical Questions

CQ1: How strong is the likelihood that the cited consequences will (may, must) occur?

CQ2: What evidence supports the claim that the cited consequences will (may, must) occur, and is it sufficient to support the strength of the claim adequately?

CQ3: Are there other opposite consequences (bad as opposed to good, for example) that should be taken into account?

References

Douglas Walton, *A Pragmatic Theory of Fallacy*. Tuscaloosa and London: University of Alabama Press, 1995, pp. 155–156.
Douglas Walton, *Scare Tactics*. Dordrecht: Kluwer, 2000, p. 123.

34. PRAGMATIC ARGUMENT FROM ALTERNATIVES

Premise 1: Either you (the respondent) must bring about *A*, or *B* will occur.
Premise 2: *B* is bad or undesirable, from your point of view.
Conclusion: Therefore, you should (ought to, practically speaking) bring about *A*.

References

Douglas Walton, *Scare Tactics*. Dordrecht: Kluwer, 2000, p. 142.

35. ARGUMENT FROM THREAT

Premise 1: If you bring about *A*, some cited bad consequences, *B*, will follow.
Premise 2: I am in position to bring about *B*.
Premise 3: I hereby assert that in fact I will see to it that *B* occurs if you bring about *A*.
Conclusion: Therefore, you had better not bring about *A*.

35.1. Argument from Disjunctive *Ad Baculum* Threat

Premise 1: You (the respondent) must bring about *A*, or I (the proponent) will undertake to see to it that *B* will occur.
Premise 2: *B* is bad or undesirable, from your point of view.
Conclusion: Therefore, you should (ought to, practically speaking) bring about *A*.

References

Douglas Walton, *A Pragmatic Theory of Fallacy*. Tuscaloosa and London: University of Alabama Press, 1995, p. 157.
Douglas Walton, *Scare Tactics*. Dordrecht: Kluwer, 2000, p. 140.

36. ARGUMENT FROM FEAR APPEAL

Premise 1: If you do not bring about *A*, then *D* will occur.
Premise 2: *D* is very bad for you.

Premise 3: Therefore, you ought to prevent *D* if possible.
Premise 4: But the only way for you to prevent *D* is bring about *A*.
Conclusion: Therefore, you ought to bring about *A*.

References

Douglas Walton, *Scare Tactics*. Dordrecht: Kluwer, 2000, p. 22.

37. ARGUMENT FROM DANGER APPEAL

Premise 1: If you (the respondent) bring about *A*, then *B* will occur.
Premise 2: *B* is a danger to you.
Conclusion: Therefore (on balance) you should not bring about *A*.

References

Douglas Walton, *Scare Tactics*. Dordrecht: Kluwer, 2000, p. 173

38. ARGUMENT FROM NEED FOR HELP

Premise 1: For all *x* and *y*, *y* ought to help *x*, if *x* is in a situation where *x* needs help, and *y* can help, and *y*'s giving help would not be too costly for *y*.
Premise 2: *x* is in a situation where some action *A* by *y* would help *x*.
Premise 3: *y* can carry out *A*.
Premise 4: *y*'s carrying out *A* would not be too costly for *y* – that is, the negative side effects would not be too great, as *y* sees it.
Conclusion: Therefore, *y* ought to carry out *A*.

Critical Questions

CQ1: Would the proposed action *A* really help *x*?
CQ2: Is it possible for *y* to carry out *A*?
CQ3: Would there be negative side effects of carrying out *A* that would be too great?

References

Douglas Walton, *Appeal to Pity*. Albany: State University of New York Press, 1997, pp. 104, 155.

39. ARGUMENT FROM DISTRESS

Premise 1: Individual *x* is in distress (is suffering).
Premise 2: If *y* brings about *A*, it will relieve or help to relieve this distress.
Conclusion: Therefore, *y* ought to bring about *A*.

Critical Questions

CQ1: Is *x* really in distress?
CQ2: Will *y*'s bringing about *A* really help or relieve this distress?
CQ3: Is it possible for *y* to bring about *A*?
CQ4: Would negative side effects of *y*'s bringing about *A* be too great?

References

Douglas Walton, *Appeal to Pity*. Albany: State University of New York Press, 1997, pp. 105, 155.

40. ARGUMENT FROM COMMITMENT

Version 1

Commitment Evidence Premise: In this case it was shown that *a* is committed to proposition *A*, according to the evidence of what he said or did.
Linkage of Commitments Premise: Generally, when an arguer is committed to *A*, it can be inferred that he is also committed to *B*.
Conclusion: In this case, *a* is committed to *B*.

Version 2

Major Premise: If arguer *a* has committed herself to proposition *A* at some point in a dialogue, then it may be inferred that she is also committed to proposition *B*, should the question of whether *B* is true become an issue later in the dialogue.
Minor Premise: Arguer *a* has committed herself to proposition *A* at some point in a dialogue.
Conclusion: At some later point in the dialogue, where the issue of *B* arises, arguer *a* may be said to be committed to proposition *B*.

Critical Questions

CQ1: What evidence in the case supports the claim that *a* is committed to *A*, and does it include contrary evidence, indicating that *a* might not be committed to *A*?
CQ2: Is there room for questioning whether there is an exception in this case to the general rule that commitment to *A* implies commitment to *B*?

References

Douglas Walton, *Fundamentals of Critical Argumentation*. Cambridge: Cambridge University Press, 2006, pp. 117–118.
Douglas Walton, "The Sunk Costs Fallacy or Argument from Waste." *Argumentation* 16 (2002), p. 490.

41. ETHOTIC ARGUMENT

Major Premise: If x is a person of good (bad) moral character, then what x says should be accepted as more plausible (rejected as less plausible).
Minor Premise: a is a person of good (bad) moral character.
Conclusion: Therefore, what x says should be accepted as more plausible (rejected as less plausible).

Critical Questions

CQ1: Is a a person of good (bad) moral character?
CQ2: Is character relevant in the dialogue?
CQ3: Is the weight of presumption claimed strongly enough warranted by the evidence given?

References

Douglas Walton, *A Pragmatic Theory of Fallacy*. Tuscaloosa and London: University of Alabama Press, 1995, p. 152.

42. GENERIC AD HOMINEM

Character Attack Premise: a is a person of bad character.
Conclusion: a's argument α should not be accepted.

Critical Questions

CQ1: How well supported by evidence is the allegation made in the character attack premise?
CQ2: Is the issue of character relevant in the type of dialogue in which the argument was used?
CQ3: Is the conclusion of the argument that α should be (absolutely) rejected, even if other evidence to support α has been presented, or is the conclusion merely (the relative claim) that α should be assigned a reduced weight of credibility as a supporter of α, relative to the total body of evidence available?

References

Douglas Walton, *Fundamentals of Critical Argumentation*. Cambridge: Cambridge University Press, 2006, p. 123.

43. PRAGMATIC INCONSISTENCY

Premise: a advocates argument α, which has proposition A as its conclusion.
Premise: a has carried out an action, or set of actions, that imply that a is personally committed to $\neg A$ (the opposite, or negation of A).
Conclusion: Therefore, a's argument α should not be accepted.

Critical Questions

CQ1: Did *a* advocate *α* in a strong way indicating her personal commitment to *A*?

CQ2: In what words was the action described, and does that description imply that *a* is personally committed to the opposite of *A*?

CQ3: Why is the pragmatic inconsistency indicated by satisfactory answers to CQ1 and CQ2 a relevant reason for not accepting argument *α*?

References

Douglas Walton, *Ad Hominem Arguments*. Tuscaloosa: University of Alabama Press, 1998, pp. 218, 251.

44. ARGUMENT FROM INCONSISTENT COMMITMENT

Initial Commitment Premise: *a* has claimed or indicated that he is committed to proposition *A* (generally, or by virtue of what he has said in the past).

Opposed Commitment Premise: Other evidence in this particular case shows that *a* is not really committed to *A*.

Conclusion: *a*'s commitments are inconsistent.

Critical Questions

CQ1: What is the evidence supposedly showing that *a* is committed to *A*?

CQ2: What further evidence in the case is alleged to show that *a* is not committed to *A*?

CQ3: How does the evidence from premise 1 and premise 2 prove that there is a conflict of commitments?

References

Douglas Walton, *Fundamentals of Critical Argumentation*. Cambridge: Cambridge University Press, 2006, pp. 120–121.

45. CIRCUMSTANTIAL AD HOMINEM

Argument Premise: *a* advocates argument *α*, which has proposition *A* as its conclusion.

Inconsistent Commitment Premise: *a* is personally committed to the opposite (negation) of *A*, as shown by commitments expressed in her/his personal actions or personal circumstances expressing such commitments.

Credibility Questioning Premise: *a*'s credibility as a sincere person who believes in his own argument has been put into question (by the two premises above).

Conclusion: The plausibility of *a*'s argument *α* is decreased or destroyed.

Critical Questions

CQ1: Is there a pair of commitments that can be identified, shown by evidence to be commitments of *a*, and taken to show that *a* is practically inconsistent?

CQ2: Once the practical inconsistency that is the focus of the attack is identified, could it be resolved or explained by further dialogue, thus preserving the consistency of the arguer's commitments in the dialogue, or showing that *a*'s inconsistent commitment does not support the claim that *a* lacks credibility?

CQ3: Is character an issue in the dialogue, and more specifically, does *a*'s argument depend on his/her credibility?

CQ4: Is the conclusion the weaker claim that *a*'s credibility is open to question or the stronger claim that the conclusion of *α* is false?

References

Douglas Walton, *Ad Hominem Arguments.* Tuscaloosa: University of Alabama Press, 1998, pp. 255–256.

Douglas Walton, *Fundamentals of Critical Argumentation.* Cambridge: Cambridge University Press, 2006, pp. 125–126.

46. ARGUMENT FROM BIAS

Major Premise: If *x* is biased, then *x* is less likely to have taken the evidence on both sides into account in arriving at conclusion *A*.

Minor Premise: Arguer *a* is biased.

Conclusion: Arguer *a* is less likely to have taken the evidence on both sides into account in arriving at conclusion *A*.

Critical Questions

CQ1: What type of dialogue are the speaker and hearer supposed to be engaged in?

CQ2: What evidence has been given to prove that the speaker is biased?

References

Douglas Walton, *A Pragmatic Theory of Fallacy.* Tuscaloosa and London: University of Alabama Press, 1995, p. 153.

47. BIAS AD HOMINEM

Premise 1: Person *a*, the proponent of argument *α*, is biased.

Premise 2: Person *a*'s bias is a failure to honestly take part in a type of dialogue *D*, that *α* is part of.

Premise 3: Therefore, *a* is a morally bad person.

Conclusion: Therefore, α should not be given as much credibility as it would have without the bias.

Critical Questions

CQ1: What is the evidence that *a* is biased?

CQ2: If *a* is biased, is it a bad bias that is detrimental to *a*'s honestly taking part in *D*, or a normal bias that is appropriate for the type of dialogue in which α was put forward?

References

Douglas Walton, *Ad Hominem Arguments*. Tuscaloosa: University of Alabama Press, 1998, p. 255.

48. ARGUMENT FROM GRADUALISM

Premise 1: Proposition *A* is true (acceptable to the respondent).

Premise 2: There is an intervening sequence of propositions, $B_1, B_2, \ldots B_{n-1}$, B_n, *C*, such that the following conditionals are true: If *A*, then B_1; if B_1, then $B_2 \ldots$; if B_{n-1}, then B_n; if B_n, then *C*.

Premise 3: The conditional "If *A*, than *C*" is not, by itself, acceptable to the respondent, nor are shorter sequences from *A* to *C* (than the one specified in the second premise) acceptable to the respondent.

Conclusion: Therefore, the proposition *C* is true (acceptable to the respondent).

References

Douglas Walton, *Argumentation Schemes for Presumptive Reasoning*. Mahwah, N.J.: Erlbaum, 1996, p. 96.

49. SLIPPERY SLOPE ARGUMENT

First Step Premise: A_0 is up for consideration as a proposal that seems initially like something that should be brought about.

Recursive Premise: Bringing up A_0 would plausibly lead (in the given circumstances, as far as we know) to A_1, which would in turn plausibly lead to A_2, and so forth, through the sequence $A_2, \ldots A_n$.

Bad Outcome Premise: A_n is a horrible (disastrous, bad) outcome.

Conclusion: A_0 should not be brought about.

Critical Questions

CQ1: What intervening propositions in the sequence linking up A_0 with A_n are actually given?

CQ2: What other steps are required to fill in the sequence of events, to make it plausible?

CQ3: What are the weakest links in the sequence, where specific critical questions should be asked on whether one event will really lead to another?

References

Douglas Walton, *Slippery Slope Arguments*. Newport News, Va.: Vale Press, 1992, pp. 93, 95.

Douglas Walton, *Fundamentals of Critical Argumentation*. Cambridge: Cambridge University Press, 2006, pp. 107, 110.

50. PRECEDENT SLIPPERY SLOPE ARGUMENT

Claim to Exceptional Status Premise: Case C_0 is claimed to be an exception to the rule R (an excusable or exceptional case).

Related Cases Premise: Case C_0 is similar to case C_1, that is, if C_0 is held to be an exception, then C_1 must be held to be an exception too (in order to be consistent in treating equal cases alike). A sequence of similar pairs of cases $\{C_i, C_j\}$ binds us by case-to-case consistency to the series $C_0, C_1, \ldots C_n$.

Intolerable Outcome Premise: Treating case C_0 as an exception to the rule R would be intolerable (for the various kinds of reasons that could be relevant).

Conclusion: Case C_0 cannot be judged to be an exception to the rule (an excusable or exceptional case).

Critical Questions

CQ1: Would C_0 set a precedent?

CQ2: What is the evidence showing why each of the precedents cited in the intervening sequence would occur?

CQ3: Is C_0 as intolerable as it is portrayed?

References

Douglas Walton, *A Pragmatic Theory of Fallacy*. Tuscaloosa and London: University of Alabama Press, 1995, p. 159.

Douglas Walton, *Slippery Slope Arguments*. Newport News, Va.: Vale Press, 1992, pp. 93, 95.

51. SORITES SLIPPERY SLOPE ARGUMENT

Initial Base Premise: It is clearly beyond contention that a_k has P.

General Inductive Premise: If a_k has P, then a_{k-1} has P.

Reapplication Sequence: A sequence of *modus ponens* subarguments, a premise set linking premises and conclusions from the clear area through the grey area.

Conclusion: a_i may have P, for all we know (or can prove).

References

Douglas Walton, "The Argument of the Beard." *Informal Logic* 18 (1996), p. 251.

52. VERBAL SLIPPERY SLOPE ARGUMENT

Premise 1: Individual a_1 has property F.
Premise 2: If a_1 has F, then a_2 has F.
Premise 3: Property F is vague, and so generally, if a_i has F, then you can't deny that the next closely neighboring individual a_n has F.
Premise 4: But quite clearly it is false that a_n has F.
Conclusion: Therefore, you can't truly say that a_1 has F.

References

Douglas Walton, *A Pragmatic Theory of Fallacy*. Tuscaloosa and London: University of Alabama Press, 1995, p. 160.

53. FULL SLIPPERY SLOPE ARGUMENT

Initial Premise: Case C_0 is tentatively acceptable as an initial presumption.
Sequential Premise: There exists a series of cases $C_0, C_1, \ldots C_{n-1}$, where each leads to the next by a combination of causal, precedent, and/or analogical steps.
Group Opinion Premise: There is a climate of social opinion such that once people come to accept each step as plausible, then they will also be led to accept the next step.
Conclusion: C_0 is not acceptable (contrary to the presumption of the initial premise).

Critical Questions

CQ1: How strong is the argument?
CQ2: How strong does the argument need to be in order to fulfill its burden of proof?
CQ3: Will people be led to accept the next step once they come to accept each step as plausible?

References

Douglas Walton, *Slippery Slope Arguments*. Newport News, Va.: Vale Press, 1992, pp. 200, 203.

54. ARGUMENT FOR CONSTITUTIVE-RULE CLAIMS

54.1. Physical World Premise Version 1

Premise: A *W* is a *D*.
Warrant: Whenever expressions are referentially coextensive, they are synonymous.
Conclusion: "*W*" means the same as "*D*."

Rebuttal Factor

1. The warrant backing does not apply if the premise is not a necessary truth.

54.2. Physical World Premise Version 2

Premise: In the official rulebook for *I* (a rule-governed institution, such as a game), it is stated that *D* counts as *W*.
Warrant: Official rulebooks correctly report the rules in force.
Conclusion: *D* counts as *W*.

Rebuttal Factor

1. The warrant backing applies unless the rule has changed since the publication of the rulebook.

54.3. Mental World Premise

Premise: Speakers of L believe that "*W*" means the same as "*D*."
Warrant: Speakers of a language are generally correct about word definitions in that language.
Conclusion: "*W*" means the same as "*D*."

Rebuttal Factor

1. Speakers are able to use the word "*W*" in the correct physical and grammatical context without being correct about the nature of its referent or property.

References

Wayne Grennan, *Informal Logic*. Montreal: McGill-Queen's University Press, 1997, p. 196.

55. ARGUMENT FROM RULES

55.1. From Established Rule

Major Premise: If carrying out types of actions including A is the established rule for x, then (unless the case is an exception), x must carry out A.

Minor Premise: Carrying out types of actions including A is the established rule for a.

Conclusion: Therefore, a must carry out A.

Critical Questions

CQ1: Does the rule require carrying out types of actions that include A as an instance?

CQ2: Are there other established rules that might conflict with or override this one?

CQ3: Is this case an exceptional one, that is, could there be extenuating circumstances or an excuse for noncompliance?

References

Douglas Walton, *A Pragmatic Theory of Fallacy*. Tuscaloosa and London: University of Alabama Press, 1995, p. 147.

55.2. From Rules

Major Premise: If A is the case, then an evaluation E is justified/conduct C is required.

Minor Premise: A is the case.

Conclusion: Therefore, evaluation E is justified/conduct C is required.

References

Peter Jan Schellens and Menno De Jong, "Argumentation Schemes in Persuasive Brochures." *Argumentation* 18, (2004), p. 311.

55.3. Regulative-Rule Premise Obligation Claim

Premise: A is prohibited (obligatory).

Warrant: One must not (must) do what is prohibited (obligatory).

Conclusion: S must (must not) do A.

Rebuttal Factor

1. S has an adequate excuse, or an overriding duty.

References

Wayne Grennan, *Informal Logic*. Montreal: McGill-Queen's University Press, 1997, p. 169.

56. ARGUMENT FOR AN EXCEPTIONAL CASE

Major Premise: If the case of x is an exception, then the established rule can be waived in the case of x.
Minor Premise: The case of a is an exception.
Conclusion: Therefore, the established rule can be waived in the case of a.

Critical Questions

CQ1: Is it the case of a a recognized type of exception?
CQ2: If it is not a recognized case, can evidence that the established rule does not apply to it be given?
CQ3: If it is a borderline case, can comparable cases be cited?

References

Douglas Walton, *A Pragmatic Theory of Fallacy*. Tuscaloosa and London: University of Alabama Press, 1995, p. 147.
Douglas Walton, *Argumentation Schemes for Presumptive Reasoning*. Mahwah, N.J.: Erlbaum, 1996, p. 93.

57. ARGUMENT FROM PRECEDENT

Major Premise: Generally, according to the established rule, if x has property F, then x also has property G.
Minor Premise: In this legitimate case, a has F but does not have G.
Conclusion: Therefore, an exception to the rule must be recognized, and the rule appropriately modified or qualified.

Critical Questions

CQ1: Does the established rule really apply to this case?
CQ2: Is the case cited legitimate, or can it be explained as only an apparent violation of the rule?
CQ3: Can the case cited be dealt with under an already recognized category of exception that does not require a change in the rule?

References

Douglas Walton, *A Pragmatic Theory of Fallacy*. Tuscaloosa and London: University of Alabama Press, 1995, p. 148.

58. ARGUMENT FROM PLEA FOR EXCUSE

Premise 1: Normally, rule *R* requires or forbids a type of action or inaction *T*, which carries with it a sanction (penalty) *S*.
Premise 2: I (the pleader) have committed *T*.
Premise 3: But I can cite special circumstances that constitute an excuse, *E*.
Conclusion: Therefore, in this instance, I ought to be exempted from *S*.

Critical Questions

CQ1: Does *E* fall under one of the recognized categories of excuses for this type of case, and, if so, can this inclusion be justified in this case?

CQ2: If *E* does not fall under a recognized category, then what about this case is special that justifies the claim to exemption?

CQ3: If *E* does not fall under a recognized category, would it set a precedent, and if so, would this pose a problem in future cases?

References

Douglas Walton, *Appeal to Pity*. Albany: State University of New York Press, 1997, pp. 154, 156.

59. ARGUMENT FROM PERCEPTION

59.1. Argument from Perception

Premise 1: Person *P* has a φ image (an image of a perceptible property).
Premise 2: To have a φ image (an image of a perceptible property) is a prima facie reason to believe that the circumstances exemplify φ.
Conclusion: It is reasonable to believe that φ is the case.

Undercutter

CQ1: Are the circumstances such that having a φ image is not a reliable indicator of φ?

References

John Pollock, *Cognitive Carpentry*. Cambridge, Mass.: MIT Press, 1995, p. 53.

59.2. Argument from Appearance

Premise: This object looks like it could be classified under verbal category *C*.
Conclusion: Therefore, this object can be classified under verbal category *C*.

Critical Questions

CQ1: Could the appearance of its looking like it could be classified under *C* be misleading for some reason?

CQ2: Although it may look like it can be classified under *C*, could there be grounds for indicating that it might be more justifiable to classify it under another category *D*?

References

Douglas Walton, "Argument from Appearance: A New Argumentation Scheme." *Logique et Analyse* 195 (2006), pp. 319–340.

60. ARGUMENT FROM MEMORY

Premise 1: Person *P* recalls ϕ.
Premise 2: Recalling ϕ is a prima facie reason to believe ϕ.
Conclusion: It is reasonable to believe ϕ.

Undercutter

CQ1: Was ϕ originally based on beliefs of which one is false?
CQ2: Is ϕ not originally believed for other reasons?
CQ3: Does the agent who recalls ϕ express doubt about ϕ ?

References

J. L. Pollock, "Defeasible Reasoning." *Cognitive Science* 11 (1987), pp. 481–518.

10

Refining the Classification of Schemes

It would be very helpful for users of the schemes to have a more refined system of classification, so that the user could search through to find a scheme applicable to her needs in a given case by searching under other, more general ones where the particular scheme being sought is known to fit. It is already fairly evident from the compendium of schemes that some schemes fit under others as subspecies of them. For example, one of the most common schemes is argument from consequences. It is closely related to practical reasoning. Other schemes, like those for the slippery slope argument, often fit under the category of argument from consequences. However, such classifications are not as straightforward as they initially seem. For example, some slippery slope arguments fit under the category of arguments from precedent, and therefore may not fit the scheme of argument from consequences, at least in any straightforward way. Another very common scheme under which many others fit as subspecies is the scheme for argument from commitment. Here we have a cluster of schemes that are closely related to each other, but in complex ways. Schemes that are very general, like those for argument from consequences and argument from ignorance, are related to many other, more specific schemes that fall under them. This chapter sets us on the road to beginning the research project of taking such clusters of schemes and investigating how they fit together with their neighboring schemes.

Of all the arguments associated with informal fallacies that have been studied, the one with the most special schemes representing subspecies that have been identified is the *argumentum ad hominem.* The most intensive studies on how to classify the critical questions matching the schemes have been conducted in the case of argument from expert opinion. In discussing the project of classification of schemes in this chapter, we will use these two types of arguments as the paradigm cases. At this point we

347

are far from being able to propose any firmly fixed or final system of classification of clusters of schemes, but we begin this chapter by proposing a
tentative system as an initial sorting device that can help to organize the
project of conducting further research on classification of the schemes
in the compendium.

1. A PROPOSED GENERAL SYSTEM FOR CLASSIFICATION OF SCHEMES

It has proved to be very difficult to classify fallacies. The multitude of logic
textbooks that treat informal fallacies typically classify the fallacies into
groups under headings like fallacies of relevance, inductive fallacies, linguistic fallacies, and so forth. Usually a distinction is drawn between formal and informal fallacies (Hamblin, 1970). Fallacies are closely related
to schemes, and indeed most of the leading informal fallacies can be
associated with misuse of schemes, or failure to pay sufficient attention
to the requirements of a scheme. For this reason, in our opinion, the
project of classifying argumentation schemes should be seen as prior to
the project of classifying informal fallacies. Thus we turn to the important
project of classifying schemes.

Not only are some schemes subspecies of others, but in many cases the
schemes appear to overlap, owing to the difficulty of defining the concepts that any classification scheme has to be based on, including concepts like knowledge, causation, inductive reasoning, expert opinion,
consequences, threat, and so forth. For these reasons, any attempt to classify schemes faces inherent conceptual difficulties, and thus, although a
tentative system for classifying schemes will be presented here, it should
be regarded as merely a hypothesis, a way of moving forward that should
be subject to improvement as each of the concepts contained in it is more
fully analyzed and defined.

That said, we have classified the schemes under three broad headings,
representing the most general categories. The first one is called reasoning.[1] Under it we include deductive reasoning, inductive reasoning,
practical reasoning,[2] abductive reasoning, and causal reasoning. Reasoning is taken in a broad sense to include different kinds of sequences in
which there is a chaining of inferences, such that the conclusion of one
local inference becomes a premise of the next one. The second category is that of source-based arguments. This category comprises all those

[1] On the distinction between reasoning and argument, see Walton, 1990.
[2] Practical reasoning is itself a scheme, but such a fundamental one that we treat it here
as a general heading. The same can be said for abductive reasoning.

arguments where the argument is dependent on a source, an agent that is in a position to know something – for example, a witness who is thought to have seen some event or to have special knowledge about it. In this kind of argumentation, evaluating the argument depends on characteristics of the source – for example, the credibility of the source or the expertise of the source. The third category, that of arguments that apply rules to particular cases, is also a very broad one. Arguments coming under this category relate to a situation in which some sort of general rule is applied to the specifics of a given case, and the argument is decided on the basis of how well the rule fits the case. Legal arguments, especially the kind used in trials, commonly fit this classification.

PROPOSED CLASSIFICATION SYSTEM FOR ARGUMENTATION SCHEMES

Reasoning

1. Deductive Reasoning
   Deductive *Modus Ponens*
   Disjunctive Syllogism
   Hypothetical Syllogism
   Reductio Ad Absurdum
   (etc.)
2. Inductive Reasoning
   Argument from a Random Sample to a Population
   (etc.)
3. Practical Reasoning
   Argument from Consequences
   Argument from Alternatives
   Argument from Waste
   Argument from Sunk Costs
   Argument from Threat
   Argument from Danger Appeal
4. Abductive Reasoning
   Argument from Sign
   Argument from Evidence to a Hypothesis
5. Causal Reasoning
   Argument from Cause to Effect
   Argument from Correlation to Cause
   Causal Slippery Slope Argument
   (see chapter on causal argumentation)

Source-Based Arguments

1. Arguments from Position to Know
   Argument from Position to Know
   Argument from Witness Testimony
   Argument from Expert Opinion
   Argument from Ignorance
2. Arguments from Commitment
   Argument from Inconsistent Commitment
3. Arguments Attacking Personal Credibility
   Argument from Allegation of Bias
   Poisoning the Well by Alleging Group Bias
   Ad Hominem Arguments (etc.)
4. Arguments from Popular Acceptance
   Argument from Popular Opinion
   Argument from Popular Practice
   (etc.)

Applying Rules to Cases

1. Arguments Based on Cases
   Argument from Example
   Argument from Analogy
   Argument from Precedent
2. Defeasible Rule-Based Arguments
   Argument from an Established Rule
   Argument from an Exceptional Case
   Argument from Plea for Excuse
3. Verbal Classification Arguments
   Argument from Verbal Classification
   Argument from Vagueness of a Verbal Classification
4. Chained Arguments Connecting Rules and Cases
   Argument from Gradualism
   Precedent Slippery Slope Argument
   Sorites Slippery Slope Argument

As noted earlier, some of the classifications are controversial. For example, argument from ignorance is classified here under source-based arguments, but a strong case could be made for classifying it as a special category of knowledge-based arguments. Studies of the argument from ignorance have shown that it has several different forms (Krabbe, 1992; Walton, 1995). Sometimes it is knowledge-based, but in other instances it is a dialectical type of argumentation that could perhaps better be classified under arguments from commitment. We make no further comment here on the issue, acknowledging that there are various different types of argument from ignorance, and that while some of them should clearly be classified as source-based arguments, many others might fit better under a knowledge-based category.

Some of the arguments have been classified under the heading of applying rules to cases. Many of these have various different forms, and thus do not fall simply into any one category. The most outstanding example is that of the slippery slope argument. In some cases slippery slopes fall under the category of argument from vagueness of a verbal classification, while in other cases they're clearly causal arguments, arguments based on precedents, or a chain of argumentation combining all of these types of arguments. The slippery slope argument is often such a complex argument, combining many other subarguments of different types, that it is impossible to classify it under any single heading. Finally, it should be noted that it is well established that ad hominem arguments and *ad populum* arguments can also be classified into many different subtypes. Thus a single argument may be properly classified under one of these general headings – as an ad hominem argument, for example – but there may be many subtypes of ad hominem that are related to other categories and other argumentation schemes and hence do not fit entirely into the ad hominem category. So we stress again that the system of classification we tentatively propose here is very broad, and much in need of refinement as the field of studying argumentation schemes begins to recognize many more special types of arguments that fit into these various categories. The preceding chapters have shown how each of these main categories of arguments, like ad hominem and *ad populum*, can have many subtypes. As research on argumentation schemes moves ahead, more and more of these subtypes will be discovered, making further refinement of this system of classification more complex.

2. CLASSIFICATION OF AD HOMINEM SCHEMES

The various argumentation schemes closely related to the *argumentum ad hominem* have been classified by Walton (1998, pp. 259–263), according to the system presented in Figure 10.1. Figure 10.1 shows which forms of argument are subspecies of others. At the top of the figure, argument from commitment is represented, and just below that, argument from inconsistent commitment. Just under that, argument from pragmatic inconsistency is represented. Related to these are the schemes for the

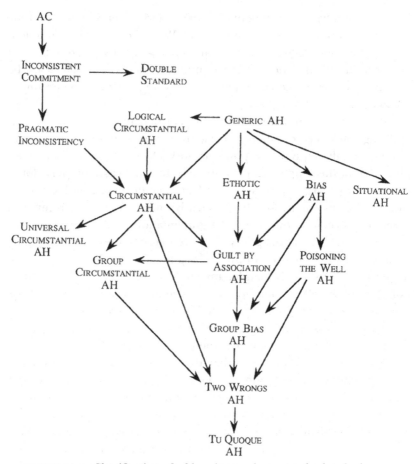

FIGURE 10.1. Classification of ad hominem schemes and related schemes.

three basic types of ad hominem argument – the generic type (sometimes called direct or abusive), the circumstantial type, and the bias type.

Here is the scheme for argument from pragmatic inconsistency, along with its matching set of critical questions. It is important to recognize that putting forward an argument fitting this scheme is not necessarily a personal attack used to attack a person by alleging that he has bad character.

ARGUMENT FROM PRAGMATIC INCONSISTENCY

Premise: *a* advocates argument α, which has proposition *A* as its conclusion.
Premise: *a* has carried out an action, or set of actions, that imply that *a* is personally committed to ¬*A* (the opposite, or negation, of *A*).
Conclusion: Therefore, *a*'s argument α should not be accepted.

Critical Questions

CQ1: Did *a* advocate *α* in a strong way indicating her personal commitment to *A*?

CQ2: In what words was the action described, and does that description imply that *a* is personally committed to the opposite of *A*?

CQ3: Why is the pragmatic inconsistency indicated by satisfactory answers to CQ1 and CQ2 a relevant reason for not accepting argument *α*?

It is important for the reader to recognize that argument from pragmatic inconsistency is not the same as the circumstantial ad hominem argument. The latter is a species of the former used to argue that the opponent is a person of bad character, and therefore that his opinion should be discounted. Of course, the circumstantial ad hominem is a species of argument from pragmatic inconsistency, but it does not follow that the two types of argument are the same in all instances.

Figure 10.1 displays the relationships among the various argumentation schemes, both for the types of ad hominem argument identified by Walton (1998) and for the closely related schemes like argument from commitment, argument from inconsistent commitment, and argument from pragmatic inconsistency. Most notable in this analysis is that none of these three forms of argument is taken to count as a genuine ad hominem argument. All the other argumentation schemes represented in Figure 10.1 are classified as genuine ad hominem arguments.

TWO WRONGS AD HOMINEM

Premise: **Proponent:** Respondent, you have committed some morally blameworthy action (and the specific action is then cited).

Conclusion: **Respondent:** You are just as bad, for you have also committed a morally blameworthy action (then cited, generally a different type of action from the one cited by the proponent but comparable with respect to being blameworthy). Therefore, you are a bad person, and your argument against me should not be accepted as having any worth.

Critical Questions

CQ1: Is there evidence to support the proponent's allegation that the respondent committed a blameworthy act?

CQ2: If the answer to CQ1 is yes, then should the respondent's counteraccusation be rated as very credible?

CQ3: Is the respondent's counteraccusation in the dialogue relevant to the proponent's original allegation?

Several of the argumentation schemes in Figure 10.1 are not in the compendium, and so the schemes for these forms of argumentation are presented here.

DOUBLE STANDARD

Premise 1: The respondent has one policy with respect to *a*.
Premise 2: The respondent has another (different) policy with respect to *b*.
Premise 3: *a* is similar to *b* (or comparable to *b* in some relevant respect).
Conclusion: Therefore, the respondent is using a double standard.

Critical Questions

CQ1: What is the respondent's policy with respect to *a*?
CQ2: What is the respondent's policy with respect to *b*?
CQ3: How is the one policy different from the other?
CQ4: How is *a* similar (or comparable) to *b*?
CQ5: Can the differences in policies be explained, or is it significant as evidence that the respondent's policies are inconsistent in some important way?

UNIVERSAL CIRCUMSTANTIAL AD HOMINEM

Premise 1: *a* advocates argument α, which has proposition *A* as its conclusion, which says that everybody should be committed to *A*.
Premise 2: *a* is bound by the 'everybody' in premise 1.
Premise 3: *a* has carried out an action, or a set of actions, that imply that *a* is personally committed to $\neg A$.
Premise 4: Therefore, *a* is a morally bad person.
Conclusion: Therefore, *a*'s argument α should not be accepted.

Critical Questions

CQ1: Does *a*'s argument conclude that everybody should be committed to *A*?
CQ2: Is there any basis for α being an exception to the commitment?
CQ3: Does the action, as described, imply that *a* is personally committed to the opposite of *A*?
CQ4: Why does it follow (if it does) that the alleged practical inconsistency shows that *a* is a bad person?
CQ5: Is *a*'s being a bad person a good reason for concluding that *a*'s argument should not be accepted?

GROUP CIRCUMSTANTIAL AD HOMINEM

Premise 1: *a* advocates argument α, which has proposition *A* as its conclusion, which says that everybody in group *G* should be committed to *A*.

Premise 2: *a* belongs to a group, *G*.

Premise 3: *a* has carried out an action, or a set of actions, that imply that *a* is personally committed to ¬*A*.

Premise 4: Therefore, *a* is a morally bad person.

Conclusion: Therefore, *a*'s argument α should not be accepted.

Critical Questions

CQ1: How exactly does the argument α state or imply that everybody in group *G* should be committed to *A*?

CQ2: Does *a* belongs to group *G*?

CQ3: Does *a* belong to other groups that would have goals affecting *a*'s commitment to *A*?

CQ4: Does *a*'s action, as described, imply that *a* is committed to the opposite of *A*?

CQ5: Why does it follow (if it does) that the alleged practical inconsistency shows that *a* is bad person?

CQ6: Is *a*'s being a bad person a good reason for concluding that *a*'s argument should not be accepted?

ARGUMENT FROM GUILT BY ASSOCIATION

Premise 1: *a* is a member of, or is associated with, group *G*, which should be morally condemned.

Premise 2: Therefore, *a* is a morally bad person.

Conclusion: Therefore, *a*'s argument α should not be accepted.

Critical Questions

CQ1: What evidence is there that *a* is a member of *G*?

CQ2: If *a* is not a member of *G*, but has been associated with *G*, how close was this association?

CQ3: Is *G* a group that should be morally condemned?

CQ4: Is it possible that even though *a* is a member of *G*, a group that ought to be condemned, *a* is not a bad person?

SITUATIONALLY DISQUALIFYING AD HOMINEM ARGUMENT

Premise 1: In dialogue *D*, *a* advocates argument α, which has proposition *A* as its conclusion.

Premise 2: *a* has certain features in his personal situation that make it inappropriate for him to make a dialectical contribution to *D*.

Premise 3: Therefore, *a* is a morally bad person.

Conclusion: Therefore, *a*'s argument α should not be accepted.

Critical Questions

CQ1: What features of *a*'s personal situation make it inappropriate for him to contribute to *D*?

CQ2: Do the features of *a*'s situation cited give any good reason to make one conclude that it is inappropriate for him to contribute to *D*?

CQ3: Could *a*'s argument be worth considering on its merits, even though there is reason to think them inappropriate for *D*?

POISONING THE WELL AD HOMINEM ARGUMENT

Premise 1: For every argument α in dialogue *D*, person *a* is biased.

Premise 2: Person *a*'s bias is a failure to honestly take part in a type of dialogue *D*, that α is part of.

Premise 3: Therefore, *a* is a morally bad person.

Conclusion: Therefore, α should not be given as much credibility as it would have without the bias.

Critical Questions

CQ1: What is the evidence that *a* has been biased with respect to every argument in the dialogue?

CQ2: Is the bias a normal partisan viewpoint that *a* has shown, or can it be shown to indicate that *a* is not honestly participating in the dialogue?

CQ3: In what respect is *a* a bad person, judging from the evidence of his participation in the dialogue that gives a reason for doubting his credibility?

LOGICAL INCONSISTENCY CIRCUMSTANTIAL
AD HOMINEM ARGUMENT

Premise 1: *a* advocates argument α, which has proposition *A* as its conclusion.

Premise 2: *a* is committed to proposition *A* (generally, or by virtue of what she has said in the past).

Premise 3: *a* is committed to proposition $\neg A$, which is opposite of the conclusion of the argument α that *a* presently advocates.

Premise 4: Therefore, *a* is a morally bad person.

Conclusion: Therefore, *a*'s argument α should not be accepted.

POISONING THE WELL BY ALLEGING GROUP BIAS

Premise 1: Person *a* has argued for a thesis *A*.

Premise 2: But *a* belongs to or is affiliated with group *G*.

Premise 3: It is known that group *G* is a special-interest partisan group that takes a biased (dogmatic, prejudiced, fanatical) quarrelling attitude in pushing exclusively for its own point of view.

Conclusion: Therefore, one cannot engage in open-minded critical discussion of an issue with any member of G, and hence the arguments of *a* for *A* are not worth listening to or paying attention to in a critical discussion.

Critical Questions

CQ1: Has *a* given any good reasons to support *A*?
CQ2: What kind of bias has *a* exhibited, and how strong is it?
CQ3: Is it the kind of bias that *a* has exhibited a good reason for concluding that she is not honestly and collaboratively taking part in the dialogue?
CQ4: Is there evidence of a dialectical shift in the case – for example, from a persuasion dialogue to a negotiation?
CQ5: Is the bias indicated in CQ2 of the very strong type that warrants the conclusion that *a* is not open to any argumentation that goes against her position (or seems to her to go against her position)?

TU QUOQUE

Premise: **Proponent**: Respondent, you are a morally bad person (because you have bad character, are circumstantially inconsistent, biased, etc.); therefore, your argument should not be accepted.
Conclusion: **Respondent**: You are just as bad, therefore your ad hominem argument against me should not be accepted as having any worth.

Critical Questions

CQ1: Is the proponent's ad hominem argument a strong one (according to the criteria for whatever type it is)?
CQ2: Is the respondent's ad hominem argument a strong one (according to the criteria for whatever type it is)?
CQ3: If the proponent's ad hominem argument is strong, how much credibility should be given to the respondent as an honest arguer who can be trusted to make such an allegation?

3. CLASSIFYING THE SUBTYPES OF AD HOMINEM ARGUMENT

The system of classification of schemes proposed in Figure 10.1 presupposes a precise knowledge of all the ad hominem argumentation schemes presented in section 2 of this chapter. Many users of the compendium who want to apply it to arguments in conversational discourse may not be familiar with the details of all these schemes, or have them ready at hand. Such users will need a more general classification of the varieties of ad hominem argument that can be used to draw distinctions between one type of ad hominem argument and another. A second classification scheme for subtypes of ad hominem argument has been presented by

Argumentation Schemes

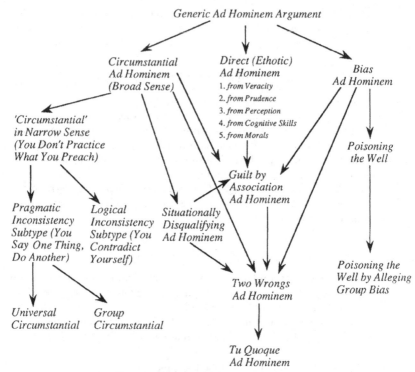

FIGURE 10.2. Classification of subtypes of ad hominem arguments.

Walton (1998, p. 261). The second classification, shown in Figure 10.2, represents a working system of classification that can be more easily used by those identifying, analyzing, or evaluating ad hominem arguments found in a text of discourse, and wishing to fit them into different classifications.

The guilt by association ad hominem argument is classified as a complex subtype involving a group attack (Walton, 1998, p. 261). The three subtypes of direct, circumstantial, and bias ad hominem are presented in Figure 10.1 as having a special place of importance. They are called the three primary subtypes.

Under the direct (ethotic) type, five special subtypes have been recognized (Walton, 1998, pp. 249–250; 2002, p. 51), as listed at the top of Figure 10.2.

Negative Ethotic Argument from Veracity

Premise 1: *a* has a bad character for veracity.
Conclusion: Therefore, *a*'s argument should not be accepted.

Negative Ethotic Argument from Prudence

Premise 1: *a* has a bad character for prudent judgment.
Conclusion: Therefore, *a*'s argument should not be accepted.

Negative Ethotic Argument from Perception

Premise 1: *a* has a bad character for realistic perception of the situation.
Conclusion: Therefore, *a*'s argument should not be accepted.

Negative Ethotic Argument from Cognitive Skills

Premise 1: *a* has a bad character for logical reasoning.
Conclusion: Therefore, *a*'s argument should not be accepted.

Negative Ethotic Argument from Morals

Premise 1: *a* has a bad character for personal moral standards.
Conclusion: Therefore, *a*'s argument should not be accepted.

Critical Questions

CQ1: Is the premise that *a* is a person of bad character true (or well supported)?

CQ2: Is the issue of character relevant in the dialogue in which the argument was used?

CQ3: Is the conclusion that the argument α should be (absolutely) rejected even if other evidence to support α has been presented, or is the conclusion merely (the relative claim) that α should be assigned a reduced weight of credibility, relative to the total body of evidence available?

Thus although identifying the direct or ethotic type of ad hominem argument may seem fairly straightforward, even it can be classified into a set of subtypes that represent special forms of it.

The general lesson here is quite important. Classifying the various subschemes of a particular type of argument helps us to identify and define it as a type of argument by drawing contrasts, not only between the different subspecies, but also between all these and other argumentation schemes that are related to them. We can see from the example of the ad hominem how many different important varieties exist as subschemes, and how it is related to other basic forms of argument, like argument from pragmatic inconsistency, with which it is often confused. The classification of these subschemes, along with related schemes, performs a function like that of a system of ontology in computing. It enables us to clearly define concepts that we need to use by constructing classification trees that display how some of the concepts in a family fall under other, more general concepts. How it works here is that, first of all, we have to construct a tree representing the precise relationships among the argumentation schemes that fall into a certain group or family. We can then use this technical device to draw up a more user-friendly classification of subtypes of a given type of argument, so that this simplified classification system can be applied without the user having to know all the precise details of the scheme tree.

4. COMPLICATIONS

Of course, this is not the first classification of forms of argument to be proposed. In antiquity as well as in modern times, both the philosophical and rhetorical traditions have developed such classifications. None, however, provides the level of detail and practical ease of use that we are after. Perelman and Olbrechts-Tyteca (1969), for example, list an extensive set of rhetorical argumentation schemes in four groups (based in part on a Quintilian model): quasi-logical, based on reality, establishing reality, and dissassociative. Beneath this four-way division, however, there is no further structure, which leaves an arguer or analyst with little to go on when trying to select a particular scheme. What we need is something like the Linnean model of arranging like species in genera, like genera in families, and so on. Linnean taxonomy supports *keys* that offer distinguishing features at each step down the classificatory tree – such keys make the job of the user of the taxonomy much easier.

Kienpointner (1986, 1992) has constructed a taxonomy based upon several fundamental distinctions. The first is the descriptive-normative distinction, whereby schemes of the former type involve no normative propositions among their premises and conclusions. The second is the real-fictitious distinction, whereby schemes of the former type involve no "fictitious" propositions (indicated by their expression in the subjunctive). The third is the pro-contra argumentation distinction, marking the use to which an argument is put in either supporting or countering a claim. Finally, drawing on Toulmin (1958), Kienpointner distinguishes warrant-using from warrant-establishing schemes, the former of which employ a warrant in arguing for a claim, while the latter have a warrant as their claim. Working very much in the same vein, Kienpointner is nonetheless much more detailed than Perelman and Olbrechts-Tyteca, but the taxonomy suffers from a number of problems.

Katzav and Reed (2004a) closely analyze Kienpointner's taxonomy in comparison to Walton's set of schemes (1996). They argue in favor in Kienpointner's "semantic" classification, relying on instrinsic features of arguments (and specifically, of premises and conclusions and their linkage and assembly) and excluding contexts and situations of argument from the classification scheme. But they then argue that many of Kienpointner's own analytical distinctions fail to be grounded in semantic properties of this sort. In place of Kienpointner's taxonomy, they propose a method of constructing a taxonomy based exclusively on the semantic properties of warrants. Katzav and Reed (2004b) construct a complete taxonomy for a similar-sized set of schemes that has several interesting properties. First, it explicitly admits of extension, so that schemes that are more specific can be slotted in beneath those that are less specific. This is an important property of a taxonomy, particularly when considering applications in artificial intelligence in which some very specific schemes are now starting to be proposed for very specific purposes (such as the VS scheme discussed in Tolchinsky et al., 2006, used to capture arguments about the viability of organs for transplant in the CARREL system). Secondly, and perhaps rather troublesomely, Katzav and Reed demonstrate how their set of schemes can be organized according to *multiple* and orthogonal hierarchical classifications. The first, and most comprehensive, hierarchy divides between relations based on intrinsic features and those based on features extrinsic to the propositions in premises and conclusions, and thence between groups of relations such as those deriving from causality, from normativity, from specificity, from identity, and so on. But then a second classificatory mechanism is described – based on the arity of the relation upon which a scheme is based. In this way, (almost) any scheme in the first hierarchy could be classified as a 2-, 3-, or *n*-arity scheme according to this second mechanism. Even more problematic is a third mechanism, based on class membership. According to this approach, the topmost division is between relations that are based on class membership and those that are not. The classification by the first mechanism is completely re-wrought by this third.

Given our experience with the task of classifying just one small subset of schemes, those related to ad hominem arguments, it is perhaps no surprise that a single all-encompassing Linnean taxonomy is deficient in the ways Katzav and Reed highlight. The success of a classification scheme depends heavily on what it is to be used for. Our focus here is squarely upon building a practical taxonomy of schemes that can be used in both teaching and research, for both producing arguments oneself and

362 Argumentation Schemes

analyzing the arguments of others. With this in mind, the classification laid out at the start of the chapter provides a natural arrangement based on semantic properties (following Kienpointner and Katzav and Reed) that aids in practical navigation. It also supports extension and refinement (following Katzav and Reed). It may be that these schemes could be reorganised (according to arity, for example), but such a reorganisation would need to demonstrate greater utility to the end users of the classification.

5. CONCLUSIONS

In arriving at the system for classification of schemes proposed in section one, we considered several systems, and tested each of the systems against the set of schemes in the compendium as well against other subschemes we know about. It turned out to be quite difficult to find a single system that could do justice to all these schemes. For this reason, we were initially very reluctant to propose any classification system at all, at this stage of research into schemes. On the other hand, logic textbooks that treat of fallacies, and the defeasible types of argument that correspond to them, already use rough systems of classification. The problem is that the systems of classification currently in use are widely different from each other and appear to exhibit little agreement. Indeed, an even worse problem is that the informal logic textbooks offer classification systems of fallacies, not realizing that any such system depends on a prior classification of the argumentation schemes underlying, and associated with, the fallacies.

Thus we seem to be caught in a dilemma. We need a working system of classification for schemes, but given that the schemes have not yet been formalized, and therefore have not yet been precisely defined in a formalistic way, any classification system will eventually have to be modified once such a formalization of the schemes has been achieved. This problem is typical of any kind of ongoing research, especially if it is still at an early stage. An ontology, or classification of basic concepts, is highly necessary, and extremely useful to help the research move forward, even though it should not be regarded as fixed in place permanently. Instead, it should be regarded as a hypothesis that is subject to refinement as new varieties and subspecies are encountered and classified.

In this chapter we used the case of the ad hominem argument to illustrate the nature of the problem of extending the proposed classification system to deal with complex types of argument that contain numerous subtypes. As research on schemes continues, more and more of these subtypes will be discovered. The tools for formalizing the schemes in the critical questions, like the Carneades system, are still at a relatively early stage, and deal only with the basic schemes and the basic critical questions matching them. Much research in argumentation theory is now needed to extend such formal models to cope with the complications of dealing with all the new varieties of schemes and critical questions that will be discovered.

11

Formalizing Schemes

The goal of this chapter is to show how to formalize the schemes by expressing each scheme as a formal inference structure in a way comparable to forms of inference we all are already familiar with in deductive logic, and to some extent in inductive reasoning. Although deductive and inductive forms of argument can be included as argumentation schemes, the most difficult part of this project is the formalization of the defeasible schemes. The defeasible schemes listed in the compendium represent the most common forms of reasoning not only in everyday discourse, but also in special contexts of use like legal and scientific reasoning. The defeasible schemes presented in the compendium look to have a rough formal structure, but there is a wide variety of them, utilizing many different kinds of variables and constants. Before the schemes can be formalized, further clarifications need to be made (Verheij, 2003).

1. THE DEFEASIBLE MODUS PONENS FORM OF SCHEMES

In order to be useful in logic, artificial intelligence, and related scientific fields, schemes must be formalized, meaning that they have to be codified in some precise way so that the coder (whether machine or human) can recognize a particular argument as fitting a scheme and then use it to derive conclusions from the given set of premises based on that identification. Once an argument is recognized as fitting a scheme, an argument markup, utilizing an argument diagram, can reconstruct the argument in a given case using the scheme as a template or pattern on which to frame the reconstruction. As part of the reconstruction, for example, the coder might identify a missing premise that needs to be inserted before the conclusion can be drawn, based on the requirements of the scheme. Such a formalization could be extremely useful for many purposes. It could be used to identify types of arguments from a natural language corpus of discourse. It could even be used to construct arguments, thus yielding a mechanized procedure for argument invention. Schemes, once formalized with constants and variables identified as occurring in their proper places in the premises and conclusion, could have a function similar to rules of deductive logic that are helpful for evaluating arguments as strong or weak, according to specify criteria.

One problem is that the schemes cited in the compendium have been defined only in a rough-and-ready way. They do have characteristic premises and conclusions that can be identified, practically speaking, but the majority of them do not conform to the notation of any known formal model of argument. Some of them look like deductive arguments; others look like inductive arguments; but the broad majority of them are defeasible arguments that don't fit deductive or inductive forms of reasoning. They are plausible arguments that give tentative grounds, subject to qualification, for accepting a conclusion as plausible. But they're also arguments that are subject to defeat when new evidence comes into a case indicating that the circumstances of the case fit one of the qualifications on which the original argument needed to be based.

As we look over the schemes in the compendium, there is a clue suggesting that at least some of them, and perhaps even most of them, have what looks to be a kind of *modus ponens* structure. Consider the scheme for argument from commitment (Walton, 1995, p. 144), where *P* is a participant in a discussion and *A* and *B* are propositions.

Scheme for Argument from Commitment

Generally, if *P* is committed to *A*, then *P* is also committed to *B*.
P is committed to *A*.
Therefore, *P* is committed to *B*.

Although this scheme looks to have a general structure comparable to *modus ponens*, the term 'generally' in the first premise indicates that the argument is best taken to be defeasible. But can there be a defeasible *modus ponens* type of argument, different from the deductive form of *modus ponens* that we are so familiar with? One development that offers a way to begin the formalization of schemes is the hypothesis that many of the schemes listed in the compendium have a form that can be reduced to a defeasible *modus ponens* structure (Walton, 1992, p. 217).

Verheij (2001, p. 232) draws a distinction between the familiar *modus ponens* form of argument from classical deductive logic and a defeasible form that he calls *modus non excipiens*. His distinction depends on the kind of rule on which the inference is based. If a rule is strict, that is, if it admits of no exceptions, the usual *modus ponens* can be applied.

Modus Ponens

As a rule, if P, then Q.
P.
Therefore, Q.

For example, if the given argument is based on a universal generalization about all triangles, without exception, *modus ponens* can be applied. But if the rule is one that admits of exceptions (a defeasible rule), an additional premise needs to be taken into account. This kind of case requires a different argumentation scheme.

Modus Non Excipiens

As a rule, if P, then Q.
P.
It is not the case that there is an exception to the rule that if P, then Q.
Therefore, Q.

Modus non excipiens should be applied in a case where both strict rules and rules admitting of exceptions might possibly come into play (Verheij, 2000, p. 5). Instead of using Latin expressions, these two forms of argument could more simply be called strict *modus ponens* (SMP) and defeasible *modus ponens* (DMP).

There is a growing consensus that all argumentation schemes can be cast in the *modus ponens* form and that this form is especially suitable for legal argumentation (Verheij, 2003; Walton, 2006). Such a hypothesis rests on the prior assumption that there can be a defeasible *modus ponens* as well as a deductive *modus ponens*. This observation in itself presents an initial problem for the classification of schemes, for if all the defeasible schemes have this form, can *modus ponens* itself be viewed as an argumentation scheme? Is DMP a general form of inference underlying most of the schemes in the compendium?

Another example of a scheme from the compendium that has the DMP form (Walton and Reed, 2002) is that of argument from popular opinion.

Scheme for Argument from Popular Opinion

If a large majority (everyone, nearly everyone, etc.) accepts *A* as true, then there exists a (defeasible) presumption in favor of *A*.
A large majority accepts *A* as true.
Therefore, there exists a presumption in favor of *A*. (Walton, 1996, p. 83)

The way this scheme is presented here shows clearly that it can be cast in a defeasible *modus ponens* form. The question now is whether many or most of the defeasible schemes in the compendium can be formulated in such a way that they explicitly show a defeasible *modus ponens* form.

This question takes us back to the related problem of classifying the schemes. Many of the schemes in the compendium appear to have a core *modus ponens* structure in general outline, but also have many special premises and variables that are also parts of the form of the argument. Consider the schemes for the slippery slope type of argument or the ad hominem type of argument. It has been shown by Walton (1992, p. 217) how *modus ponens* has a defeasible form where the major premise is a defeasible conditional, and how the structure of slippery slope argumentation is based on a forward chaining of such arguments (p. 225). Many of the other complex schemes listed in the compendium might be fitted under argument from commitment, which has a defeasible *modus ponens* form, but also has many other components that are not easily formalized, and do not easily yield to some kind of analysis using only *modus ponens* reasoning. Therefore the task of seeing how these complex forms are based on an underlying *modus ponens* structure involves considerable analysis and study. Still, the conjecture that many of the schemes in the compendium have the underlying structure of a defeasible *modus ponens* form of argument is a good hypothesis for moving forward and investigating the project of formalization of schemes.

2. SCHEMES IN AML

Araucaria, explored in more detail in Chapter 12, provides a software tool for creating argument analyses and, most importantly for our task here, for creating analyses that are founded upon argumentation schemes. In order to build a software system, it is necessary to have at least some sort of formal characterization of the analytical components that the software will need to store and manipulate. This is a "weak" kind of formalization that is purely descriptive.

Araucaria's underlying representation language is an XML language, the Argument Markup Language. AML is defined using a DTD, a simple and straightforward language-design mechanism. One of the basic components of arguments from Araucaria's point of view is a proposition or PROP – loosely, a box in examples such as Figure 1.2 above. The definition of this component is as follows:

```
<!ELEMENT PROP (PROPTEXT, OWNER*, INSCHEME*)>
```

The PROPTEXT component details the text or, roughly, the propositional content of a given PROP. The OWNERs of a PROP allow analysts to distinguish between viewpoints in an argument (and to lay a foundation for marking up argumentative dialogue, which is currently work in progress). Finally, the INSCHEME component allows the analyst to indicate that a PROP belongs to a given scheme. Notice that the Kleene star in the definition allows multiple INSCHEME tags for a given PROP – that is, a given proposition can be in more than one argumentation scheme.

The following definition of the (empty) INSCHEME tag, includes two references, one to a unique scheme name, the *scheme* attribute, and one to a unique identifier, *schid*. It is important to include both so that any given PROP can be marked as belonging not only to a scheme of a particular type, but also to a particular instance of that scheme within the current text (so that multiple instances of a given scheme can be identified uniquely).

```
<!ATTLIST INSCHEME scheme   CDATA #REQUIRED
                   schid CDATA #REQUIRED>
```

Finally, the *scheme* attribute in the definition corresponds (in processing, not in AML definition) to an element in the SCHEMESET tag of the AML file. For ease of exchange and independence, each AML analysis includes the complete set of scheme definitions that are used in the analyzed text. The SCHEMESET (which can also be saved separately, and thereby adopted in different analyses) is composed of a series of SCHEME elements.

```
<!ELEMENT SCHEME (NAME, FORM, CQ*)>
```

Thus each scheme has a unique name (e.g., "Argument from Expert Opinion" in the schemeset corresponding to Walton, 1996). The CQ elements allow specification of critical questions, and the FORM element supports specification of the formal structure of a scheme, thus,

```
<!ELEMENT FORM (PREMISE*, CONCLUSION)>
```

where both PREMISEs and CONCLUSIONs are ultimately just propositions expressed in text.

In this way, AML supports the specification of argumentation schemes in a machine-readable format. It is flexible enough to capture various types of argumentation schemes, including examples from Kienpointner (1986), Walton (1996), Grennan (1997), and Katzav and Reed (2004). Similarly, it is flexible enough to handle and match other types of argumentation analysis in diverse domains, including Wigmore charts in reasoning about legal evidence (Prakken et al., 2003) and Pollock-style inference graphs (Pollock, 1995). At the same time, the language is simple enough to support manipulation by a number of systems, tools, and utilities, including, of course, Araucaria. But AML is also used by several other utilities, and its schemes are being employed in the construction of a large online corpus[1] of natural argumentation, available online at <http://arg.computing.dundee.ac.uk>.

AML, however, does not provide the "strong" type of formalization required for reasoning about, and with, the structure of schemes. For this, more detail is required. The next three sections review three complimentary approaches that work toward such strong formalization.

3. ELEMENTS OF A LOGICAL FORMALIZATION OF SCHEMES

A simple starting point for formalization is propositional logic, *PL*, from which we take our propositions (*Props*) and propositional variables, and all the usual operators. Next, we define a set of attributes, T. This set contains any number of arbitrary tokens. Attributes are associated with propositions by the typing relation, τ, thus: $\tau : Props \rightarrow P(T)$. That is, the typing relation associates with every proposition a set of attributes, or "type."

The next step is to define scheme structures formally. The approach presented here is based on the implementation of the Argument Markup Language DTD (Reed and Rowe, 2001), and is designed to facilitate practical and reusable implementation. The set, Ξ, of schemes in a particular system is comprised of a set of tuples of the following form: <*SName, SConclusion, SPremises*>, where *SName* is some arbitrary token, *SConclusion* $\in T$, and *SPremises* $\subset T$. If $\exists \xi \in \Xi$ such that $\xi = <\sigma_0, \sigma_1,$

---

[1] Clearly, the use of a markup language and the presentation here is suggestive of other work in corpus linguistics. There is not space here to explore the relationships between AML and corpus research; the interested reader is directed to the website for further details.

$\sigma_2 >$ then $\neg \exists \xi' \in \Xi / \{\xi\}$, such that $\xi' = <\sigma_0, \sigma_3, \sigma_4 >$ or $\xi' = <\sigma_5, \sigma_1, \sigma_2 >$, for any σ_3, σ_4, σ_5. In this way, a scheme is uniquely named and is associated with a conclusion type and a set of premise types.

Finally, an instantiation is an argument based upon one of the schemes. An instantiation is thus a tuple, $<$*Name, Conclusion, Premises*$>$ such that for some $< SName, t, SPremises> \in \Xi$, where $SName = Name$,

$$Conclusion \in Props \quad \wedge \quad \tau(Conclusion) = t, \quad \text{and}$$

$$\forall p \in Premises, \; p \in Props \wedge \text{ the set } \{\pi | \pi = \tau(p)\} = SPremises^2$$

In this way, an instantiation of a scheme named *SName* must have a conclusion of the right type, and all the premises, each of which is also of the right type. (Note that this latter requirement is actually a little too strong for most natural models of scheme usage, as schemes often involve some premises being left implicit, to form enthymematic arguments. The simplification is useful at this stage of development and does not preclude more sophisticated handling later.)

This model supports a straightforward mechanism for representation of schemes. It does not, as it stands, give an agent a mechanism for reasoning with schemes and for building (that is to say, chaining) arguments using schemes. Through structures such as critical questions (Walton, 1996), argumentation schemes offer the potential for a sophisticated model of dialectical argument-based nonmonotonic reasoning. Such a model is currently under development (see Prakken et al., 2003, for some preliminary steps in this direction.) In the meantime, a simple solution suffices to support development of both theory and implementation.

To sketch how this works, we define a new operator, $\rightarrow$, that corresponds to implication extended to schemes. That is, in this system, if $\alpha \supset \beta$, then $\alpha \rightarrow \beta$, but also, if there exists an instantiation of an argument scheme $<N, C, P>$ in which $\beta = C$ and $\alpha \in P$, then $\alpha \rightarrow \beta$. Dung-style definitions of acceptability and admissibility are then formed using deductive closure on $\rightarrow$ rather than $\supset$, and everything else remains as before. Thus, the representation of argumentation schemes is brought into the standard models of defeasible argumentation of Dung, Prakken, Vreeswijk, Verheij, and others.

There are two distinct facets to implementation that can handle schemes. The first is the ability to represent and manipulate scheme-based structures in the one-agent setting in a flexible and scalable way. The second is the ability to utilize that representation in the multi-agent case, and to exploit representational structure in communication design.

---

[2] Set equivalence here is taken to mean identical membership.

4. FORMALIZATION OF SCHEMES IN THE CARNEADES SYSTEM

Carneades is not a formalization of argument in the manner of a deductive formal system of logic. It is a computational model that builds on ontologies from the semantic web to build a system for using argumentation schemes in legal reasoning. In effect, the model is an abstract functional specification of a computer program and can be implemented in any programming language. Arguments modeled in the Carneades system can be visualized using an argument diagram because the basic structure Carneades uses, following the model of the semantic web, is that of the directed labelled graph. The nodes represent objects and the arrows represent binary relations.[3] It is especially important for our purpose to look carefully at how Carneades models the notion of argument defeat. What are called defeaters or rebuttals in Pollock's language are modeled as arguments that go in the opposite direction from that of another argument at issue. For example, if one argument at issue is *pro*, its rebuttal would be another argument *con* that argument. Premise defeat is modeled by an argument *con* an antecedent or assumption, or *pro* an exception (Gordon, 2005, p. 56). In Carneades, a Pollock-style undercutter of an argument *n* is modeled as an argument against the implicit applicability presumption of a scheme, that is, an argument *con* the atom.

The Carneades system defines three kinds of defeat relations or argument rebuttals. The first type consists of arguments *con* the consequent of a *pro* argument. The second type consists of arguments *pro* the consequent of a *con* argument. The third type consists of arguments *pro* alternative positions on the same issue. Undercutters correspond to arguments *con* applicability presumptions. An example is the assumption that the generalization "Birds fly" is applicable to Tweety.

An output representing an argument in Carneades can be visualized using an Araucaria argument diagram. The basic units of the system, called atoms, as in the example (`asserts Gloria (killed joe sam)`), are made up of subjects, objects, and binary predicates. Atoms are defined as RDF triples of the following kind (Gordon, 2005, 54).

```
type atom =
{predicate: symbol,
subject: symbol,
object: datum}
```

---

[3] The resource description framework (RDF) of the semantic web provides an XML syntax representing a graph.

Argumentation is viewed as a model construction process that tracks issues about which atoms should be included in a domain model. An *issue* is defined as a record for keeping track of the arguments *pro* and *con* each position (Gordon, 2005, p. 55). A *position,* which can be accepted or rejected or can remain undecided, is a proposed or claimed value of an atom. An argument is defined as a single atom called a consequent and another set of atoms called premises. The distinctive feature of how Carneades models arguments is that it recognizes three different types of premises called antecedents, assumptions, and exceptions. Antecedents are normal premises that are assumed to be acceptable, and that must be justifiable to make an argument acceptable. Assumptions are assumed to be acceptable unless called into question. Exceptions are premises that are not assumed to be acceptable, but are taken for granted for the sake of argument unless they are challenged. This distinction can be clarified by comparing the three kinds of premises to critical questions matching an argumentation scheme. They can be classified into three categories, depending on whether they are treated as antecedents, assumptions, or exceptions. Antecedents are like premises that are already present as required premises in a scheme, and so critical questions questioning them can be seen as redundant. Assumptions are premises that are assumed to be true, and exceptions are premises that are assumed to be false, even though they may later be shown to be true.

It is important to note that an exception is different from the negation of a presumption. A presumption is an additional premise in the generalization (conditional) that represents the structure of the argument. Consider the conditional in the Tweety example, shown as an argument in Carneades.

```
(if and (isa tweety bird) (applies arg-1 true))
(flies tweety true))
```

If we try to model an exception as the negation of an assumption, we get the conditional:

```
if (and (isa tweety bird) (not (isa tweety penguin))
(applies arg-1 true)) (flies tweety true))
```

The proper way to model an exception is by a conditional of this form:

```
if (and (isa tweety penguin) (applies arg-1 true))
bottom)
```

An exception, as modeled in Carneades, expresses a constraint on an argument. It cannot be treated in the same way as an application of an argument.

Each argument is provided with an identifier called an *id*. The following definition (Gordon, 2005, p. 56) shows how all the various elements are combined in the definition of the data type for an argument. Arguments are provided with an identifier allowing propositions to be about arguments, because the identifier of an argument can be used as the subject or object of an atom. Using this feature, an applicability presumption of the form (`applies <argument> id true`) is added to every argument.

```
type argument   =
{id: id,
   direction: {pro, con},
   consequent: atom,
   antecedent: atom list,
   assumptions: atom list,
   exceptions: atom list}
```

As noted earlier, premise defeat is modeled by an argument *con* an antecedent or presumption, or *pro* an exception (Gordon, 2005, p. 56). Also as noted earlier, Carneades follows Verheij's lead in modeling the most familiar sorts of argumentation schemes as having the DMP form. However, it also differs from Verheij's system in several respects. One of the most significant of these is that instead of representing an undercutter of an argument in the way that the argument diagramming system *ArguMed* does, it models argument attack and defeat notions in relation to how the pair of arguments relates to the underlying issue at dispute. Carneades diagrams arguments in a way comparable to Araucaria. But Araucaria represents refutation by using a function on the toolbar that displays an implicit proposition shown as the conclusion of an opposed argument that negates the conclusion of the argument attacked. This difference is a significant one when it comes to the project of formalizing the schemes for the ad hominem type of argument, essentially a type of argument that is used to attack another given argument.

5. FORMALLY MODELING THE CRITICAL QUESTIONS

Verheij has shown, using the example of ad hominem argumentation, that the critical questions matching a scheme are of four different types. Some are used to question whether a premise of a scheme holds. Some point to exceptional situations in which a scheme should not be used. Some set conditions for the proper use of a scheme. Some point to other arguments (rebuttals) that might be used to attack the scheme. Citing the first type, Verheij argues that critical questions that criticize the premises of a scheme are redundant because they merely ask whether the premise is true. We routinely ask anyway, in evaluating any argument, whether the premises are true. Such critical questions merely restate a premise, and are therefore redundant, and can be ignored. For example, consider the field question in the list of critical questions matching argument from expert opinion in Chapter 1 (also shown in Chapter 9). It can be seen as redundant, because the major premise already says that E is an expert in field F containing proposition A. As another example, consider the first critical question matching the scheme for the direct ad hominem argument that asks whether the arguer is a bad person. Such a critical question may be useful for the practical purpose of teaching students to think more critically, but is not necessary in a formalized system in which arguments fitting schemes are to be evaluated.

The original motivation of the Carneades system was to accommodate two different theories about what happens when a respondent asks a critical question (Walton and Gordon, 2005). On the one theory, when a critical question is asked, the burden of proof shifts to the proponent's side to answer it. On the other theory, merely asking the question does not defeat the proponent's argument until the respondent offers some evidence to back it up. Carneades approaches this distinction by distinguishing three types of premises, called ordinary premises, assumptions, and exceptions. Ordinary premises behave like assumptions at issue. An assumption holds if it is undisputed or accepted, but not if it is rejected. An undisputed ordinary assumption holds if its statement is acceptable, given its proof standard, or if it has been accepted, but not if it has been rejected. Exceptions hold unless the statement of the exception has been proven acceptable.

The following syntax suggests how the scheme for argument from expert opinion can be modeled in Carneades, where the conclusion is statement a.

```
(argument arg-1
  (scheme argument-from-expert-opinion)
  (premises
    (domain e d)
    (asserted e a)
    (within a d)
    (assuming (credible e))
```

```
(assuming (based-on-evidence (asserted e a)))
(unless (not (trustworthy e)))
(unless (not (consistent-with-other-experts e))))
(conclusion (pro a)))
```

The three ordinary premises are that the expert is an expert in the subject domain of the claim, that she asserted the claim in question, and that the claim is in the subject domain in which she is an expert. The two assumptions are that the expert is credible as an expert and that what she says is based on evidence, and they are taken to hold until called into question by the opponent. The additional two premises, that the expert is not trustworthy and that what she says is not consistent with what other experts say, are assumed not to hold, until such time as new evidence comes in showing that they are acceptable. There are like exceptions to a rule in defeasible reasoning. Carneades is a further development of Verheij's system of formalization, because it portrays every scheme as based on a set of premises, one of which is a defeasible conditional statement, thus viewing presumptive schemes as having the DMP form. The advantage of Carneades is that it provides a way of formalizing the critical questions as propositional components of a scheme.

6. THE ARGUMENT INTERCHANGE FORMAT

Most computational research on argumentation has focused on one of three areas: analysis of arguments, generation of arguments, and reasoning with arguments. These three areas place different demands and different emphases on the underlying representation of argumentative structures, and as a result, the techniques used by Araucaria and Carneades – to take just two examples – are very different. This is a shame, because although analysis, generation, and reasoning work in different ways, there are inevitably strong similarities between their substrates – there is no a priori reason why arguments that have been autonomously generated might not be subjected to computational analysis, or why computational analyses might not be used as the basis of autonomous reasoning. To tackle this challenge, an international group of researchers has come together to design a common language that supports the needs of each of the three communities. This *argument interchange format* includes schemes as a central part of its vision (Chesñevar et al., 2006).

The argument interchange format (AIF) is constructed in three layers. At the most abstract layer, the AIF provides an "ontology" of concepts that can be used to talk about argument structure. This ontology describes an argument by conceiving of it as a network of connected nodes that are of two types: information nodes that capture data (such as datum and claim nodes in a Toulmin analysis, or premises and conclusions in a traditional analysis), and scheme nodes that describe passages between information nodes (similar to warrants or rules of inference). Scheme nodes, in turn, come in several different guises, including scheme nodes that correspond to support or inference (or "rule application nodes"), scheme nodes that correspond to conflict or refutation (or "conflict application nodes"), and scheme nodes that correspond to value judgments or preference orderings (or "preference application nodes"). At this top-most layer, there are various constraints on how components interact: information nodes, for example, can be connected to other information nodes only via scheme nodes of one sort or another. Scheme nodes, on the other hand, can be connected to other scheme nodes directly (in cases, for example, of arguments that have inferential components as conclusions, e.g., in patterns such as Kienpointner's (1992) "warrant-establishing arguments").

A second, intermediate layer provides a set of specific argumentation schemes (and value hierarchies, and conflict patterns). Such a set of devices is a development of the basic concept of a "schemeset" that was available in the AML representation language used in Araucaria. Thus, the uppermost layer in the AIF ontology lays out that presumptive argumentation schemes are types of rule application nodes, but it is the intermediate layer that cashes those presumptive argumentation schemes out into argument from consequences, argument from cause to effect, and so on. At this layer, the form of specific argumentation schemes is defined: each will have a conclusion description (such as "A may plausibly be taken to be true") and one or more premise descriptions (such as "E is an expert in domain D").

It is also at this layer that, as Rahwan and colleagues (2007) have shown, the AIF supports a sophisticated representation of schemes and their critical questions. In addition to descriptions of premises and conclusions, each presumptive inference scheme also specifies descriptions of its presumptions and exceptions. Presumptions are represented explicitly as information nodes, but, as some schemes have premise descriptions that entail certain presumptions, the scheme definitions also support entailment relations between premises and presumptions. Finally, presumptive inference schemes also represent exceptions (or, rather, potential exception descriptions). Rather than handling these too as information nodes, the conflict handling mechanism built in to the AIF ontology is harnessed, in effect, to see (potential) exceptions as (potential) conflicts. Thus, for example, an exception to the argument from expert opinion scheme is that the expert is biased. The corresponding exception description is constructed from a conflict scheme (the conflict scheme from bias). The definition of conflict schemes follows that of presumptive argumentation schemes in specifying premises and conclusions, and so the conflict scheme from bias used here involves a premise descriptor (in this case, "Speaker is biased"). Between them, these representations of presumptions and exceptions represent growth points for an argument – ways in which they can be questioned and further supported. They correspond to critical questions.

In this way, the representation of the presumptive inference scheme argument from expert opinion (a specific scheme node) is linked to many premise, conclusion, and presumption nodes (all information nodes), and also to several conflict nodes (also scheme nodes) – the constraints specified at the most abstract layer (about how scheme and information nodes may interact) are inherited at this level, and new constraints are introduced.

Finally, the third and most concrete level supports the integration of actual fragments of argument, with individual argument components (such as strings of text) instantiating elements of the layer above. At this third layer, an actual instance of a given scheme is represented as a rule application node – the terminology now becomes clearer. This rule application node is said to *fulfill* one of the presumptive argumentation scheme descriptors at the level above. As a result of this fulfillment relation, premises of the rule application node fulfill the premise descriptors; the conclusion fulfils the conclusion descriptor; presumptions can fulfill presumption descriptors; and conflicts can be instantiated via instances of conflict schemes that fulfill the conflict scheme descriptors at the level above. Again, all the constraints at the intermediate layer are inherited, and new constraints are introduced by virtue of the structure of the argument at hand. Figure 11.1 shows diagrammatically how all of these pieces fit together (but it should be borne in mind that Fig. 11.1 aims to diagram how the AIF works – it is a poor diagram of the argument that is represented).

The "fulfills" relationships are indicated by dotted lines, and inference between object layer components (i.e., the arguments themselves) by thick lines. Remaining lines show "is a" relationships. Note that many details have been omitted for clarity, including the way in which scheme descriptions are constructed from forms.

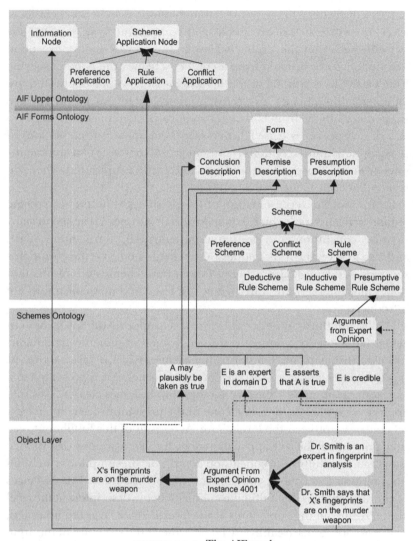

FIGURE 11.1. The AIF stack.

7. THE RESEARCH PROJECT FOR DEVELOPING
A FORMAL SYSTEM

So far we have the outlines of a research project for developing the formalization of schemes in which schemes can be modeled as identifiable types of arguments leading from premises to a conclusion. The three systems of formalization of schemes described here use different notation, different argument diagramming methods, and different formal and computational methods, but they also have much in common, including a common philosophical viewpoint. When you put them together, they form a platform for further research on the formal modeling of schemes for their computational implementation. They take us much deeper into the investigation of how schemes can be modeled in a formal structure of the kind that would be useful in helping us to identify arguments in a natural language text discourse, and for other purposes as well.

Before we can take the program for classification and formalization of schemes much further, we need to ask the following question. How can the results of these efforts at formalization that have been developed so far be applied to the most important argumentation schemes represented in the compendium? We began in Chapter 1 with a particular scheme that is often taken as the leading example of efforts at analysis and formalization of schemes, namely, the argument from expert opinion. Let's go back to that scheme to illustrate how the research efforts described here can be applied to the most important of the schemes in the compendium, taking us deeper into the analysis of schemes.

As indicated earlier in this chapter, the most difficult obstacle to moving ahead, given the present state of the art, is the problem of formalizing the critical questions in such a way that they could be adapted to the current technology of artificial intelligence. The six basic questions corresponding to the scheme for argument from expert opinion, for example, can be analyzed in Carneades by classifying these critical questions as antecedents, assumptions, or exceptions. Here we make no attempt to provide anything like final answers, but will discuss each question in turn.

The expertise question asks how strong the expert's mastery of the field is. Should it be assumed that an expert in a field has some mastery of that field? If so, failure to give a specific answer should not be enough to make the argument default.

The field question is already a premise in the argument from expert opinion. Since it is a premise, failure to back it up adequately means that the argument should default. If in response to the opinion question, the proponent fails to supply a proposition representing what the expert claimed, her argument should default. But the discussion on the issue of how to judge whether the proposition needs to match exactly what the expert said may remain open. However, if the proponent gives no answer at all to the question, the original argument should default. When the trustworthiness question is asked, unless the respondent gives positive evidence indicating that the expert is untrustworthy, the proponent could simply reply, "There is no evidence of that at all," shifting the burden to the respondent's side. To make the consistency question have any force, some evidence to support it seems to be required. In relation to the backup evidence question, what an expert claims needs to be backed up by some evidence in the domain of knowledge of the expert in order to make the argument hold. Thus it is reasonable to rule that some evidence backing up the question is required, or the original argument will default. If these rulings are reasonable, Carneades can then be applied to the scheme for argument from expert opinion, enabling each of the critical questions to be represented as an antecedent, assumption, or exception. The project of formalization can then move ahead, providing a formal model of the argument from expert opinion that can be implemented in an automated system of argumentation schemes. But this is not the end of the project.

To move forward with a deeper analysis of the structure of argument from expert opinion – and the same could be said for all of the schemes in the compendium – one needs to realize that in addition to the basic critical questions matching each scheme, there are also subquestions under each basic question. Hence a theoretical problem is posed. If it is possible to go on asking critical questions indefinitely, is there any end to the sequence of such questioning? In other words, can an argument matching a scheme ever be finally evaluated so that the process of questioning ends, and we can say definitively whether the argument is strong or weak, or meets whatever criteria of success we have adopted in a particular case? Such a possibility of a chain of argumentation being an unfinished infinite sequence of this kind is not so worrisome once we recall that we are dealing with defeasible arguments. Such arguments need to be seen as potentially open to defeat as new evidence enters a case. Of course, closure will arrive at some point as we declare that a discussion or investigation has reached the concluding stage. But the problem remains: if the critical questions can always have some questions that lie beneath them, how can any formal model of argumentation schemes deal with a complex layers of critical questioning that can arise even in the case of a basic scheme like argument from expert opinion, never mind more complex forms of argument like those for slippery slope?

In Chapter 1, and in the compendium (Chapter 9), the basic critical questions for argument for expert opinion were shown. But investigations of this scheme (Walton, 1997, pp. 211–225) have shown that under each basic question is nested a set of subquestions. Here is the argumentation scheme for argument from expert opinion along with a set of such sub-questions as presented by Walton (1997, pp. 211–225).

Argument from Expert Opinion

Major Premise: Source E is an expert in subject domain S containing proposition A.
Minor Premise: E asserts that proposition A is true (false).
Conclusion: A is true (false).

Critical Questions

CQ1: *Expertise Question*. How credible is E as an expert source?
  1.1. What is E's name, job or official capacity, location, and employer?
  1.2. What degrees, professional qualifications, or certification by licensing agencies does E hold?
  1.3. Can testimony of peer experts in the same field be given to support E's competence?
  1.4. What is E's record of experience, or other indications of practiced skill in S?
  1.5. What is E's record of peer-reviewed publications or contributions to knowledge in S?
CQ2: *Field Question*. Is E an expert in the field that A is in?
  2.1. Is the field of expertise cited in the appeal a genuine area of knowledge, or an area of technical skill that supports a claim to knowledge?
  2.2. If E is an expert in a field closely related to the field cited in the appeal, how close is the relationship between the expertise in the two fields?
  2.3. Is the issue one where expert knowledge in *any* field is directly relevant to deciding the issue?
  2.4. Is the field of expertise cited an area in which there are changes in techniques or rapid developments in new knowledge, and, if so, is the expert up to date in these developments?
CQ3: *Opinion Question*. What did E assert that implies A?
  3.1. Was E quoted as asserting A? Was a reference to the source of the quote given, and can it be verified that E actually said A?
  3.2. If E did not say A exactly, then what did E assert, and how was A inferred?

3.3. If the inference to *A* was based on more than one premise, could one premise have come from *E* and the other from a different expert? If so, is there evidence of disagreement between what the two experts (separately) asserted?

3.4. Is what *E* asserted clear? If not, was the process of interpretation of what *E* said by the respondent who used *E*'s opinion justified? Are other interpretations plausible? Could important qualifications have been left out?

CQ4: *Trustworthiness Question.* Is *E* personally reliable as a source?

    4.1. Is *E* biased?

    4.2. Is *E* honest?

    4.3. Is *E* conscientious?

CQ5: *Consistency Question.* Is *A* consistent with what other experts assert?

    5.1. Does *A* have general acceptance in *S*?

    5.2. If not, can *E* explain why not, and give reasons why there is good evidence for *A*?

CQ6: *Backup Evidence Question.* Is *E*'s assertion based on evidence?

    6.1. What is the internal evidence the expert herself used to arrive at this opinion as her conclusion?

    6.2. If there is external evidence – for example, physical evidence reported independently of the expert – can the expert deal with this adequately?

    6.3. Can it be shown that the opinion given is not one that is scientifically unverifiable?

Thus we can see there is much work left to be done, not only in developing systems for formalization and computational modeling of argumentation schemes, such as Carneades, but also in applying such formal systems to the schemes in the compendium. The kind of further research work needed should be carried on at two levels. At the first level, the task is to go through each of the basic questions and show how each question can be accommodated to the formal model by classifying it as an antecedent, assumption, or exception to the scheme. Then, at the second level, the task is to go through the subquestions classified under each of the basic questions and classify them as antecedents, assumptions, or exceptions.

8. SCHEMES IN DIALOGUE

Though Carneades and the current version of the AIF both make significant contributions to our understanding of how schemes work in monologic structures, they say little about how schemes fit into dialogues. In a sense, it is surprising that little investigation has been carried out of how schemes are used in dialogue, because they have an inherently dialogical nature. Critical questions, after all, are questions that an opponent can pose in order to test and probe inferential links, or to shift the burden of proof (Prakken et al., 2005): and these are all dialogical activities.

One technique for investigating dialogic structures that is equally familiar to both computer scientists and philosophers is the formal dialogue game. These games lay out the locutions that can be uttered by participants, the ways in which those locutions modify the state of the game (often through manipulating commitment stores), and the rules that govern how games start and finish. They have a long history in philosophy and, because of their similarity to concepts such as finite state automata, are increasingly popular in computer science and artificial intelligence (McBurney and Parsons, 2002; Wells, 2007). Formal dialogue games offer a good starting point for exploring how argumentation schemes might be understood dialogically.

Our aim is to define an Argumentation Scheme Dialogue, or ASD. We take as our starting point Walton's (1984) game, CB. As a subset of Mackenzie's (1979) DC game and Hamblin's (1970) H, it is both familiar and simple. CB has just four locutions, five commitment rules, and three further dialogue rules and, with such a small set, allows us to explore what is unique to dialogue with schemes. We hope in future to apply scheme-based extensions to much richer accounts of argumentative dialogue such as PPD (Walton and Krabbe, 1995). We do not here want to defend CB as a model of (even idealized) dialogue, but rather use it as a tool to explore how a dialogue game can be extended to encompass argumentation schemes.

Walton's CB is as defined as follows (Walton, 1984, pp. 133–135):

Locution Rules

i. *Statements*: Statement letters, S, T, U, . . . , are permissible locutions, and truth-functional compounds of statement letters.
ii. *Withdrawals*: "No commitment S" is the locution or withdrawal (retraction) of a statement.
iii. *Questions*: The question "S?" asks "Is it the case that S is true?"
iv. *Challenges*: The challenge "Why S?" requests some statement that can serve as a basis in (a possibly defeasible) proof for S.

Commitment Rules

   i. After a player makes a statement, S, it is included in his commitment store.

  ii. After the withdrawal of S, the statement S is deleted from the speaker's commitment store.

 iii. "Why S?" places S in the hearer's commitment store unless it is already there or unless the hearer immediately retracts his commitment to S.

 iv. Every statement that is shown by the speaker to be an immediate consequence of statements that are commitments of the hearer then becomes a commitment of the hearer's and is included in his commitment store.

  v. No commitment that is shown by the speaker to be an immediate consequence of statements that are previous commitments of the hearer may be withdrawn by the hearer.

Dialogue Rules

  R1. Each speaker takes his turn to move by advancing one locution at each turn. A no-commitment locution, however, may accompany a why-locution as one turn.

  R2. A question "S?" must be followed by (i) a statement "S," (ii) a statement "Not-S," or (iii) "No commitment S."

  R3. "Why S?" must be followed by (i) "No commitment S" or (ii) some statement "T," where S is a consequence of T.

Walton goes on to describe further strategic rules that are not of interest here and are invariant to the extensions for argumentation schemes.

Like Mackenzie, Walton requires his dialogue game to have available to it a set of rules of inference. Walton (1984) assumes that these are selected from the rules of inference of propositional logic, but here we relax that assumption and instead expand the set of rules to include argumentation schemes. Walton's (1984) definitions of immediate consequence and consequence hold for schemes just as they do for deductive rules of inference:

immediate consequence. A statement T is an *immediate consequence* of a set of statements S0, S1,... Sn if and only if 'S0, S1,..., Sn therefore T' is a substitution-instance of some rule of the game.

consequence. A statement T is a *consequence* of a set of statements So, S1,... Sn if and only if T is derived by a finite number of immediate consequence steps from immediate consequences of So, S1,... Sn. (Walton, 1984, pp. 132–133)

The analogy between deductive and inductive reasoning is strong here: where deductively, if we know A and A ⊃ B, then B is an immediate consequence (assuming that *modus ponens* is among the rules of inference we accept); where inductively, if we know E is an expert (in the domain of X) and E asserts X, then X is an immediate consequence (assuming that argument from expert opinion is among the rules of inference we accept).

CB supports what we might call 'reasoning elicitation' (i.e., the support of a conclusion by a premise) along a single specific path: one player states S; the other player challenges, "Why S?"; and the original player can respond with T, of which S is a consequence. It is this reasoning elicitation step that is the focus of the extension of CB to form ASD.

We introduce a new dialogue rule that is applicable after the S–Why S?–T pattern:

(R4) After a statement T has been offered in response to a challenge locution, "Why S?," then if (S, T) is a substitution instance A of some argumentation scheme of the game, the locution pose(C) is a legal move, where C is a critical question of scheme A appropriately instantiated.

There are several things to note about rule R4. First, CB does not deal explicitly with enthymemes. It is not particularly instructive to extend it in order to do so, but it is important to note that argumentation schemes have an arbitrary number of premises, and only some of these premises may be introduced explicitly in a dialogue. So (S, T) in rule R4 may be a *partial* substitution instance. We may, for example, say:

My doctor has said I need to eat less salt, so I probably should try and cut down.

This is an enthymeme in which the doctor's expertise in dietary matters is left implicit – we have one of the premises of the scheme, plus its conclusion:

Minor Premise: My doctor has said I should eat less salt.
Major Premise: My doctor is an expert on my diet.
Conclusion: I should eat less salt.

The instantiation (My doctor has said I should eat less salt, I should eat less salt) thus counts as a partial substitution instance of the general form of the scheme, (E says A, E is an expert, A), and would license a pose(C) move under rule R4.

The rule R4 introduces a new locution move, pose(C), which can be added as a fifth in the list of CB's locutions:

(v) *Critical Attacks*: The attack "Pose C" poses the critical question C associated with an argumentation scheme.

Though there is a single locution type, it has two distinct effects on the dialogue depending on whether the critical question being posed is an exception or a assumption. Broadly, we want to capture the intuition that assumptions are required in order for an argument to go through, that they function much like missing or implied premises, and that questioning them requires the proponent of the argument to justify. Exceptions, by contrast, are potential attacks in which the barrier is set higher: an interlocutor must not just ask the question but must provide some evidence for the potential counter. This distinction is discussed using legal examples by Prakken and colleagues (2005) and with computational ramifications by Gordon and Walton (2006). We assume that the different types of critical questions are explicitly marked – both Gordon and Walton (2006) and Rahwan and colleagues (2007) suggest that this is a reasonable expectation in computational practice as well as in theory. We can then formulate a new dialogue rule to handle the pose move:

(R5) After a "Pose C" move, then either
    (a) if C is an assumption of its argumentation scheme, the move is followed by
       (i) a statement "C"
       (ii) a statement "not-C"
       (iii) "No commitment C"
    (b) if C is an exception to its argumentation scheme, the move is followed by
       (i) a statement "C"
       (ii) a statement "not-C"
       (iii) "No commitment C"
       (iv) "Why not-C?"

Part (a) of rule R5 is of course a rehearsal of the rule for the question move: questioning an assumption is analogous (as Gordon and Walton [2006] and Verheij [2005] have discussed) to questioning an implicit premise (in these earlier versions of the theory, what are now called assumptions were referred to as *presumptions*). Part (b) extends the permitted responses to include a reciprocal challenge move, requiring the other party to justify his critical questioning.

So, for example, an assumption of the expert opinion scheme is that E is indeed an expert in the right field (an assumption associated with the second critical question of that scheme). If this is questioned, the speaker must state that it is the case, or that it is not, or withdraw commitment to it. Similarly, the fourth critical question of the expert opinion scheme allows a critic to probe the expert's reliability. The question that can be posed is "Is the expert reliable?," to which the proponent can respond not only with statement or withdrawal moves but also with a challenge, "Why is the expert not reliable?"

But what happens if critical questions are not posed? Clearly, with respect to exceptions, there is nothing to do: only if an interlocutor takes exception is there any work to be done – exceptions function as potential "growth points" of an argument, but the opportunity for growth lapses if not taken up. But for assumptions, it seems natural to account for the assuming somehow. This is an ideal candidate for the notion of *mere concession* introduced in Walton and Krabbe's (1995) PPD. Mere concessions provide a way of accounting for propositions that are agreed to (typically by a hearer) but in a weaker way than commitments. Specifically, commitments are usually required to be defended on demand, while mere concessions are not. Unfortunately, CB does not have such a subcategory of commitment – so we instead sketch how bare commitment to assumptions can be introduced.

(iv) Every statement that is shown by the speaker to be an immediate consequence of statements that are commitments of the hearer via some rule of inference or argumentation scheme A, then becomes a commitment of the hearer's, along with all the assumptions of A.

This encompasses the previous version of this commitment rule because the definition encompasses rules of inference that have no assumptions, such as *modus ponens*, as well as argumentation schemes that may have critical questions, some of which will be marked as assumptions.

The new pose move also introduces commitment directly, but it does so in exactly the same way as the challenge move, and so can be accommodated by a change to CB's third commitment rule:

(iii) Both "Why S?" and "Pose S" place S in the hearer's commitment store unless it is already there or unless the hearer immediately retracts his commitment to S.

In combination, these changes to the rules of commitment and the rules of dialogue ensure that inference over argumentation schemes can contribute to a dialogue in just the same way as inference over deductive rules, and furthermore that the burden of proof is distributed appropriately across the different critical questions: the proponent takes the burden of proof for assumptions, the questioner for exceptions.

9. SUMMARY OF THE DIALECTICAL SYSTEM ASD

To summarise, our new dialectical game, ASD, is set out as follows:

Locution Rules

    i. *Statements*: Statement letters, S, T, U, . . . , are permissible locutions, and truth-functional compounds of statement letters.
    ii. *Withdrawals*: "No commitment S" is the locution or withdrawal (retraction) of a statement.
    iii. *Questions*: The question "S?" asks "Is it the case that S is true?"
    iv. *Challenges*: The challenge "Why S?" requests some statement that can serve as a basis in (a possibly defeasible) proof for S.
    v. *Critical Attacks*: The attack "Pose C" poses the critical question C associated with an argumentation scheme.

Commitment Rules

    i. After a player makes a statement, S, it is included in his commitment store.
    ii. After the withdrawal of S, the statement S is deleted from the speaker's commitment store.
    iii. "Why S?" places S in the hearer's commitment store unless it is already there or unless the hearer immediately retracts his commitment to S.
    iv. Every statement that is shown by the speaker to be an immediate consequence of statements that are commitments of the hearer via some rule of inference or argumentation scheme A, then becomes a commitment of the hearer's and is included in his commitment store along with all the assumptions of A.
    v. No commitment that is shown by the speaker to be an immediate consequence of statements that are previous commitments of the hearer may be withdrawn by the hearer.

Dialogue Rules

R1. Each speaker takes his turn to move by advancing once locution at each turn. A no-commitment locution, however, may accompany a why-locution as one turn.

R2. A question "S?" must be followed by (i) a statement "S," (ii) a statement "Not-S," or (iii) "No commitment S."

R3. "Why S?" must be followed by (i) "No commitment S" or (ii) some statement "T," where S is a consequence of T.

R4. After a statement T has been offered in response to a challenge locution, "Why S?," then if (S, T) is a substitution instance A of some argumentation scheme of the game, the locution pose (C) is a legal move, where C is a critical question of scheme A appropriately instantiated.

R5. After a "Pose C" move, then either

(a) if C is a *assumption* of its argumentation scheme, the move is followed by
   (i)   a statement "C"
   (ii) a statement "not-C"
   (iii) "No commitment C"

(b) if C is an *exception* to its argumentation scheme, the move is followed by
   (i)   a statement "C"
   (ii) a statement "not-C"
   (iii) "No commitment C"
   (iv) "Why not-C?"

10. A WORKED EXAMPLE OF A DIALOGUE IN ASD

As an example, let us take the following dialogue fragment between Bob and Wilma. Wilma isn't sure about Bob's assertion that the conference program is perfect, and tries to convince Bob that it has a mistake in it.

(L1) Bob: OSSA's great: all the experts go. And they're brilliant at doing the program – they never make a mistake.

(L2) Wilma: Hmm. Look at the program: either Alf is staying home, or they've made a mistake on this one.

(L3) Bob: Yes, I suppose so.

(L4) Wilma: Well, do you remember that "expert" piece that Alf wrote in *South Western Ontario Philosophy Monthly* that said that most Canadian philosophers go to OSSA?

(L5) Bob: Yes, I remember.

(L6) Wilma: Well Alf should know, so we can take it that most Canadian philosophers do indeed go.

(L7) Bob: Yes, but he'd have a biased opinion.
(L8) Wilma: Why do you think he's biased?
(L9) Bob: Er, not sure – OK, so what if he wasn't biased? So what?
(L10) Wilma: Well, if we know that Canadian philosophers are going, then, given that Alf is a Canadian philosopher, he'll certainly be going.
(L11) Bob: No, not necessarily – he only said *most* philosophers.
(L12) Wilma: Well, you must agree that if all experts are going (as you said) and Alf is an expert (as you agreed), then Alf must be going anyway.
(L13) Bob: Yes, I suppose so.

We can reconstruct the dialogue propositionally, tracking the changes to the commitment stores thus:

Locution	Speaker	Content	B's Commitments	W's Commitments
L1	B	a, b	a, b	
L2	W	c ∨ not-a ?		
L3	B	c ∨ not-a	c ∨ not-a	
L4	W	e ?		
L5	B	e	e	
L6	W	d	d	d
		by AS1 that has e as premise and f as an assumption and g as potential exception	f	f
L7	B	Pose(g)		
L8	W	Why not-g		
L9	B	g	g	
L10	W	e ⊃ not-c?		
L11	B	No commitment e ⊃ not-c		
L12	W	(f ∧ b) ⊃ not-c?		
L13	B	(f ∧ b) ⊃ not-c	(f ∧ b) ⊃ not-c not-a	

Key List

a: They never make a mistake doing the program at OSSA.
b: All the experts go to OSSA.
c: Alf is staying home.
d: Most Canadian philosophers go to OSSA.
e: Alf said most Canadian philosophers go to OSSA.
f: Alf is an expert.
g: Alf is unbiased.

At L13, Wilma wins – one can see how this is so with the following proof through Bob's commitments to Wilma's thesis T_W that not-a (we assume that all the rules of propositional logic are available to this dialogue; they are indicated in the proof by "PL"):

(1) $(f \wedge b) \supset$ not-c   (established at L13)
(2)   b   (established at L1)
(3)   f   (established at L6)
(4) $f \wedge b$ (2, 3, PL)
(5)   not-c   (1, 4, PL)
(6) $c \wedge$ not-a   (established at L3)
(7)   not-a   (5, 6, PL) ∎

The trick to Wilma's success lies in using the assumption implicit in the argumentation scheme at L6 to formulate the final implication at L12, having failed with the implicative gambit (or "corner" in the language of Walton [1984]) of L10. In more detail, the game starts with Bob offering his position, which commits him to two propositions (and, indeed, several others omitted here for the sake of clarity). In order to try and have Bob retract his commitment to *a* (that they never make a mistake in the conference program), Wilma adopts a strategy based on disjunctive syllogism, pointing out that (in the face of some evidence), either Alf isn't going (*c*), or else a mistake has been made (*not-a*). At L3, Bob accepts the disjunction, so Wilma proceeds to disprove the first disjunct. She starts by introducing the claim that Alf wrote an article saying that most Canadian philosophers attend OSSA (*e*). At L5, Bob agrees to this claim. At L6, Wilma uses an argumentation scheme (we assume it is rather like the scheme from expert opinion presented in the first section) based on the premise *e*, and with the conclusion *d*, that most Canadian philosophers do indeed attend OSSA. The use of this scheme involves an assumption – that Alf is an expert (*f*), which gets tacitly added as another commitment under commitment rule (iv). The scheme also has a potential exception – that the expert may be biased – here referred to as *g*. (Note that the scheme carries a number of other assumptions and exceptions omitted from this analysis for the sake of clarity.) Bob decides to pose the critical question associated with this exception – not because he has any particular reason to doubt the honesty of the expert, but rather as a strategic ploy, in an attempt to shift the burden of proof to Wilma (Prakken et al., 2005). Unfortunately for Bob, at L8, Wilma rejects the ploy by demanding justification for the critical question, as she is permitted to do with critical questions associated with exceptions, under locution rule (R5b). With no further evidence, Bob is compelled to concede that his question has no further justification, and so agrees that the exception does not hold (g). At L10, Wilma attempts to win her "corner" – the premise that will guarantee success, namely, the conditional that if most Canadian philosophers are going, then Alf is going. Bob rejects this conditional, however. So, at L12, Wilma adopts a different strategy, this time using the implicit assumption from the argumentation scheme used earlier at L6. She again aims for agreement on a conditional, this time that if Alf is an expert, and all experts are attending OSSA, then Alf is attending OSSA. This time, Bob agrees, so Wilma wins her corner – she has one of the disjuncts in her disjunctive syllogism shown to be false, so the other – her ultimate thesis – must be true.

11. CONCLUSIONS

We have shown how, with an understanding of defeasibility in schemes, various techniques can be used to formally describe argumentation schemes. These different techniques have various strengths and weaknesses, but all provide insight into the internal structure and function of schemes. We used the case of the argument from expert opinion to illustrate the problem of how to achieve a deeper formal analysis of argumentation schemes by formally modeling the critical questions matching each scheme. Just as the ad hominem case illustrates how many subtypes we have to deal with, the case of the argument from the expert opinion illustrates how many subquestions we have to deal with when trying to formalize the set of critical questions matching any given scheme.

Dealing with critical questions naturally leads to wanting an adequate explanation of how those questions can actually be asked – that is, how argumentation scheme structures can be slotted into dialogical systems. We used a simple example of a formal dialogue to which schemes and their critical questions can be added in order to sketch how this can be achieved more generally.

We also described briefly the Argument Interchange Format, a new direction for formalizing schemes in a reusable way. One of the key reasons for exploring these different characterisations (logical formalization, the representations in Araucaria and Carneades, and the AIF) is to provide a stepping stone toward computational systems. It is to these we turn in the next chapter.

12

Schemes in Computer Systems

Argumentation theory has laid foundations for and has had influence upon a wide variety of computational systems (Reed and Norman, 2003). This chapter explores four distinct areas, reviewing the ways in which argumentation schemes have been put to work: in natural language generation, in interagent communication, in automated reasoning, and in various specific computational applications. To start, however, we look at the tools that are being used to support the development of these applications and that allow the creation, analysis, and manipulation of the raw computational resources that involve argumentation schemes.

1. SCHEMES IN ARAUCARIA

Following work examining the diagramming of natural argument – an important topic from the practical, pedagogic point of view (van Gelder and Rizzo, 2001), but also a driver of theoretical development in informal logic (Walton & Reed, 2004) – Reed and Rowe (2004) developed Araucaria, a system for aiding human analysts and students in marking up argument. Araucaria adopts the "standard treatment" (Freeman, 1991) for argument analysis, based on identification of propositions (as vertices in a diagram) and the relationships of support and attack holding between them (edges in a diagram). It is thus similar to a range of argument visualization tools (see Kirschner et al., 2003, for an overview), and familiar from AI techniques such as Pollock's (1995) inference graphs and even Bayesian nets and qualitative probabilistic networks. As well as having a number of features that make it particularly well suited to teaching and research in argumentation, it is also unique in having explicit support for argumentation schemes.

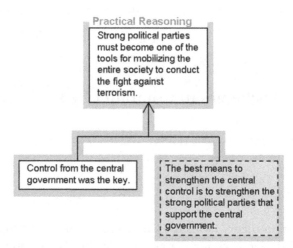

FIGURE 12.1. An example of the practical reasoning scheme.

In Chapter 11, we have seen how Araucaria stores argumentation schemes using an XML-based language. But Araucaria's main job, of course, is to visualise. The visual presentation of schemes is problematic, and often reflects deep-seated assumptions about the way in which inference works. In some conceptions of inference, it is important to "reify" the inference, allowing it to be manipulated directly. Pollock (1995), for example, talks of undercutters of an inference: such undercutters attack the inference itself (and inference must therefore be visualised in such a way that undercutters can be shown to attack). Toulmin (1958), on the other hand, sees inferences as being warranted, and it is those warrants that take primary place in both conception and visualization. This is a particular challenge for Araucaria, which aims to support diagrammatic analysis according to multiple styles of analysis (including, for example, Toulmin and Wigmore).

In its standard analysis of argumentation, Araucaria shows argumentation schemes as shaded regions that encapsulate parts of the argument tree. So, for example, in Figure 12.1 the argumentation scheme from practical reasoning encapsulates the premises and conclusion in the context of an argument extracted from a British local newspaper about Putin's political strategy for the Duma.[1]

---

[1] This example is taken from the *Evening Telegraph* (Lancashire), September 13, 2004, and forms a part of the AraucariaDB online corpus available at <http://araucaria.computing.dundee.ac.uk>.

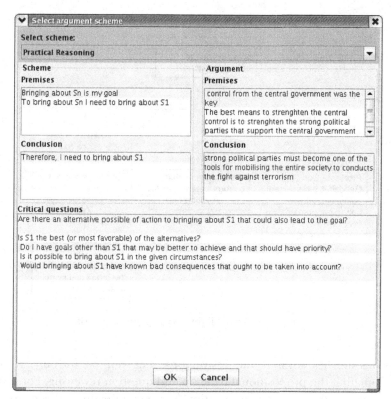

FIGURE 12.2. Araucaria software dialog box for scheme marking.

As mentioned in Chapter 11, Araucaria adopts a "weak" approach to formalizing schemes that focuses solely on representation. In usage, this means that the software can do nothing but suggest and guide. So, for example, if a user identifies premises and conclusions of an argument and decides to associate with them the practical reasoning scheme, the software offers the dialogue box shown in Figure 12.2. On the left are the definitions of premises and conclusions available in the template offered by the scheme; on the right are the selected argument components; and at the bottom are the critical questions, associated with the scheme. The tying up of actual premises and conclusion with the template premises and conclusion, and the task of reviewing the critical questions, are all jobs for the analyst that are unsupported by the software. This is good for rapid analysis, and for pedagogical purposes (the two primary foci of Araucaria), but it can lead to problems when used for generating argument structures, because it is easy to introduce errors and

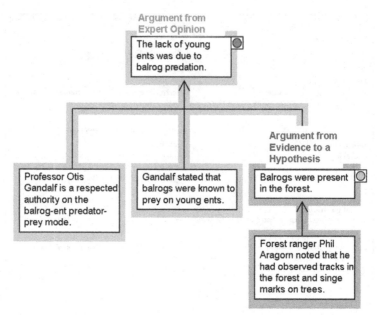

FIGURE 12.3. Traffic lights example in Araucaria.

inconsistencies without the restrictions that can be introduced by automating the propositional tying-up and question reviewing. For software that introduces this automation, see section eight of this chapter for a discussion of ArgDF.

Finally, for very rapid checking of scheme structure and, in particular, for swift review of critical questions, a version of Araucaria has been produced in which critical questioning is achieved through "traffic lights" in the main diagram. In the example shown, in Figure 12.3, the little circles in boxes at the top left of claims indicate traffic light status. Araucaria's visualization is in color, and shows, in this example, the top traffic light to be green. This shows that all the critical questions of the argument from expert opinion have been checked by the analyst (through a simple check-box interface shown in Figure 12.4). The lower traffic light, however, is amber, indicating that only some of the critical questions of the argument from evidence to hypothesis have been answered. In cases where Araucaria is used as a part of an analytical process (e.g., in constructing judicial summaries), these traffic lights act as a handy and obvious indicator that all appropriate paths and challenges have been checked.

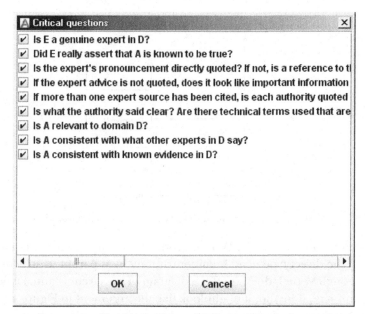

FIGURE 12.4. Checking off the critical questions in Araucaria.

2. SCHEMES IN ARGUMED

Verheij (2003) has proposed a method for formalizing argumentation schemes, basing his analysis (p. 176) on the minimal assumption that any argumentation scheme can be expressed in the following general format:

Premise 1. Premise 2.... Premise n. Therefore, *Conclusion.*

Verheij (p. 177) uses an argument diagramming method, ArguMed, to represent schemes as they apply to argumentation. An ArguMed diagram like the one in Figure 12.5, which is comparable to one of Verheij's diagrams (2003, p. 177), shows how arguments are represented in ArguMed.

To show how his proposed system of formalization works, Verheij has studied the problem of how to formally model the various schemes for ad hominem argumentation presented by Walton (1998). His diagram (Verheij, 2003, p. 177) of the argumentation scheme for the direct type of ad hominem argument discussed by Walton (1998) is like the one shown in Figure 12.6.

Verheij presents a comparable diagram (Verheij, 2003, p. 178) representing the type of argument called the guilt by association type of

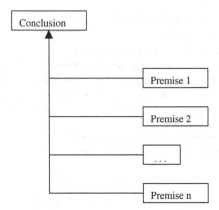

FIGURE 12.5. Argument diagram structure in ArguMed.

ad hominem argument by Walton (1998). The latter is shown to incorporate the direct type of ad hominem argument as part of its structure.

Although Verheij doesn't offer a diagram of the circumstantial type of ad hominem, if he did, it might look like the argument in Figure 12.7.

In ArguMed, undercutting moves, like asking a critical question, are modeled by a concept called entanglement. The question, or other rebuttal, attacks the inferential link between the premises and conclusion of the original argument, and thereby requires the retraction of the original conclusion. On a diagram, entanglement is represented as a line that meets another line at a junction marked by an X. Entanglement is shown in Figure 12.8 as applied to an instance of a circumstantial ad hominem argument.

The bottom three boxes of Figure 12.8 show how the given argument fits the form of the circumstantial ad hominem. The top box, when inserted to represent new evidence in the case, makes the ad hominem argument default. It is a typical defeasible argument.

This example shows how asking one of the critical questions matching a given scheme can make the original *argumentum ad hominem* default, leading to an important general question about how to model schemes formally. Can the critical questions be modeled as implicit premises that,

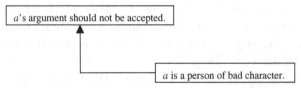

FIGURE 12.6. ArguMed diagram for the direct ad hominem argument.

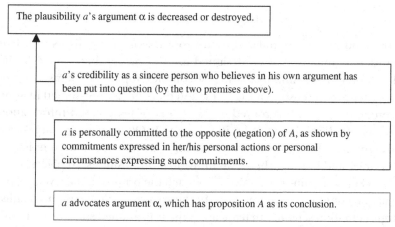

FIGURE 12.7. ArguMed diagram for the circumstantial ad hominem argument.

when added to the original premises of an argument fitting a scheme, can make it default? Formalizing argumentation schemes would potentially be easier if the critical questions matching the scheme could be represented as propositions that function as additional premises of the argument. If this could be done, we wouldn't have to confront the problem of how to formalize questions. We could simply model the critical questions as additional premises of the scheme.

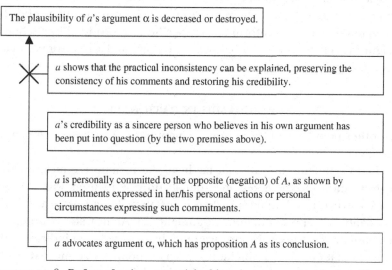

FIGURE 12.8. Defeat of a circumstantial ad hominem argument in ArguMed.

3. SCHEMES IN COMPENDIUM

Compendium is a popular software tool for mapping discussion. It is a wide-ranging application designed to offer practical assistance in the development and understanding of complex topics, and aims to cover more types of discussion and interaction than are usually studied in argumentation theory. It is rooted in the theory of Issue Based Information Systems (IBIS) originated by Rittel (1970). Often, it is the large-scale structures of discussion that form atomic units in Compendium: issues, critiques, and so on, rather than individual propositions. In places, however, Compendium-style analysis can reach the propositional level, particularly in contentious interactions. In these cases, it can be useful to enlist argumentation schemes to tease apart the fine-grained structures of positions and claims. Buckingham Shum (2007) has developed a translation mechanism that allows the XML schemesets developed for Araucaria to be imported into Compendium as *patterns* that guide the Compendium user in developing detail for an argument. Figure 12.9 shows argument from expert opinion as a Compendium map. Figure 12.10 shows multiple schemesets in Compendium.

By virtue of the XML specification, Compendium can inherit from Araucaria the multiple schemesets reflecting the different approaches to the characterization of argumentation schemes proposed by different authors.

Like Araucaria, Compendium uses schemes as a guide, providing a visual aide-mémoire for the analyst as they construct graphs that describe an unfolding dialogue. There are no in-built constraints that inflexibly demand that a user conform to the specification provided by a scheme. Figure 12.11 shows how both scheme structure and actual argument are available simultaneously.

4. SCHEMES IN RATIONALE

Another popular tool for argument mapping is Austhink's commercial product, Rationale.

Like Compendium, Rationale has been able to import Araucaria's AML specification of schemesets in order to provide its users with scheme-based structures. A difference between Rationale and Araucaria is the modality of representing the argument. In Araucaria, an argument instantiating a scheme is portrayed in a highlighted area indicating the argumentation scheme, including a set of premises connected to the conclusion. In Rationale, on the other hand, it is not possible to classify

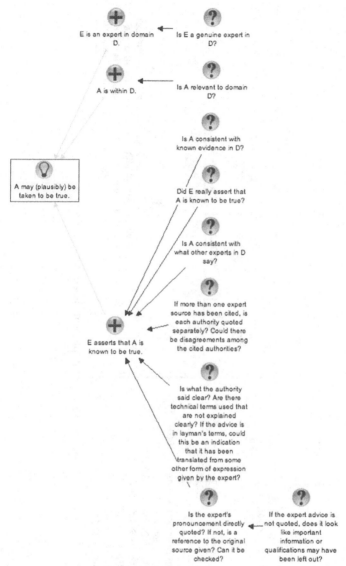

FIGURE 12.9. Argument from expert opinion as a compendium map.

an argument under a particular pattern of inference, so only the abstract model of the scheme is given, to be filled in by the user with the real premises and conclusion. For instance, the pattern of the argument from expert opinion is represented as shown in Figure 12.12.[2]

[2] Available at <http://wiki.austhink.com/Walton>.

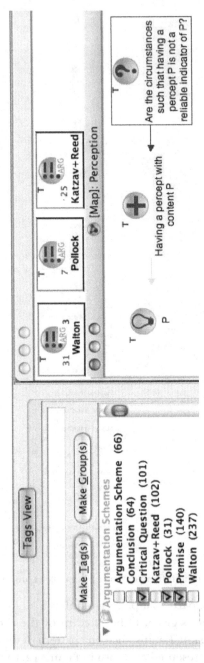

FIGURE 12.10. Multiple schemesets in Compendium.

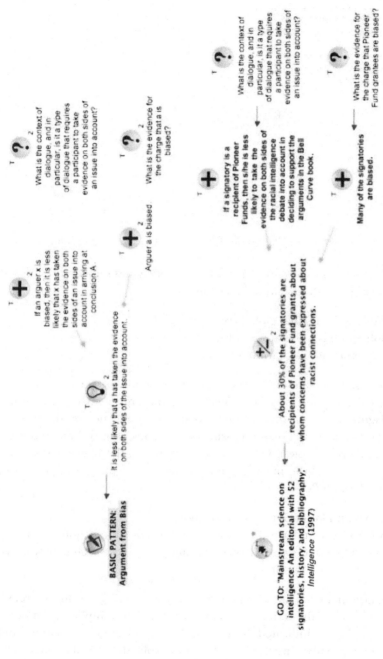

FIGURE 12.11. Using an argumentation scheme in Compendium.

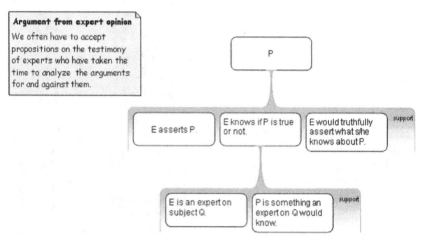

FIGURE 12.12. Pattern for the argument from expert opinion.

The pattern can be filled in with the explicit and implicit premises of a real argument, such as the following one:

> Bob Smith says that global warming is real.
> Bob Smith is a climate scientist.
> Therefore, global warming is real.

If we want to use Rationale for analyzing this argument, we need to reconstruct the missing premises constituting the argument pattern shown in Figure 12.12. In this abstract model, we should notice, the missing premises represent not only missing assumptions, as in Walton's schemes, but also critical questions, such as, for instance, the field question. We can reconstruct the missing premises needed to match the pattern in Rationale as follows:

> Bob Smith knows if this is true or not
> Bob Smith would truthfully assert what he knows about global warming.
> The nature and risks of global warming are subjects a climate scientist would know about.

The argument can be diagrammed as in Figure 12.13. In this figure, we should notice, the difference between explicit and implicit premises is not graphically indicated. Moreover, the type of inference pattern and the set of premises instantiating it are not explicitly indicated, but can be drawn from the typology of the premises and the manner in which they are connected.

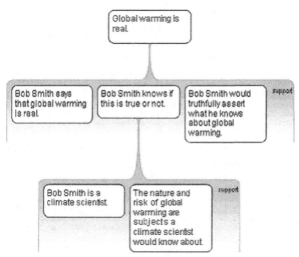

FIGURE 12.13. Using Walton's argument from expert opinion in Rationale.

Rationale allows the user to represent some inference patterns in another way. Premises may be marked with an icon, indicating the type of support they provide to the conclusion. For instance, a premise may support the conclusion because the argument instantiates argument from expert opinion, or argument by definition, or argument by example. For instance, in Figure 12.14, the example analyzed in Figure 12.13 has been diagrammed using the latter technique. In this diagram, the image of the argument has been emphasized in order to show the relation between premises, conclusion, and scheme.

Having discussed Araucaria, ArguMed, Compendium, and Rationale, we now go on to some general considerations.

5. SCHEMES IN NATURAL LANGUAGE GENERATION

Previous work (Reed, 1998; 1999) has shown how deductive and nondeductive reasoning patterns can be operationalized in an AI planner. This implementation can then be used to generate the textual structure of a monologic argument. Thus an argument structured around *modus ponens* (a, $a \rightarrow b$, b) is characterized as a single planning *operator*, with a postcondition that the hearer believes b, and preconditions that the hearer believes both minor (a) and major ($a \rightarrow b$) premises. Similar operators are constructed for a range of deductive and nondeductive forms. The effect of such a characterization is that arguments can be built up by a

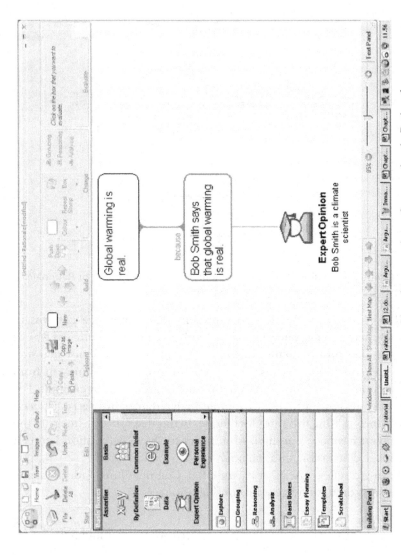

FIGURE 12.14. Argumentation schemes as icons for premises in Rationale.

process of means-ends analysis: planning to bring the hearer to believe a conclusion by bringing him to believe the premises, which in turn can be achieved by the application of operators, which again have preconditions, and so on. There are a number of shortcomings to this work, including its reliance upon belief (rather than upon the more accessible notions of commitment and acceptability) and the requirements of the knowledge base from which arguments are constructed (the implementation focuses upon the classical '*divisio*' rather than '*inventio*'). Nevertheless, the model provides a basis from which to develop an implementation of argumentation schemes.

Each argumentation scheme has a goal. That is, an argument's author is using a particular scheme to some specific effect. In many cases, the effect is one of lending support to a claim in the hope that some audience will accept the claim, though in others it is the determination of a course of action or the apportionment of resources that is the aim (cf. Walton and Krabbe's [1995] analysis of dialogue types). The claim-supporting (or, perhaps, standpoint-supporting) type of goal can be crudely captured for the purposes of automatically generating an argument in natural language through the mechanism described earlier, *viz.* as $BEL(h, p)$, where BEL is a predicate capturing belief, h refers to the audience (or an individual stereotype of that audience), and p is the claim supported by the argumentation scheme. A goal of this form then constitutes the postcondition of an argumentation scheme operator. The preconditions of the operator include, in the first place, the premises of a particular scheme, similarly construed. So, for example, in argument from consequences, the scheme is constructed from a premise and a conclusion thus (Walton, 1996, p. 76):

(P1) If A is brought about, then good (bad) consequences will (may plausibly) occur.
(C) Therefore, A should (not) be brought about.

In addition, there are the critical questions:

(CQ1) How strong is the likelihood that these cited consequences will (may, must, etc.) occur?
(CQ2) If A is brought about, will (or might) these consequences occur, and what evidence supports this claim?
(CQ3) Are there other consequences of the opposite value that should be taken into account?

An implementation of this scheme, following the structure provided by the Rhetorica system employed by Reed (1999), would be constructed from the postcondition

$Bel(h, do(A))$

and preconditions

$Bel(h, leads\_to(do(A), good\_consequences))$
$Bel(h, not(leads\_to(do(A), bad\_consequences)))$

Like many other schemes described by Walton (1996), this characterization of argument from consequences involves critical questions that probe for evidence. Rather than requiring additional machinery, and explicit mention within a scheme operator, this evidential function is handled naturally as part of the system's processing. In order to gather evidence for any statement in an argument – including statements that play the role of premises in argumentation schemes – all that is required is another round of processing. Thus the first precondition goal, $Bel(h, leads\_to(do(A), good\_consequences))$, can in its turn be considered as a goal to be fulfilled by the application of further operators. Consideration of the evidence for this claim (as well as its linked sibling premise) is thus a natural part of the functioning of the system.

Although it is clear from the brief example just given that this approach places heavy demands on the knowledge representation system, it does suffice to demonstrate that defeasible argumentation schemes – replete with their critical questions – can be characterized quite formally in an existing implemented system. With a sufficiently rich knowledge base, textual arguments that involve schemes can thus be generated automatically, significantly extending the range of the arguments that can be produced by Reed's (1998) implementation.

6. SCHEMES IN INTERAGENT COMMUNICATION

Research into multi-agent systems (MAS) is a rapidly expanding area of AI and computer science (Wooldridge, 2002), and as such it has not yet achieved universally accepted definitions or delimitations. The concept of an agent employed here follows the trend typified by the work of Wooldridge and Jennings (1995), in which an agent has several defining properties. First, an agent is autonomous and persistent. When an agent is started, it typically persists for an extended period and, crucially, has no further direct manipulation by either a user or another computer process. It may interact with either a human or another software agent, but its internal state is not under the direct influence of either. Second, an agent is proactive. An agent will typically be characterized through a mentalistic definition, employing logics of beliefs and goals. Endowed with more or less sophisticated reasoning capabilities, an agent will determine a course of action that furthers its own goals. The demands of autonomy and proactivity typically produce agents that are ultimately "selfish." Finally, an agent is communicative. The power of multi-agent systems lies not in the capabilities of individual agents, but rather in the abilities of heterogeneous agents to form teams and to make joint plans that exploit individual capabilities. The crux of designing agent systems that work effectively lies in building mechanisms by which agents can reach agreements. Examples of such mechanisms include auctions, rules of information exchange, and contract specification.

Many approaches to the design of interagent communication protocols have enjoyed considerable success through application of traditional deductive logics in supporting the distributed reasoning that is required for reaching agreements. The work of Cohen and Levesque (1990) provides a good example of this approach. By defining complex communicative acts in terms of the mental states of the agents involved in a communicative encounter, they have developed mechanisms for the formation of close-knit teams and the design of plans distributed across members of those teams.

This approach, however, has important limitations. In particular, a number of works (Parsons and Jennings, 1996; Parsons et al., 1998; Reed, 1998; McBurney and Parsons, 2002) argue that the inherent incompleteness and uncertainty of data available to individual agents can effectively stymie monotonic, deductive reasoning, and that a more flexible alternative is required. Argumentation has been proposed and implemented as a means of providing this flexibility, in two distinct ways.

In the first place, traditional defeasible logics have been extended (e.g., by Dung [1995]) to capture formal abstractions of arguments. These systems typically borrow terminology from Toulmin (1958), referring to undercutters and rebutters, and then distinguish sets of propositions and their closures on the basis of "acceptability classes." (Note that the use of the term "acceptability" here is only very distantly related to its meaning in argumentation theory.) An argument belongs to the highest acceptability class if its closure is consistent with current knowledge, or to some lower class depending upon the availability of undercutting propositions, rebutting propositions, and counters to those propositions. This work then served as the foundation for a multi-agent interpretation according to which sets of these formal structures are exchanged between agents and then evaluated (Parsons and Jennings, 1996).

The second use of argumentation focuses upon the dialogical aspects, exploiting work such as Walton and Krabbe's dialogue typology (1995) and Mackenzie's (1990) dialogue games. In these situations, it is the structure of the communication language itself that avails itself of argumentation theoretic concepts. This is typified by the work on dialogue frames described by Reed (1998) and the work on the *ludens* language done by McBurney and Parsons (2002).

Though successful and the focus of ongoing research, these two classes of applications of argumentation to multi-agent communication both suffer from a further limitation. Consider again the argument from expert opinion. If such an argument were to be employed by an agent capable of producing defeasible arguments, then the example situation in Ex2 might be rendered between two agents as something along the lines of Ex3:

(Ex2) Bob and Elaine are working on a computer science assignment and come across a question asking whether or not a particular machine possesses an accumulator. They discuss it for a moment, and it turns out that Elaine thinks the answer is "No." "But wait," says Bob, remembering his expert friend, "Mary said that this computer did have an accumulator."

(Ex3) *agent1 sends agent2 the message*
  (C) *has_accumulator*(*computer1*) $\wedge$
  $(P1)$ *said*(*mary,has_accumulator*(*computer1*)) $\wedge$
  $(P2)$ *defeasibly_implies*($P1, C$)

There are two problems that occur with a conventional approach to handling Ex3. The first is that sending the message in this form misses a crucial piece of information. This is an example of an argument from expert opinion. Such arguments are characteristically different from other types of argument, and are to be evaluated in a characteristically different way. By reducing all argument types to "plain vanilla" defeasible logic, this valuable distinguishing feature is lost. One key difficulty arising from this loss manifests itself as the second problem. The agent receiving the message has to perform processing and evaluation before updating its beliefs. The predominant approach would focus upon propositions that deductively or defeasibly rebut or undercut any of C, P1, or P2. It is quite possible that the receiving agent has stored (*not expert_in*(*mary, computer_science*)), but there is no mechanism by which this can be brought to bear in evaluating the message received. The agent has no means of determining that this information is relevant, because there is no recognition that the message fits the pattern of an argument from expert opinion.

By explicitly handling argumentation schemes – either in the message itself, or as a component of agent reasoning – it becomes possible for agents to at once broaden the scope of relevant information and narrow down selection on the basis of the argument scheme detected. Argumentation schemes may thus offer at least one part of an attack upon the thorny problem of relevance – which becomes critical as the size of agents' knowledge bases grows past the trivial.

7. SCHEMES IN AUTOMATED REASONING

In a series of works (see, e.g., 1989 and, particularly, 1995), Pollock describes the development of OSCAR, an automated theorem prover that on many classes of problems is at least as efficient as the best, and in some cases does very much better. In its later incarnations, OSCAR was built upon argumentation schemes. Pollock uses schemes to characterize the inferencing mechanisms in OSCAR. He calls them *prima facie reason schemes*, and has a set of seven that can be paraphrased thus:

(i) *Perception*: Having a percept with content p is a prima facie reason for believing p.

(ii) *Memory*: Recalling that p is a prima facie reason for believing p.

(iii) *Statistical Syllogism*: c is an F, and F's are usually G's, together form a prima facie reason for believing p.

(iv) *Induction*: Most observed F's are G's is a prima facie reason for believing that F's are usually G's.

(v) *Temporal Persistence*: p is believed to be true at t_1, and t_1 is before t_2, together form a prima facie reason for believing p is true at time t_2.

Together, these five schemes form the core of OSCAR's defeasible reasoning system, by which arguments can be formulated in defense of particular propositional claims, and attacks against them can be constructed. All inferencing based on prima facie reasons can be *rebutted* (by an argument with the opposite conclusion) or *undercut* (by an argument that demonstrates why the prima facie reason fails to hold in a given case). These latter forms of attack correspond closely to critical questions, and Pollock's prima facie reasons have associated with them schematic forms of undercutters. So, for example, his description of the perception scheme is:

> Where φ is a perceptible property, an agent's having a φ image constitutes a prima facie reason for the agent to believe ⌈My circumstances exemplify φ⌉. (Pollock, 1995, p. 55)

Associated with this scheme is the following example of a schematic undercutter, which captures how epistemic processes can interact with and override percepts:

> Where θ is a perceptible property, an agent's having a θ image and believing that if his circumstances exemplify θ then they do not exemplify φ constitutes an undercutting defeater. (Pollock, 1995, p. 56)

Pollock also has a generic form for another class of schematic undercutters:

> If belief in P is a prima facie reason for belief in Q, the belief:
>
> the present circumstances are of a general type C such that the probability is not high of Q's being true given that P is true under circumstances of type C
>
> is an undercutting defeater for this prima facie reason. (Pollock, 1995, p. 56)

Pollock's approach has also been extended to include further prima facie reasons – for example, for evidential reasoning (Prakken et al., 2003) – and such extension of schemes is a common feature among computational applications of schemes in general.

8. SCHEMES IN COMPUTATIONAL APPLICATIONS

In addition to software systems built specifically to represent or reason with argument, argumentation schemes have also been finding a wide variety of roles in applied computer systems in a range of different domains. Typically, these applications have been harnessing individual schemes for particular purposes. A good example is the PARMENIDES system, a proof-of-concept application that is aimed at supporting e-democracy (Atkinson et al., 2006). PARMENIDES focuses upon practical reasoning, and aims to encourage users to be critical in their assessment of arguments for a particular course of action. It uses a refined version of the argumentation scheme for practical reasoning in which there are thirty-four distinct potential types of attack (these attack types correspond closely to Walton's critical questions). These thirty-four attack types are used to provide the means by which a user can analyze and critique an existing argument using the scheme (in the sample case from Atkinson et al., 2006, the argument concerns the war in Iraq). Following such analysis and critique, users can construct their own versions of the scheme instantiated with their own premises. Structuring the process of critique in this way is argued to have benefits both for public understanding of complex policy decisions and for engagement of the public in policy formation.

Using schemes in software to help formulate the process of decision making is a common technique. CARREL is a prototype system that aims to improve the efficiency of the organ transplant systems in Western Europe by allowing software agents to negotiate autonomously in order to produce good matches for available organs quickly (Tolchinsky et al., 2006). The basic system has been extended with a model called ProCLAIM that, in the context of CARREL, employs a set of very specific argumentation schemes. One example is their Donor Disease Transfer Contraindication Scheme:

> When transplanting organ O from donor D with condition C_1 to a recipient R, R may result in having condition C_2, where C_2 is harmful. Hence, C_1 is a contraindication for donating O. (Tolchinsky et al., 2006, p. 77)

Though clearly very specialized, this scheme has its roots in the scheme from negative consequences, and indeed it is usually the case that schemes developed for specific application domains are more refined (and more heavily parameterized) versions of more generic scheme forms. This suggests that taxonomic hierarchies such as those proposed earlier in Chapter 10 and by Katzav and Reed (2004b) can be easily expanded to accommodate these new schemes.

With this potential proliferation of schemes that can work together harmoniously, the power of the Argument Interchange Format (Chesñevar et al., 2006) described in Chapter 11 becomes more important. Computational systems that use the AIF might be envisaged in which arguments from different sources and different authors are combined using combinations of schemes that might not have been anticipated. Thus two arguments that happen to make use of the same proposition (as a

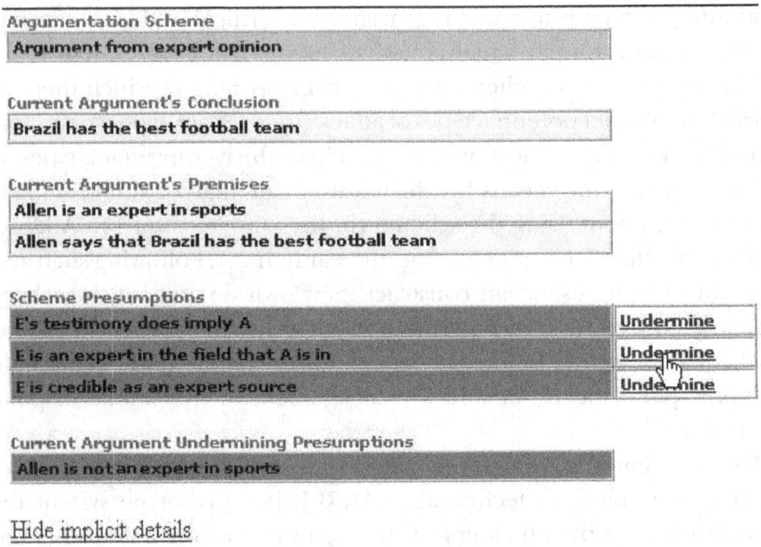

Argumentation Scheme
Argument from expert opinion

Current Argument's Conclusion
Brazil has the best football team

Current Argument's Premises
Allen is an expert in sports
Allen says that Brazil has the best football team

Scheme Presumptions

E's testimony does imply A	Undermine
E is an expert in the field that A is in	Undermine
E is credible as an expert source	Undermine

Current Argument Undermining Presumptions
Allen is not an expert in sports

Hide implicit details

FIGURE 12.15. Part of the ArgDF interface.

premise or as a conclusion, say) are in some manner connected. These connections might be represented, or displayed, or navigated by computational systems. In this way, a whole ocean of arguments, contributed by different people, using different argumentation schemes and practices, might be freely navigable and extensible. This is the vision behind the World Wide Argumentation Web (Rahwan et al., 2007). One of the first building blocks of this web to be implemented is a system that allows users to access arguments online and to critique them (using the critical questions associated with schemes), challenge them, support them, or extend them. The system, ArgDF, shows that it is possible to use the AIF to represent and manipulate these arguments while retaining a rich scheme-based structure that forms the links between the components. As an example, Figure 12.15 shows part of one of the interfaces to ArgDF: someone has entered an argument from expert opinion in which the conclusion that "Brazil has the best football team" is supported by a claim from Allen, an expert. The new user is now about to attack this argument by challenging one of the presumptions of the scheme.

9. CONCLUSIONS

Two things are clear from this brief survey of the uses of argumentation schemes in computational systems. The first is the breadth of application.

Artificial intelligence is concerned with autonomous reasoning, distributed reasoning, and the expression of reasoning, so perhaps it should come as no surprise that a fundamental aspect of reasoning – the use of defeasible schemes – should be ubiquitous across AI. What is more surprising is that there is appreciable impact in many of those diverse areas already. Argumentation schemes are simply turning out to be a good fit for the sorts of problems that many AI systems are trying to tackle. The second thing that is striking, however, is that this work is as yet very underdeveloped. There is a bootstrapping process that needs to be completed so that we have the basic representational devices (that might be provided by a language such as the AIF), which can in turn be employed by the tools (such as Araucaria, Compendium, and Rationale) that can support, the construction of systems that perform autonomous reasoning, language generation, decision support, and so on. At the moment, representational devices, tools, and systems are being developed in parallel, each making use of the latest advances of the others. Though hectic, this activity is stimulating enormous interest. Argumentation schemes have enormous potential to contribute to solutions to AI problems, and the immediate future is going to be an exciting time in this field.

Bibliography

Abelard, Petrus (1970). Dialectica. In Lambertus Marie de Rijk (Ed.), *Petrus Abelardus: Dialectica.* Assen: Van Gorcum.

Agricola Rodolphus (1976). *De inventione dialectica libri tres.* Hildesheim: Georg Olms.

Alberts, Laurie (2001). Causation in toxic tort litigation. *Villanova Environmental Law Journal* **12**: 33–63.

Amgoud, Leila, and Cayrol, Claudette (2002). A model of reasoning based on the production of acceptable arguments. *Annals of Mathematics and Artificial Intelligence* **34**: 197–216.

Anderson, Terence, Schum, David, and Twining, William (2005). *Analysis of Evidence*, 2nd edition. Cambridge: Cambridge University Press.

Aristotle (1851). *On Rhetoric.* Translated by T. Buckley. London: Henry G. Bohn.

Aristotle (1937). *The Art of Rhetoric.* Translated by John Henry Freese. Loeb Classical Library. Cambridge, Mass.: Harvard University Press.

Aristotle (1939). *Topica.* Translated by E. S. Forster. Loeb Classical Library. Cambridge, Mass.: Harvard University Press.

Aristotle (1964). *Prior and Posterior Analitics.* Translated by John Warrington. New York: Everyman.

Aristotle (1984). Prior analytics. Translated by Jonathan Barnes. In Jonathan Barnes (ed.), *The Complete Works of Aristotle*, vol. I. Princeton, N.J.: Princeton University Press.

Arnaud, Antoine, and Nicole, Pierre (1964). *The Art of Thinking.* Indianapolis: The Bobbs-Merril.

Ashley, Kevin D., and Rissland, Edwina L. (2003). Law, learning and representation. *Artificial Intelligence* **150**: 17–58.

Atkinson, Katie, Bench-Capon, Trevor, and McBurney, Peter (2006). PARMENIDES: Facilitating deliberation in democracies. *Artificial Intelligence and Law* **14** (4): 261–275.

Audi, Robert (1989). *Practical Reasoning.* London: Routledge.

Barker, Evelyn (1989). Beardsley's theory of analogy. *Informal Logic* **11** (3): 185–194.

Beardsley, Monroe (1950). *Practical Logic.* New York: Prentice-Hall.

Beardsley, Monroe (1956). *Thinking Straight.* Englewood Cliffs, N.J.: Prentice-Hall.

Bench-Capon, Trevor (1998). Specifying the interaction between information sources. *Proceedings of DEXA,* Vienna, Austria, August 24–8. Berlin: Springer, 425–434.

Bench-Capon, Trevor (2003a). Persuasion in practical argument using value-based argumentation frameworks. *Journal of Logic and Computation* **13**: 429–448.

Bench-Capon, Trevor (2003b). Agreeing to differ: Modelling persuasive dialogue between parties without a consensus about values. *Informal Logic* **22**: 231–245.

Bench-Capon, Trevor, and Prakken, Henry (2005). Argumentation. In Arno Lodder and Anja Oskamp (eds.), *Information Technology and Lawyers: Advanced Technology in the Legal Domain, from Challenges to Daily Routine.* Berlin: Springer Verlag, 61–80.

Bench-Capon, Trevor, and Sartor, Giovanni (2003). A model of legal reasoning with cases incorporating theories and values. *Artificial Intelligence* **97**: 97–143.

Best, Joel (2001). *Damned Lies and Statistics.* Berkeley: University of California Press.

Bex, Floris, Prakken, Henry, Reed, Chris, and Walton, Douglas (2003). Towards a formal account of reasoning about evidence: argumentation schemes and generalizations. *Artificial Intelligence and Law* **11**: 125–165.

Boh, Ivan (1984). Epistemic and alethic iteration in later medieval logic. *Philosophia Naturalis* **21**: 492–506.

Boller, Paul F. (1967). *Quotemanship. The Use and Abuse of Quotations for Polemical and Other Purposes.* Dallas: Southern Methodist University Press.

Braet, Antoine C. (2004). The oldest typology of argumentation schemes. *Argumentation* **18**: 127–148.

Bratman, Michael (1987). *Intentions, Plans, and Practical Reason.* Cambridge, Mass.: Harvard University Press.

Brewer, Scott (1996). Exemplary reasoning: Semantics, pragmatics, and the rational force of legal argument by analogy. *Harvard Law Review* **109**: 923–1038.

Brown, William (1989). Two traditions of analogy. *Informal Logic* **11** (3): 161–172.

Buckingham Shum, S. (2007). Mapping dialogue and argumentation in international development: The case of Compendium and OpenLearn LabSpace. *Workshop on User Centered Design and International Development,* ACM Computer-Human Interaction Conference, April 28, 2007, San Jose. <http://www-static.cc.gatech.edu/~mikeb/UCDandIDWorkshop/papers/shum.pdf>.

Burbridge, John (1990). *Within Reason: A Guide to Non Deductive Reasoning.* Peterborough, Ont.: Broadview Press.

Burke, Michael (1985). Unstated premises. *Informal Logic* **7**: 107–118.

Burnyeat, Myles F. (1994). Enthymeme: Aristotle on the logic of persuasion. In David J. Furley and Alexander Nehemas (eds.), *Aristotle's Rhetoric: Philosophical Essays.* Princeton, N.J.: Princeton University Press, 3–55.

Carberry, Sandra (1990). *Plan Recognition in Natural Language Dialogue.* Cambridge, Mass.: MIT Press.

Carbogim, Daniela V., Robertson, David S, and Lee, John R. (2000). Argument-based applications to knowledge engineering. *The Knowledge Engineering Review* **15** (2): 119–149.

Chesnevar, Carlos, McGinnis, Jarred, Modgil, Sanjay, Rahwan, Iyad, Reed, Chris, Simari, Guillermo, South, Matthew, Vreeswijk, Gerard, and Willmott, Steven (2006). Towards an argument interchange format. *Knowledge Engineering Review* **21** (**4**): 293–316.

Chorley Alison, and Bench-Capon, Trevor (2004). AGATHA: Automation of the construction of theories in case law domains. In Tom Gordon (ed.), *Legal Knowledge and Information Systems. Jurix 2004: The Seventeenth Annual Conference.* Amsterdam: IOS Press, 89–98.

Cicero, Marcus Tullius (1949). *De Inventione, De optimo genere oraturum, Topica.* Translated by H. Hubbell. Cambridge, Mass.: Harvard University Press.

Cicero, Marcus Tullius (1965). *Rhetorica ad Herennium.* Translated by Harry Caplan. Cambridge, Mass.: Harvard University Press.

Clark, Keith (1978). Negation as failure. In Hervé Gallaire and Jack Minker (eds.), *Logic and Data Bases.* New York: Plenum Press, 293–322.

Clarke, David S., Jr. (1985). *Practical Inferences.* London: Routledge.

Cohen, Philip R., and Levesque, Hector J. (1990). Intention is choice with commitment. *Artificial Intelligence* **42**: 213–261.

Console, Luca, and Torasso, Pietro (1990). Hypothetical reasoning in causal models. *International Journal of Intelligent Systems* **5**: 83–124.

Copi, Irving (1986). *Informal Logic.* London: Collier Macmillan.

Copi, Irving, and Burgess-Jackson, Keith (1992). *Informal Logic.* New York: Macmillan.

Copi, Irving, and Cohen, Carl (1998). *Introduction to Logic.* Upper Saddle River, N.J.: Prentice-Hall.

Crenshaw, Kimberle (1998). Demarginalizing the intersection of race and sex: a black feminist critique of antidiscrimination doctrine, feminist theory and antiracist politics. *Chicago Legal Forum*: 139–167.

De Pater, Wilhelmus (1965). *Les Topiques d'Aristote et la Dialectique Platonicienne.* Fribourg, Germany: Éditions de St. Paul.

Dijkstra, Pieter, Bex, Floris, Prakken, Henry, and de Vey Mestdagh, Kees (2005). Towards a multi-agent system for regulated information exchange in crime investigations. *Artificial Intelligence and Law* **13**: 133–151.

Doyle, Conan Arthur (1932). *The Complete Sherlock Holmes.* Garden City, N.Y.: Doubleday.

Dung, Phan Minh (1995). On the acceptability of arguments and its fundamental role in nonmonotonic reasoning, logic programming, and *n*-person games. *Artificial Intelligence* **77**: 321–357.

Eemeren, Frans H. van, and Grootendorst, Rob (1984). *Speech Acts in Communicative Discussions.* Dordrecht: Foris.

Eemeren, Frans H. van, and Grootendorst, Rob (1992). *Argumentation, Communication and Fallacies.* Hillsdale, N.J.: Erlbaum.

Eemeren, Frans H. van, and Kruiger, Tjark (1987). Identifying argumentation schemes. In Frans Van Eemeren, Rob Grootendorst, Anthony Blair, and Charles Willard (eds.), *Argumentation: Perspectives and Approaches.* Dordrecht: Foris, 70–81.

Engel, Morris S. (1980). *Analyzing Informal Fallacies.* Englewood Cliffs, N.J.: Prentice-Hall.

Ennis, Robert H. (1982). Identifying implicit assumptions. *Synthese* **51**: 61–86.
Ennis, Robert H. (2001). Argument appraisal strategy: A comprehensive approach. *Informal Logic* **21** (**2**): 97–140.
Farrell, Thomas B. (2000). Aristotle's enthymeme as tacit reference. In Alan Gross and Arthur Walzer (eds.), *Rereading Aristotle's Rhetoric*. Carbondale: Southern Illinois University Press, 93–106.
Fox, John, and Das, Subrata (2000). *Safe and Sound: Artificial Intelligence in Hazardous Applications*. Cambridge, Mass.: MIT Press.
Freeman, James B. (1991). *Dialectics and the Macrostructure of Argument*. Dordrecht: Foris.
Freeman, James B. (1995). The appeal to popularity and presumption by common knowledge. In Hans V. Hansen and Robert C. Pinto (eds.), *Fallacies: Classical and Contemporary Readings*. University Park: Pennsylvania State University Press, 263–273.
Garssen, Bart (2001). Argumentation schemes. In Frans van Eemeren (ed.), *Crucial Concepts in Argumentation Theory*. Amsterdam: Amsterdam University Press, 81–99.
Gerritsen, Susanne (2001). Unexpressed premises. In Frans van Eemeren (Ed.), *Crucial Concepts in Argumentation Theory*. Amsterdam: Amsterdam University Press, 51–79.
Gilbert, Michael (1991). The enthymeme buster. *Informal Logic* **13**: 159–166.
Gilbert, Michael A. (1997). *Coalescent Argumentation*. Mahwah, N.J.: Erlbaum.
Goldschmidt, Victor (1947). *Le Paradigme dans la Dialectique Platonicienne*. Paris: Presses Universitaires de France.
Gordon, Thomas F. (1995). *The Pleadings Game: An Artificial Intelligence Model of Procedural Justice*. Dordrecht: Kluwer.
Gordon, Thomas F. (2005). A computational model of argument for legal reasoning support systems. In Paul Dunne and Trevor Bench-Capon (eds.), *Argumentation in Artificial Intelligence and Law*, IAAIL Workshop Series. Nijmegen: Wolf Legal Publishers, 53–64.
Gordon, Thomas F., and Walton, Douglas (2006). Pierson vs. Post revisted – a reconstruction using the Carneades Argumentation Framework. In Paul Dunne and Trevor Bench-Capon (eds.), *Proceedings of the First International Conference on Computational Models of Argument (COMMA 06)*. Liverpool: IOS Press. 208–219.
Gough, James, and Tindale, Christopher (1985). Hidden or missing premises. *Informal Logic* **7**: 99–106.
Govier, Trudy (1989). Analogies and missing premises. *Informal Logic* **11** (**3**): 141–152.
Govier, Trudy ([1992], 2005). *A Practical Study of Argument*. Belmont, Calif.: Wadsworth.
Grasso, Floriana, Cawsey, Alison, and Jones, Ray (2000). Dialectical argumentation to solve conflicts in advice giving: A case study in the promotion of healthy nutrition. *International Journal of Human-Computer Studies* **53** (**6**): 1077–1115.
Green-Pedersen, Niels J. (1984). *The Tradition of Topics in the Middle Age*. Munich: Philosophia Verlag.

Grennan, Wayne (1997). *Informal Logic.* Montreal: McGill-Queen's University Press.

Groarke, Leo (1999). Deductivism within pragma-dialectics. *Argumentation* **13**: 1–16.

Groarke, Leo (2001). Argumentation schemes in pedagogy and AI. In Hans Hansen and Christopher Tindale (eds.), *Proceedings of the OSSA '2001 Conference on Argument and its Applications.* Windsor, Ontario: Society for the Study of Argumentation.

Guarini, Marcello (2004). A defense of non-deductive reconstructions of analogical arguments. *Informal Logic* **24**: 153–168.

Hage, Jaap (2005). The logic of analogy in the law. *Argumentation* **19**: 401–415.

Hamblin, Charles. L. (1970). *Fallacies,* London: Methuen.

Hart, H. L. A. (1957–58). Positivism and the Separation of Law and Morals. *Harvard Law Review* **71**: 593–629.

Hart, H. L. A., and Honoré, Tony (1962). *Causation in the Law.* Oxford: Oxford University Press.

Hastings, Arthur. C. (1963). "A Reformulation of the Modes of Reasoning in Argumentation." Ph.D. dissertation, Northwestern University, Evanston, Ill.

Hintikka, Jaakko (1979). Information-seeking dialogues: A model. *Erkenntnis* **38**: 355–368.

Hintikka, Jaakko (1992). The interrogative model of inquiry as a general theory of argumentation. *Communication and Cognition* **25**: 221–242.

Hintikka, Jaakko (1993). Socratic questioning, logic and rhetoric. *Revue Internationale de Philosophie* **184**: 5–30.

Hintikka, Jaakko (1995). The games of logic and the games of inquiry. *Dialectica* **49**: 229–249.

Hitchcock, David (1981). Deduction, induction, and conduction. *Informal Logic Newsletter* **3 (2)**: 7–15.

Hitchcock, David (1985). Enthymematic arguments. *Informal Logic* **7**: 83–97.

Hitchcock, David (2005). Good reasoning on the Toulmin model. *Argumentation* **19**: 373–391.

Huber, Peter W. (1991). *Galileo's Revenge: Junk Science in the Courtroom.* New York: Basic Books.

Hurley, Patrick J. (2000). *A Concise Introduction to Logic,* 7th ed. Belmont, Calif.: Wadsworth.

Jackson, Sally, and Jacobs, Scott (1980). Structure of conversational argument: Pragmatic bases for the enthymeme. *Quarterly Journal of Speech* **66**: 251–165.

Johnson, Ralph H. (2000). *Manifest Rationality: A Pragmatic Theory of Argument.* Mahwah, N.J.: Erlbaum.

Johnson, Ralph, and Blair, Anthony ([1983], 1994). *Logical Self Defence.* Toronto: McGraw-Hill.

Josephson, John R., and Josephson, Susan G. (1994). *Abductive Inference: Computation, Philosophy, Technology.* New York: Cambridge University Press.

Jovicic, Taeda (2002). "Authority-based Argumentative Strategies." Doctoral dissertation in the Department of Theoretical Philosophy, Uppsala University. Uppsala, Sweden.

Juthe, André (2005). Argument by analogy. *Argumentation* **19**: 1–27.

Katzav, Joel, and Reed, Chris (2004). On argumentation schemes and the natural classification of argument. *Argumentation* **18**: 239–259.

Katzav, Joel, and Reed, Chris (2004a). A Classification system for argument. Department of Applied Computing Technical Report, University of Dundee, Dundee, Scotland.

Kienpointner, Manfred (1986). Towards a typology of argument schemes. In Frans H. van Eemeren et al. (eds.), *Argumentation: Across the Lines of Discipline.* Dordrecht: Foris, 275–287.

Kienpointner, Manfred (1992). *Alltagslogik: Struktur und Funktion von Argumentationsmustern.* Stuttgart: Fromman-Holzboog.

Kienpointner, Manfred (1992a). How to classify arguments. In F. H. van Eemeren, R. Grootendorst, J. A. Blair and C. A. Willard (eds.), *Argumentation Illuminated.* Amsterdam: Amsterdam University Press, 178–188.

Kienpointner, Manfred (2002). Perelman on causal arguments: The argument from waste. In Frans van Eemeren et al. (eds.), *Proceedings of the 5th International Conference on Argumentation.* Amsterdam: SicSat, 611–616.

Kirschner, Paul A., Buckingham Shum, Simon J., and Carr, Chad S. (eds.) (2003). *Visualizing Argumentation.* London: Springer-Verlag.

Kneale, William, and Kneale, Martha (1962). *The Development of Logic.* Oxford: Clarendon Press.

Kozinski, Alex (2001). How I narrowly escaped insanity. *U.C.L.A. Law Review* **48**: 1293–1304.

Krabbe, Erik (1992). So what? Profiles for relevance criticism in persuasion dialogues. *Argumentation* **6**: 271–283.

Krabbe, Erik C. W. (1999). Profiles of dialogue. In Jelle Gerbrandy, Maarten Marx, Maarten de Rijke, and Yde Venema (eds.), *JFAK: Essays Dedicated to Johan van Benthem on the Occasion of his 50th Birthday.* Amsterdam: Amsterdam University Press, 25–36.

Kupperman, Joel (1991). *Character.* Oxford: Oxford University Press.

Lenat, Douglas (1995). CYC: A large-scale investment in knowledge infrastructure. *Communications of the ACM* **38** (**11**): 33–38.

Lloyd, Geoffrey E. R. (1966). *Polarity and Analogy.* Cambridge: Cambridge University Press.

Mackenzie, James D. (1979). Question-begging in non-cumulative systems. *Journal of Philosophical Logic* **8**: 117–133.

Mackenzie, James D. (1990). Four dialogue systems. *Studia Logica* **49**: 567–583.

Mackie, John L. (1965). Causes and conditions. *American Philosophical Quarterly* **2**: 245–264.

Mann, William, and Thompson, Sandra (1987). Rhetorical structure theory. *Text* **8**: 243–281.

Maruyama Magoroh (1968). The second cybernetics: Deviation-amplifying mutual causal processes. In Walter Buckley (ed.), *Modern Systems Research for the Behavioral Scientist.* Chicago: Aldine, 304–313.

Mathews, Nieves (1996). *Francis Bacon: The History of a Character Assassination.* New Haven and London: Yale University Press.

McBurney, James (1936). Some recent interpretations of the Aristotelian enthymeme. *Papers of the Michigan Academy of Sciences, Arts, and Letters* 21: 489–500.

McBurney, Peter, and Parsons, Simon (2002). Games that agents play. *Journal of Logic, Language and Information* 11 (3): 315–334.

Mitchell, Melanie (2001). Analogy-making as a complex adaptive system. In Lee Segel and Irun Cohen (eds.), *Design Principles for the Immune System and Other Distributed Autonomous Systems.* New York: Oxford University Press.

Mitchell, Melanie (2002). *Perception and Analogy-making in Complex Adaptive Systems.* Available at <http://www.jsmf.org/grants/cs/essays/2002/mitchell. htm>.

Norman, Timothy, Carbogim, Daniela V., Krabbe, Erik C, and Walton, Douglas N. (2003). Argument and multi-agent systems. In Chris Reed and Timothy Norman (eds.), *Argumentation Machines: New Frontiers in Argument and Computation.* Dordrecht: Kluwer, 15–54.

Parsons, Simon, and Jennings, Nicholas R. (1996). Negotiation through argumentation – a preliminary report. In *Proceedings of the 2nd International Conference on Multi-Agent Systems.* Menlo Park, Calif.: AAAI Press, 267–274.

Parsons, Simon, Sierra, Carles, and Jennings, Nicholas R. (1998). Agents that reason and negotiate by arguing. *Journal of Logic and Computation* 8 (3): 261–292.

Patry, William (forthcoming). *Patry on Copyright.* Rochester, N.Y.: West Publishing.

Pearl, Judea (2000). *Causality: Models, Reasoning and Inference.* Cambridge: Cambridge University Press.

Pearl, Judea (2001). Causal inference in the health sciences: A conceptual introduction. *Health Services and Outcomes Research Methodology* 2: 189–220.

Pearl, Judea (2002). Bayesianism and causality, or, why I am only a half-Bayesian. In David Corfield and Jan Williamson (eds.), *Foundations of Bayesianism,* Applied Logic Series Volume 24. Dordrecht: Kluwer, 19–36. Document available on the personal web page of Judea Pearl.

Peirce, Charles S. (1965). *Collected Papers of Charles S. Peirce.* Edited by Charles Hartshorne and Paul Weiss. Cambridge, Mass.: Harvard University Press.

Perelman, Chaim, and Olbrechts-Tyteca, Lucie (1969). *The New Rhetoric: A Treatise on Argumentation.* Translated by J. Wilkinson and P. Weaver. Notre Dame, Ind.: University of Notre Dame Press.

Peter of Spain (1980). *Language in Dispute.* Translated by Francis Dinneen. Amsterdam/Philadelphia: John Benjamins.

Pinto, Robert C., Blair, Anthony J, and Parr, Katherine E. (1993). *Reasoning: A Practical Guide for Canadian Students.* Scarborough, Ont.: Prentice Hall Canada.

Plato (1835). *Phaedo.* Translated by C. Stanford. New York: Hurst and Company.

Plato (1990). *Sophist.* Translated by William Cobb. Savage, Md.: Rowman and Littlefield.

Pollock, John L. (1974). *Knowledge and Justification.* Princeton, N.J.: Princeton University Press.

Pollock, John L. (1987). Defeasible reasoning. *Cognitive Science* 11: 481–518.

Pollock, John L. (1989). *How to Build a Person.* Cambridge, Mass.: MIT Press.
Pollock, John L. (1995). *Cognitive Carpentry.* Cambridge, Mass.: MIT Press.
Prakken, Henry (1993). An argumentation framework in default logic. *Annals of Mathematics and Artificial Intelligence* 9: 91–132.
Prakken, Henry (1997). *Logical Tools for Modelling Legal Argument: A Study of Defeasible Reasoning in Law.* Dordrecht: Kluwer.
Prakken, Henry (2001). Modelling defeasibility in law: Logic or procedure? *Fundamenta Informaticae* 48: 253–271.
Prakken, Henry (2002). Incomplete arguments in legal discourse: A case study. In Trevor Bench-Capon, Aspassia Daskalopulu, and Radboud Winkels (eds.), *Legal Knowledge and Information Systems. JURIX 2002: The Fifteenth Annual Conference.* Amsterdam: IOS Press, 93–102.
Prakken, Henry (2005). AI and law, logic and argumentation schemes. *Argumentation* 19: 303–320. Available at <http://www.cs.uu.nl/people/henry>.
Prakken, Henry, and Renooij, Silja (2001). Reconstructing causal reasoning about evidence: A case study. In Bart Verheij, Arno R. Lodder, Ronald P. Loui, and A. Muntjewerjj (eds.), *Legal Knowledge and Information Systems.* Amsterdam: IOS Press, 131–142.
Prakken, Henry, and Sartor, Giovanni (1996). A dialectical model of assessing conflicting arguments in legal reasoning. *Artificial Intelligence and Law* 4: 331–368.
Prakken, Henry, and Sartor, Giovanni (1997). Argument-based logic programming with defeasible priorities. *Journal of Applied Non-classical Logics* 7: 25–75.
Prakken, Henry, and Sartor, Giovanni (2004). The three faces of defeasibility in the law. *Ratio Juris* 17 (1): 118–139.
Prakken, Henry, and Vreeswijk, Gerard (2002). Logics for defeasible argumentation. In Dov Gabbay and Franz Guenther (eds.), *Handbook of Philosophical Logic*, vol. 4, Dordercht: Kluwer, 218–319.
Quintilian, Maximus Fabius (1996). *Institutio Oratoria.* Translated by H. E. Butler. Cambridge, Mass.: Harvard University Press.
Rahwan, Iyad, Zablith, Fouad, and Reed, Chris (2007). Laying the foundations for a World Wide Argument Web. *Artificial Intelligenc* 171: 897–921.
Ramus, Petrus (1969). *Dialecticae Libri Duo.* Amsterdam: Theatrum Orbis Terrarum; New York: Da Capo Press.
Reed, Chris, and Norman, Timothy J. (eds.) (2003). *Argumentation Machines.* Dordrecht,: Kluwer.
Reed, Chris, and Rowe, Glenn (2004). Araucaria: software for argument analysis, diagramming and representation. *International Journal of AI Tools*, 13 (4): 961–980.
Reed, Chris, and Rowe, Glenn (2005). *Araucaria, Version 3.* Available free at <http://www.computing.dundee.ac.uk/staff/creed/araucaria>.
Reed, Chris, and Rowe, Glenn (2001). Araucaria: Software for Puzzles in Argument Diagramming and XML. Department of Applied Computing, University of Dundee Technical Report. Available at <http://www.computing.dundee.ac.uk/staff/creed>.

Reed, Chris, and Walton, Douglas (2005). Towards a formal and implemented model of argumentation schemes in agent communication. *Autonomous Agents and Multi-Agent Systems* **11**: 173–188.

Reed, Chris (1997). Representing and applying knowledge for argumentation in a social context. *AI & Society* **11 (3–4)**: 138–154.

Reed, Chris (1998). Dialogue frames in agent communication. In Yves Demazeau (ed.), *Proceedings of the 3rd International Conference on Multi Agent Systems (ICMAS'98)*. Paris: IEEE Press, 246–253.

Reed, Chris (1998). "Generating Arguments in Natural Language." Ph.D. thesis, University College, London.

Reed, Chris (1999). The role of saliency in generating natural language arguments. In Thomas Dean (ed.), *Proceedings of the 16th International Joint Conference on Artificial Intelligence (IJCAI'99)*. Stockholm: Morgan Kaufmann, 876–883.

Reed, Chris, Norman, Timothy J., and Gabbay, Dov (eds). "Handbook of Argument and Computation." Unpublished manuscript.

Reiter, Raymond (1980). A logic for default reasoning. *Artificial Intelligence* **13**: 81–132.

Rigotti, Eddo, and Rocci, Andrea (2001). Sens – non-sens – contresens. *Studies in Communication Sciences* **1**: 45–80.

Rigotti, Eddo (1997). La retorica classica come una prima teoria della comunicazione. In Elena Bussi, Marina Bondi, and Francesca Gatta (eds.), *Understanding Argument: The Informal Logic of Discourse*. Bologna: CLUEB, 1–8.

Rigotti, Eddo. "Elementi di Topica." Unpublished manuscript.

Rigotti, Eddo (2006a). Can classical topics be revived within the contemporary theory of argumentation? Paper presented at the ISSA Conference on Argumentation Theory, Amsterdam, June.

Rissland, Edwina L. (1990). Artificial intelligence and law: stepping stones to a model of legal reasoning. *Yale Law Journal* **99**: 1957–1980.

Robinson, Richard (1962). *Plato's Earlier Dialectic*, 2nd ed. Oxford: Clarendon Press.

Rowe, Glenn, Reed, Chris, Macagno, Fabrizio, and Walton, Douglas (2006). Araucaria as a tool for diagramming arguments in teaching and studying philosophy. *Teaching Philosophy* **29 (2)**: 111–124.

Sadock, Jerrold M. (1997). Modus brevis: The truncated argument. In Woodford Beach, Samuel Fox, and Shulamith Philosoph (eds.), *Papers from the 13th Regional Meeting of the Chicago Linguistics Society*. Chicago: Chicago Linguistics Society, 545–554.

Schauer, Frederick (1995). *Playing by the Rules: A Philosophical Examination of Rule-Based Decision-Making in Law and Life*. Oxford: Oxford University Press.

Schellens, Peter Jan, and Menno DeJong, (2004). Argumentation schemes in persuasive brochures. *Argumentation* **18**: 295–323.

Schupp, Franz (1988). *Logical Problems of the Medieval Theory of Consequences*. Napoli, Italy: Bibliopolis.

Scriven, Michael (1964). Critical study of "The Structure of Science." *Review of Metaphysics* **17**: 403–424.

Scriven, Michael (1976). *Reasoning.* New York: McGraw-Hill.
Sextus Empiricus (1933). *Outlines of Pyrrhonism.* Translated by R. G. Bury. Cambridge, Mass.: Harvard University Press.
Sherwin, Emily (1999). A defense of analogical reasoning in law. *University of Chicago Law Review* **66**: 1179–1198.
Simmons, Reid (1992). The roles of associational and causal reasoning in problem solving. *Artificial Intelligence* **53 (2–3)**: 159–207.
Snoeck Henkemans, A. Francisca (2001). Argumentation structures. In Frans H. van Eemeren (ed.), *Crucial Concepts in Argumentation Theory.* Amsterdam: Amsterdam University Press, 101–134.
Stump, Eleonore (1982). Topics: Their development and absorption into the consequences. In Norman Kretzmann, Anthony Kenny, and Jan Pinborg (eds.), *Cambridge History of Later Medieval Philosophy.* Cambridge: Cambridge University Press, 315–334.
Stump, Eleonore (1989). *Dialectic and Its Place in the Development of Medieval Logic.* Ithaca, N.Y.: Cornell University Press.
Stump, Eleonore (trans.) (1978). *Boethius' "De topicis differentiis."* Ithaca, N.Y.: Cornell University Press.
Stump, Eleonore (trans.) (1988). *In Ciceronis Topica.* Ithaca, N.Y.: Cornell University Press.
Tardini, Stefano (2005). Endoxa and communities: Grounding enthymematic arguments. *Studies in Communication Sciences. Special Issue, "Argumentation in Dialogic Interaction"*: 279–294.
Thomson, Judith J. (1971). A defense of abortion. *Philosophy and Public Affairs* **1**: 47–66.
Tillers, Peter (2002). Making sense of the process of proof in litigation. In Peter Tillers and Marilyn MacCrimmon (eds.), *The Dynamics of Judicial Proof.* Heidelberg: Springer, 3–11.
Tolchinsky, Pancho, Modgil, Sanjay, Cortés, Ulises, and Sànchez-Marrè, Miquel (2006). CBR and argument schemes for collaborative decision making. In Paul Dunne and Trevor Bench-Capon (eds.), *Computational Models of Argument: Proceedings of COMMA 2006.* Amsterdam: IOS Press, 71–82.
Toulmin, Stephen E. (1958). *The Uses of Argument.* Cambridge: Cambridge University Press.
Toulmin, Stephen E., Richard, Rieke, and Janik, Allan (1979). *An Introduction to Reasoning.* New York: Macmillan.
Van Gelder, Tim, and Rizzo, Alberto (2001). Reason!Able across the curriculum. In *Is IT an Odyssey in Learning? Proceedings of the 2001 Conference of the Computing in Education Group of Victoria.* Victoria, Australia.
Verheij, Bart, and Hage, Jaap (1994). Reasoning by analogy: A formal reconstruction. In H. Prakken, A. J. Muntjewerff, and A. Soeteman (eds.), *Legal Knowledge Based Systems.* Lelystad: Koninklijke Vermande, 65–78.
Verheij, Bart (1996). *Rules, Reasons and Arguments: Formal Studies of Argumentation and Defeat.* Doctoral dissertation, University of Maastricht.
Verheij, Bart (1999a). Logic, context and valid inference. Or: Can there be a logic of law? In Jaap van den Herik, Marie-Francine Moens, Jon Bing, Bea van Buggenhout, John Zeleznikow, and Carolus Grütters (eds.), *Legal*

Knowledge Based Systems. JURIX 1999: The Twelfth Conference. Available at <http://citeseer.jst.psu.edu/verheij99logic.html>.

Verheij, Bart (1999). Automated argument assistance for lawyers. In *The Seventh International Conference on Artificial Intelligence and Law: Proceedings of the Conference* (June 14–17, Oslo, Norway). New York: ACM, 43–52. Available at <bart.verheij@metajur.unimaas.nl, http://www.metajur.unimaas.nl/~bart>.

Verheij, Bart (2001). Legal decision making as dialectical theory construction with argumentation schemes. In *ICAIL 2001: The Eighth International Conference on Artificial Intelligence and Law.* New York: ACM, 225–226. The full paper is available at <http://www.ai.rug.nl/~verheij/publications/pdf/argsch.pdf>.

Verheij, Bart (2003). Dialectical argumentation with argumentation schemes: Towards a methodology for the investigation of argumentation schemes. In Frans van Eemeren, Anthony Blair, Charles. Willard, and Francisca Snoeck Henkemans (eds.), *Proceedings of the Fifth Conference of the International Society for the Study of Argumentation (ISSA 2002).* Amsterdam: Sic Sat, 1033–1037.

Verheij, Bart (2003a). DefLog: On the logical interpretation of prima facie justified assumptions. *Journal of Logic and Computation* 13: 319–346. Available at <http://www.ai.rug.nl/~verheij/publications.htm>.

Verheij, Bart (2003b). Dialectical argumentation with argumentation schemes: An approach to legal logic. *Artificial Intelligence and Law* 11: 167–195.

Verheij, Bart (2005). *Virtual Arguments: On the Design of Argument Assistants for Lawyers and Other Arguers.* The Hague: T. M. C. Asser Press.

Waller, Bruce N. (2001). Classifying and analyzing analogies. *Informal Logic* 21: 199–218.

Walton, Douglas (1982). *Argument: The Logic of Fallacies.* Toronto: McGraw-Hill Ryerson.

Walton, Douglas (1984). *Logical Dialogue-Games and Fallacies.* Lanham, Md.: University Press of America.

Walton, Douglas (1987). The ad hominem argument as an informal fallacy. *Argumentation* 1: 317–331.

Walton, Douglas (1989). *Informal Logic.* New York: Cambridge University Press.

Walton, Douglas (1990). *Practical Reasoning.* Savage, Md.: Rowman and Littlefield.

Walton, Douglas (1991). Bias, critical doubt and fallacies. *Argumentation and Advocacy* 28: 1–22.

Walton, Douglas (1992). Nonfallacious arguments from ignorance. *American Philosophical Quarterly* 29: 381–387.

Walton, Douglas (1992a). *Slippery Slope Arguments.* Newport News, Va.: Vale Press.

Walton, Douglas (1992b). *The Place of Emotion in Argument.* University Park: Pennsylvania State University Press.

Walton, Douglas (1995). *A Pragmatic Theory of Fallacy.* Tuscaloosa and London: University of Alabama Press.

Walton, Douglas (1996). *Argumentation Schemes for Presumptive Reasoning.* Mahwah, N.J.: Erlbaum.

Walton, Douglas (1996a). The argument of the beard. *Informal Logic* 18: 235–251.

Walton, Douglas (1996b). *Arguments from Ignorance.* University Park: Pennsylvania State University Press.

Walton, Douglas (1997). *Appeal to Expert Opinion.* University Park: Pennsylvania State University.

Walton, Douglas (1997a). Actions and inconsistency: The closure problem of practical reasoning. In Ghita Holmstrom-Hintikka and Raimo Tuomela (eds.), *Contemporary Action Theory,* vol. 1. Dordrecht: Kluwer, 159–175.

Walton, Douglas (1997b). *Appeal to Pity.* Albany: State University of New York Press.

Walton, Douglas (1998). *Ad Hominem Arguments.* Tuscaloosa: University of Alabama Press.

Walton, Douglas (1998a). *The New Dialectic: Conversational Contexts of Argument.* Toronto: University of Toronto Press.

Walton, Douglas (1999). *Appeal to Popular Opinion.* University Park: Pennsylvania State University Press.

Walton, Douglas (2000). *Scare Tactics.* Dordrecht: Kluwer.

Walton, Douglas (2000a). The place of dialogue theory in logic, computer science and communication studies. *Synthese* **123**: 327–346.

Walton, Douglas (2001). Abductive, presumptive and plausible arguments. *Informal Logic* **21**: 141–169.

Walton, Douglas (2001a). Enthymemes, common knowledge and plausible inference. *Philosophy and Rhetoric* **34**: 93–112.

Walton, Douglas (2001b). Abductive, presumptive and plausible arguments. *Informal Logic* **21**: 141–169.

Walton, Douglas (2002). *Legal Argumentation and Evidence.* University Park: Pennsylvania State University Press.

Walton, Douglas (2002a). The sunk costs fallacy or argument from waste. *Argumentation* **16**: 473–503.

Walton, Douglas (2002b). Are some *modus ponens* arguments deductively invalid? *Informal Logic* **22**: 19–46.

Walton, Douglas (2006). *Fundamentals of Critical Argumentation.* New York: Cambridge University Press.

Walton, Douglas (2006a). *Character Evidence: An Abductive Theory.* Dordrecht: Springer.

Walton, Douglas (2006b). Argument from appearance: A new dialectical scheme. *Logique et Analyse* **195**: 319–340.

Walton, Douglas (2007). *Witness Testimony Evidence.* New York: Cambridge University Press.

Walton, Douglas, and Reed, Chris (2002). Argumentation schemes and defeasible inferences. In Giuseppe Carenini, Floriana Grasso, and Chris Reed (eds.), *Working Notes of the ECAI'2002 Workshop on Computational Models of Natural Argument.* Lyon, France, 45–55.

Walton, Douglas, and Gordon, Thomas F. (2005). Critical questions in computational models of legal argument. In Paul Dunne and Trevor Bench-Capon (eds.), *Argumentation in Artificial Intelligence and Law,* IAAIL Workshop Series. Nijmegen: Wolf Legal Publishers, 103–111.

Walton, Douglas, and Krabbe, Erik C. W. (1995). *Commitment in Dialogue.* Albany: State University of New York Press.

Walton, Douglas, and Reed, Chris (2003). Diagramming, argumentation schemes and critical questions. In Frans H. van Eemeren, J. Anthony Blair, Charles A. Willard, and A. Francisca Snoek Henkemans (eds.), *Anyone Who Has a View: Theoretical Contributions to the Study of Argumentation.* Dordrecht: Kluwer, 195–212.

Walton, Douglas, and Reed, Chris (2005). Argumentation schemes and enthymemes. *Synthese* **145**, 2005, 339–370.

Walton, Douglas, Prakken, Henry, and Reed, Chris (2003). Argumentation schemes and generalizations in reasoning about evidence. In *Proceedings of the 9th International Conference on Artificial Intelligence and Law, Edinburgh, 2003.* New York: ACM, 32–41.

Warnick, Barbara (2000). Two systems of invention: The topics in the *Rhetoric* and *The New Rhetoric.* In Alan G. Gross and Arthur E. Walzer (eds.), *Rereading Aristotle's "Rhetoric."* Carbondale: Southern Illinois University Press, 107–129.

Weinreb, Lloyd L. (2005). *Legal Reason: The Use of Analogy in Legal Argument.* New York: Cambridge University Press.

Weitzenfeld, Julian (1984). Valid reasoning from analogy. *Philosophy of Science* **51** (1): 137–149.

Wells, Simon (2007). "Formal Dialectical Games in Multiagent Argumentation," Ph.D. thesis, University of Dundee, Dundee, Scotland.

William of Ockham (1966). *Introduction to Logic.* Translated by Norman Kretzmann. Minneapolis: University of Minnesota Press.

Windes, Russel R., and Hastings, Arthur (1965). *Argumentation and Advocacy.* New York: Random House.

Woods, John, and Walton, Douglas. (1982). *Argument: The Logic of Fallacies.* Toronto: McGraw-Hill Ryerson.

Wooldridge, Michael, and Jennings, Nicholas R. (1995). Intelligent agents: Theory and practice. *The Knowledge Engineering Review* **10** (2): 115–152.

Wooldridge, Michael (2000). *Reasoning about Rational Agents.* Cambridge, Mass.: MIT Press.

Wooldridge, Michael (2002). *An Introduction to MultiAgent Systems.* Chichester: Wiley.

Wright, Richard W. (1985). Causation in tort law. *California Law Review* **73**: 1735–1828.

Yu, Bin, and Singh, Munindar P. (2000). A social mechanism of reputation management in electronic communities. In Matthias Klusch and Larry Kerschberg (eds.), *Proceedings of the 4th International Workshop on Cooperative Information Agents,* Lecture Notes in Computer Science, vol. 1860. London: Springer, 154–165.

Index

Page numbers in square brackets [ ] indicate pages with an argumentation scheme listed with no discussion.
Page numbers in cursive brackets { } indicate pages with critical questions listed with no discussion.
Page numbers in parentheses ( ) indicate pages with diagrams of argumentation models.

Printed in the United States
By Bookmasters